THE ROUTLEDGE HANDBOOK OF HOUSING AND WELFARE

This handbook takes on one of the most pressing issues of today's society – the question of housing. It is a cutting-edge edited volume about the disputed interrelationship between housing and the wider welfare state. Although housing scholars generally agree that housing should be regarded as part of such a wider welfare system, it has proven hard to pinpoint and operationalize its position within it. Moreover, the relationship became considerably more complex as a result of the period of intense globalization and the integration of national housing finance systems into world finance markets. Furthermore, welfare systems reflect economic as well as social models and these, too, have changed as countries have responded to globalization, and traditional ideological frameworks have become less distinct.

Thus, there is a need to redefine the connection between housing and welfare in light of changes in both welfare and housing systems. By investigating the current situation and historical development of housing provision and welfare distribution in different contexts worldwide, this book aims to contribute to an expanded understanding of housing and welfare.

The book brings together 25 international housing researchers covering 15 countries worldwide. With such a global approach, the book aims to provide an updated empirical picture and analysis of different housing systems and their connection to the welfare regime in different national contexts. The book moves beyond the usual focus on affordable housing provision in the context of well-developed welfare regimes and includes countries from the global south, incorporating regions where it is debatable whether there are welfare systems present at all. Thus, the book aims to provide the reader with an insight into the large differences in housing provision in international contexts with large differences regarding how the welfare state is comprised. From these insights, we reflect on whether regime approaches continue to provide a suitable theoretical framework for understanding the relationship between housing and the wider welfare state.

This handbook is essential reading for researchers, students, policymakers, and other professionals in the fields of housing studies, welfare studies, economics, urban studies, social work, social and public policy, and sociology.

Martin Grander is Associate Senior Lecturer in Urban Studies at Malmö University and Director of the Interdisciplinary Research Environment Studies in Housing and Welfare.

Mark Stephens is Ian Mactaggart Chair in Land, Property and Urban Studies at the University of Glasgow. He also leads the international theme in the UK Collaborative Centre for Housing Evidence (CaCHE). Much of his contribution to this book was undertaken during his appointment as Visiting Professor in the School of Architecture, Planning and Design, The University of Sydney.

THE ROUTLEDGE HANDBOOK OF HOUSING AND WELFARE

Edited by Martin Grander and Mark Stephens

Routledge
Taylor & Francis Group
LONDON AND NEW YORK

Designed cover image: © Mark Stephens

First published 2024
by Routledge
4 Park Square, Milton Park, Abingdon, Oxon OX14 4RN

and by Routledge
605 Third Avenue, New York, NY 10158

Routledge is an imprint of the Taylor & Francis Group, an informa business

British Library Cataloguing-in-Publication Data
A catalogue record for this book is available from the British Library

Library of Congress Cataloging-in-Publication Data
Names: Grander, Martin, editor. | Stephens, Mark, 1963–editor.
Title: The Routledge handbook of housing and welfare / edited by Martin Grander and Mark Stephens.
Description: New York, NY: Routledge, 2024. |
Includes bibliographical references and index.
Identifiers: LCCN 2023037787 (print) | LCCN 2023037788 (ebook) |
ISBN 9781032074337 (hardback) | ISBN 9781032080529 (paperback) |
ISBN 9781003212690 (ebook)
Subjects: LCSH: Housing policy. | Welfare state.
Classification: LCC HD7287.3 .R688 2024 (print) | LCC HD7287.3 (ebook) |
DDC 363.5/561—dc23/eng/20230824
LC record available at https://lccn.loc.gov/2023037787
LC ebook record available at https://lccn.loc.gov/2023037788

ISBN: 978-1-032-07433-7 (hbk)
ISBN: 978-1-032-08052-9 (pbk)
ISBN: 978-1-003-21269-0 (ebk)

DOI: 10.1201/9781003212690

Typeset in Times New Roman
by codeMantra

In memory of David Donnison (1926–2018) and Jim Kemeny (1942–2020)

CONTENTS

ACKNOWLEDGEMENTS

Martin Grander is grateful to Ed Needle at Routledge for contacting him and presenting him with the opportunity to write this book.

Mark Stephens is grateful to the University of Glasgow for granting him a period of study leave that greatly assisted with the timely completion of this project. He is also indebted to The University of Sydney for hosting him during this period.

TABLES AND FIGURES

Tables

Figures

FOREWORD

Comparing the unique – on the comparative study of housing and welfare regimes

Bo Bengtsson

This handbook is based on housing regimes and welfare state regimes and the relationship between them. The character of this relationship is not as obvious as one might think at the first glance, which is indicated by Ulf Torgersen's much-quoted description of housing as "the wobbly pillar under the welfare state" (Torgersen, 1987). Of course, housing is generally seen as crucial to people's welfare, which is also emphasized in a number of international documents on human rights. However, unlike education, health care and several other welfare services, which are typically provided directly to citizens by the state, housing provision is normally implemented via markets, and the role of the state is to set up correctives that help make the aggregate outcomes of housing market transactions socially acceptable. Although in recent decades, with the introduction of different aspects of "new public management", huge steps towards marketization have been taken in these other welfare sectors, the provision of schooling and healthcare to citizens, unlike housing, remains a responsibility of the central and local state. Consequently, the role of market mechanisms is limited, at least in principle, to securing competition and users' choice between different welfare producers. So housing remains the "wobbly pillar under the welfare state".

Housing as the wobbly pillar of welfare research

This "wobbly" relationship between housing and general welfare has also left its trace on the research on housing and welfare state regimes. As Jim Kemeny pointed out more than 20 years ago, housing is often omitted from comparative studies of welfare state policy (Kemeny, 1995), even though it is politically seen as a key welfare sector. Harold Wilensky, one such classic student of comparative welfare, justified his omission of housing from his analysis by reference to the lack of data – "housing belongs but data are thin" (Wilensky, 1975, p. 7). The obvious reason for such measurement problems is the central role of markets in the implementation of housing policy. This means that housing quality and housing costs are to a large extent affected by demand and supply in the housing market, and in consequence, the policy outcomes are difficult to measure in budgetary terms.

The same can be said about another leading welfare state scholar, Gøsta Esping-Andersen, who plays a central role in the present volume. For some rather obscure reason, Esping-Andersen does not include housing provision in his seminal work *The Three Worlds of Welfare Capitalism*

(Esping-Andersen, 1990). In fact, Esping-Andersen did analyse housing in previous work, comparing Denmark and Sweden (Esping-Andersen, 1980) and later also Norway (Esping-Andersen, 1985). Unfortunately, however, as observed by Norwegian historian and housing specialist Erling Annaniassen, Esping-Andersen's understanding of housing institutions in all three countries had evident shortcomings (Annaniassen, 2000). Another leading scholar in comparative housing, Michael Harloe, even speculates that the inability of Scandinavian social democracy to achieve a solidaristic housing policy may account for Esping-Andersen's decision to omit any consideration of housing policies from his later work (Harloe, 1995, pp. 539f.).

Over the years, Esping-Andersen's distinction between three types of welfare state regimes (liberal, conservative and social democratic) has had, and still has, an immense following among scholars in the field. He has also been criticized for ignoring important aspects of welfare, as well as for excluding some relevant welfare states in his analysis. Consequently, several additions and amendments to his ideal-type model have been suggested by other researchers.

This is, of course, only a natural expression of scientific discourse and progress. Perhaps it is more remarkable that Esping-Andersen's general welfare types have also been applied – at least as a starting point – in several comparative and single-country studies of housing regimes, where the analysis often has to start with an attempt to transform his model to the specific field of housing (Kemeny, 1995, 2006; Hoekstra, 2003; Stamsø, 2009 are some examples). Interestingly, as it seems, no such transformations have been based on Esping-Andersen's own analysis of Scandinavian housing regimes. So – perhaps not surprisingly considering Annaniassen's criticism – Esping-Andersen's previous work on (Scandinavian) housing appears to have been difficult to integrate with his "three worlds of welfare capitalism".

The challenge of comparing housing regimes

Comparing housing regimes – and drawing more general conclusions from the results – is a challenge that several leading housing scholars have taken on over the years, with varying degrees of success. National housing regimes are extremely complex, involving the state, markets and civil society, housing production, finance and management and the academic disciplines of economics, law, sociology and political science. Moreover, and perhaps even more importantly, all national housing regimes are in some way unique due to the strong path dependence of housing provision (e.g., Kleinman, 1996; Kemp, 2000; Lowe, 2004, Chapter 6; Kay, 2005; Holt-Jensen & Pollock, 2009; Nielsen, 2010; Malpass, 2011; Kohl, 2017; Bengtsson & Kohl, 2020; Blackwell & Bengtsson, 2021). In short, the common historical pattern is that more or less contingent solutions to acute housing problems, typically in the first half of the 20th century, have tended to open up institutional trajectories that have made system change at some later point difficult due to the obstructing "path dependence mechanisms" of efficiency, legitimacy and power (Bengtsson & Ruonavaara, 2010).

Several years ago, I was part of a project that may in itself serve to illuminate the uniqueness of housing regimes. In the project, we compared the housing regimes of the five Nordic countries Denmark, Sweden, Norway, Finland and Iceland, of which at least three – and arguably all five – belong to Esping-Andersen's social democratic welfare state category. Despite this, and despite geographical proximity and partly common history, as well as cultural, social and political similarities, each of the five countries represents one specific housing regime, and this could largely be explained by path dependence arising from differences between national and local solutions to earlier housing problems. It was no coincidence that the title of the project, and the book summarizing the results, was "Why so Different?" (Bengtsson et al., 2013).

Comparative politics and comparative method

The ambition of the present book is to integrate the analysis of housing regimes with the analysis of other welfare state regimes and to carry this integration beyond the Western focus that has traditionally dominated comparative research on welfare and housing. Both the outline and general logic, and the way that the work has been organized – with chapters on different countries and regions written by specialists following a common, but not overly strict and detailed template – allows for a multitude of fruitful comparisons.

One subfield of the academic discipline of political science has traditionally been labelled "comparative politics". According to a well-recognized article by Arendt Lijphart (1971, p. 682) "comparative politics" carries a methodological and not a substantive label. As a matter of fact, the label is more misleading than methodological, since it is not exclusively used to describe studies applying comparative *methods*, but more generally to studies with an interest in policy and politics in more than one country, regardless of methodological approach. The book review section of the leading journal *American Political Science Review* takes this even further by making the rather ethnocentric distinction between the headings "American politics" and "Comparative politics". The first part reviews books about US policy and politics, whilst the second part shifts the focus away from the US and towards one or more other countries. This organization of the review section implicitly suggests that readers should use the US as the benchmark against which other countries are compared.

So of course, for the sake of clarity, a distinction should be made between "comparative method" and "comparative politics" in political science. Analogously, in the specific field of housing studies, a distinction should be made between "comparative method" and the more substantive "comparative housing" or "comparative housing research" (cf. Stephens & Norris, 2011; Stephens & Hick, forthcoming). Applying this distinction, the present book is certainly one of comparative housing, i.e. containing substantive studies with the ambition to help us understand in comparative terms (mainly) national housing regimes. The handbook has obvious comparative elements in the editors' innovative theoretical analysis in the concluding part, in the common template for the country and regional chapters and sometimes within individual chapters comprising more than one country. Yet the book does not explicitly and unambiguously make use of comparative methods, at least not in any strict sense. (This is certainly not to be expected either, since it is a handbook and not a research project.)

What can we learn from comparative housing studies?

The classical logic of choosing and arranging comparing case studies is in accordance with some "quasi-experimental" design, which dates back to John Stuart Mill and his "method of difference" and "method of agreement" (Mill [1843] 1967). The "method of difference" means studying cases with different values on the dependent variable and trying to find an important independent variable that varies between the cases in a similar way to the independent variable, thus trying to explain the difference. In contrast, the 'method of agreement' means choosing cases with the same values on the dependent variable, which rules out an explanation all independent variables with different values between the cases. Other related frameworks are what Przeworski and Teune (1970, pp. 31f.) have called "most similar systems design" and "most different system design" respectively. Analogously to medical and behavioural experiments, the quasi-experimentally inclined case study researcher chooses a "treatment case" and a "control case" in the hope of obtaining an outcome that either confirms or falsifies the hypothesis. One or two easily identifiable background variables like population size may also be added to such a design.

The quasi-experimental comparative case study design has a symmetrical logic and neatness that can be seductive. Nevertheless, as Stanley Lieberson convincingly argues, it is pointless to draw any causal generalizations from comparative case studies as long as you cannot assume a deterministic relation – which is always risky in the open-ended social sciences where interaction between actors with intentions, preferences and norms is a basic ingredient. "Put bluntly, application of Mill's methods to small-*N* situations does not allow for probabilistic theories, interaction effects, measurement errors or even the presence of more than one cause" (Lieberson, 1992, p. 177).

Accepting Lieberson's harsh argument, we understand that quasi-experimentally ordered comparative case studies are insufficient to provide evidence for *testing and developing theory* in the form of empirical generalizations based on necessary and sufficient conditions or on statistically significant interrelations. Furthermore, the complexity of housing and welfare regimes also makes it difficult to establish strong statistical relations between variables in a convincing way.

Nonetheless, comparative case studies are an essential component of social science research, but it is not necessary – and sometimes it may even be misleading – to maintain a strict distinction between single case- and comparative case studies when we appreciate the limitations of quasi-experimental designs. What can be learned from two-case or multi-case studies can also be learned from one well-designed and well-implemented single-case study – only with the additional advantage of having more cases at your disposal. Instead of quasi-experimentally trying to test empirical generalizations both single-case and comparative case studies are better used to identify *ideal-type social mechanisms*, which may to some extent be generalized to other situations (cf. Bengtsson & Hertting, 2014). Bengtsson and Ruonavaara (2017) have suggested a logic of "comparative process tracing" for identifying such social mechanisms drawing on historical processes, e.g., of national housing regimes in different countries.

An equally important strength of comparative case studies is the potential to provide useful *counterfactual contrasts* between two or more countries or other cases.

A fourth, more political, aim of comparative housing studies may be to offer concrete *solutions to practical and political housing problems*. An important general lesson from history is, however, that it is almost impossible – due to the complexity and ever-present path dependence of housing provision – to import housing systems and institutions directly from one country to another. Nevertheless, comparison with other countries is often useful in providing ideas about various micro solutions at the same time demonstrating that things can be done differently.

In sum, the problems with Mill's methods pointed out by Lieberson and others should not discourage researchers from applying comparative analysis, but rather encourage a more relaxed and pragmatic attitude to (non-perfect) comparative approaches. Even if we can rarely assume determinism in the social sciences, and we are seldom in a position to estimate probabilities of complex outcomes either, several observations pointing in the same direction can help establish trends and expectations. At the same time exceptions to such trends and expectations always deserve special attention if we are looking for social mechanisms. I end by highlighting three examples of this from the Swedish housing regime, all further discussed in the Handbook. What are the mechanisms behind these trends – and behind these exceptions? (The solutions to these puzzles – both the trends and the Swedish exceptions – are also discussed, for example, in Bengtsson, 1992; Grander, 2017; Bengtsson & Grander, 2023; Bengtsson, forthcoming.)

1. Today, Sweden is perhaps the only country holding on to some elements of the universal logic of housing provision, a system much discussed by Jim Kemeny, for example. Why have so many countries abandoned that system? And why does Sweden remain an exception?

2. While the rental system of most countries is based on either individual market rents or rent regulation or, in most cases, a combination of both, Sweden has its own unique system based on collective bargaining and judicial estimations of use value. What has made that possible? And why have no other countries followed suit?
3. While cooperative housing is struggling in most countries, housing cooperatives represent no less than 25 per cent of the housing stock in Sweden. Why has cooperative housing been such a successful part of the Swedish housing market? And why is this the case in so few other countries?

I invite the reader to make use of the Handbook not only for general reference but also as a source for comparison in order to identify social mechanisms, to provide counterfactual contrasts and to identify new practical solutions to national housing problems – and by extension even to develop theory.

References

Annaniassen, E. (2000). *Housing Policy in the Scandinavian Countries: One Nordic Model or Different Approaches*. NOVA, paper presented at the ENHR Conference in Gävle 2000.

Bengtsson, B. (1992). Not the Middle Way but Both Ways – Cooperative Housing in Sweden. *Scandinavian Housing and Planning Research, 9*(sup. 2), 87–104. https://doi.org/10.1080/02815737.1992.10801445

Bengtsson, B. (forthcoming). The Changing Role of Cooperatives in the Swedish Housing Regime – A Path Dependence Analysis. *Housing Studies*.

Bengtsson, B. (Ed.) Annaniassen, E., Jensen, L., Ruonavaara, H., & Sveinsson, J. R. (2013). *Varför så olika? Nordisk Bostadspolitik i Jämförande Historiskt ljus* [Why so Different? Nordic Housing Policy in Comparative Light]. Égalité.

Bengtsson, B., & Grander, M. (2023). *Bostadsfrågan som Politik och Intressekamp* [Housing as Policy and Politics]. Égalité.

Bengtsson B., & Hertting,N. (2014). Generalization by Mechanism. Thin Rationality and Ideal–type Analysis in Case Study Research. *Philosophy of the Social Sciences, 44*(6), 707–732. https://doi.org/10.1177/0048393113506495

Bengtsson, B., & Kohl, S. (2020). Incremental Change in Housing Regimes: Some Theoretical Propositions with Empirical Illustrations. *Critical Housing Analysis, 7*(1), 15–24. https://www.housing-critical.com/viewfile.asp?file=2654

Bengtsson, B., & Ruonavaara, H. (2010). Introduction to the Special Issue: Path Dependence in Housing. *Housing, Theory and Society, 27*(3), 192–203. https://doi.org/10.1080/14036090903326411

Bengtsson, B., & Ruonavaara, H. (2017). Comparative Process Tracing: Making Historical Comparison Structured and Focused. *Philosophy of the Social Sciences, 47*(1), 44–66. https://doi.org/10.1177/0048393116658549

Blackwell, T., & Bengtsson, B. (2021). The Resilience of Social Rental Housing in the United Kingdom, Sweden and Denmark. How Institutions Matter. *Housing Studies, 38*(2), 269–289. https://doi.org/10.1080/02673037.2021.1879996.

Esping-Andersen, G. (1980). *Social Class, Social Democracy, and State Politics.* Nyt Fra Samfundsvitenskaberne.

Esping-Andersen, G. (1985). *Politics Against Markets. The Social Democratic Road to Power*. Princeton University Press.

Esping-Andersen, G. (1990). *The Three Worlds of Welfare Capitalism*. Polity Press.

Grander, M. (2017). New Public Housing: A Selective Model Disguised as Universal? Implications of the Market Adaptation of Swedish Public Housing. *International Journal of Housing Policy, 17*(3), 335–352. https://doi.org/10.1080/19491247.2016.1265266

Harloe, M. (1995). *The People's Home? Social Rented Housing in Europe and America*. Blackwell.

Holt-Jensen, A., & Pollock, E. (Eds.) (2009). *Urban Sustainability and Governance: New Challenges in Nordic-Baltic Housing Policies.* Nova Science.

Hoekstra, J. (2003). Housing and the Welfare State in the Netherlands: An Application of Esping-Andersen's Typology. *Housing, Theory and Society, 20*(2), 58–71. https://doi.org/10.1080/14036090310000634

Kay, A. (2005). A Critique of the Use of Path Dependency in Policy Studies. *Public Administration, 83*(3), 553–571. https://doi.org/10.1111/j.0033-3298.2005.00462.x

Kemeny, J. (1995). *From Public Housing to the Social Market: Rental Policy Strategies in Comparative Perspective*. Routledge.

Kemeny, J. (2006). Corporatism and Housing Regimes. *Housing, Theory and Society, 23*(1), 1–18. https://doi.org/10.1080/14036090500375423

Kemp, P. E. (2000). Housing Benefit and Welfare Retrenchment in Britain. *Journal of Social Policy, 29*(2), 263–279. https://doi.org/10.1017/S0047279400005912

Kleinman, M. (1996). Housing, Welfare and the State in Europe. Edward Elgar.

Kohl, S. (2017). *Homeownership, Renting and Society: Historical and Comparative Perspectives*. Routledge.

Lieberson, S. (1992). Small *N*'s and Big Conclusions: An Examination of the Reasoning in Comparative Studies Based on a Small Number of Cases. In C. C. Ragin & H. S. Becker (Eds.), *What Is a Case? Exploring the Foundations of Social Inquiry* (pp. 105–118). Cambridge University Press.

Lijphart, A. (1971). Comparative Politics and the Comparative Method. *American Political Science Review, 65*(3), 682–693. https://doi.org/10.2307/1955513

Lowe, S. (2004). *Housing Policy Analysis. British Housing in Cultural and Comparative Context*. Palgrave MacMillan.

Malpass, P. (2011). Path Dependence and the Measurement of Change in Housing Policy. *Housing, Theory and* Society, *28*(4), 305–319. https://doi.org/10.1080/14036096.2011.554852

Mill, J. S. ([1843] 1967). *A System of Logic. Ratiocinative and Inductive*. University of Toronto Press.

Nielsen, B. G. (2010). T*he Hidden Politics of a Haunted Sector. Retrenchment in Danish Housing Policy 2001–2009* (Doctoral dissertation, Copenhagen University).

Przeworski, A., & Teune, H. (1970). *The Logic of Comparative Social Inquiry*. John Wiley & Sons.

Stamsø, M. A. (2009). Housing and the Welfare State in Norway. *Scandinavian Political Studies, 32*(2), 195–220. https://doi.org/10.1111/j.1467-9477.2008.00223.x

Stephens, M., & Hick, R. (forthcoming). Comparative Housing Research. In K. Jacobs, K. Flanagan, J. Verdouw, & J. de Vries (Eds.), *Research Handbook on Housing and Society*. [(16) (PDF) Comparative housing research (researchgate.net) (Accessed: 26 May 2023].

Stephens, M., & Norris, M. (2011). Introduction to Special Issue: Comparative Housing Research. *International Journal of Housing Policy, 11*(4), 333–336. https://doi.org/10.1080/14616718.2011.626596

Torgersen, U. (1987). Housing: The Wobbly Pillar under the Welfare State. *Scandinavian Housing and Planning Research, 4*(sup. 1), 116–126. https://www.tandfonline.com/doi/abs/10.1080/02815737.1987.10801428

Wilensky, H. L. (1975). *The Welfare State and Equality*. University of California Press.

CONTRIBUTORS

Valentina Abufhele is a Postdoctoral Researcher at the Institute of Urban and Territorial Studies (IEUT), Pontificia Universidad Católica de Chile (ANID/Fondecyt N°3200926). Her research focuses on urban studies and political sociology, specifically on urban poverty and inequality, State housing policies, and the political practices and subjectivities of low-income groups.

Nicolás Angelcos is a Professor in the Sociology Department at the University of Chile and an Associate Researcher at the Center for the Study of Conflict and Social Cohesion (COES). His research focuses on political sociology and urban studies. Until 2022, he led the research project "From Protagonism to Abstention? A Study on Political Participation in Low-Income Neighborhoods of Santiago." With Miguel Pérez, he edited the book "Vivir con dignidad: transformaciones sociales y políticas de los sectores populares en Chile" (Fondo de Cultura Económica, 2023).

Melissa García-Lamarca is a Marie Skłodowska-Curie Research Fellow at the Polytechnic of Turin and member of the Beyond Inhabitation Lab and the Barcelona Lab for Urban Environmental Justice and Sustainability. Her research untangles lived experiences of housing financialization and collective struggles towards housing justice, as well as the financial- and real-estate-related dimensions of urban green inequalities.

Job Gbadegesin is a Postdoctoral Fellow at the Centre for Development Support, University of the Free State (UFS). He works on housing and urban issues in South Africa and Nigeria.

Zoë Goodall is a Research Associate and PhD candidate at the Centre for Urban Transitions, Swinburne University of Technology, Australia. Zoë's research interests include shared housing in the private rental sector, tenancy law, housing justice for renters, and gendered housing issues.

Patrick Grogan-Myers is a PhD student in the Department of Urban Studies, Levin College of Urban Affairs. He also works for a local government in planning and development. His professional work and research focus on the nexus of affordable housing development, local government policy leadership, and public transportation access.

József Hegedüs is a founding member of the Metropolitan Research Institute (MRI), a think tank working on urban and housing policy since 1989. He has an MA in Economics (1975) and PhD in Sociology (1994) and worked five years at the Planning Institute of Budapest and ten years at the Institute of Sociology of The Hungarian Academy of Sciences before joining MRI. His field of research is housing and urban policy. He has widely published in journals, and he is a co-editor of five books dealing with housing regimes in post-socialist countries.

Stephen Hincks is a Reader in the Department of Urban Studies and Planning at the University of Sheffield, UK. His research lies within the broad scope of urban and regional studies, coalescing around three interrelated themes of housing, neighbourhoods, and inequality, applied spatial analysis and GIS, and urban-regional policy and planning.

Kath Hulse is an Emeritus Professor at the Centre for Urban Transitions, Swinburne University of Technology (Melbourne). Her research investigates housing system changes in the shifting landscape of the political economy of welfare. She is particularly interested in how these changes intersect to shape social and economic inequalities and spatial advantage/disadvantage.

Alexander Kalyukin is a Postdoctoral Researcher at the Institute for Urban Research, Malmö University. He holds a PhD in Social and Economic Geography from Uppsala University. His research interests revolve around questions of housing and everyday life, modernist residential architecture, historical development of housing regimes, gentrification, public space, and heritage in urban redevelopment.

Daniela Leonardi is a Postdoctoral Researcher in Sociology at the University of Turin. Her research focuses on housing as a key focus for analysing inequalities, welfare, poverty and social exclusion, homelessness, public policy analysis, Street-level Bureaucracy Theory. Currently, she is studying the interaction between lower classes and institutions.

Carina Listerborn is a Professor in Urban Planning at Malmö University and the Vice-Director of the Institute for Urban Research (IUR). Her research focuses on feminist urban perspectives on neoliberal planning, critical housing studies, and smart housing.

Lochner Marais is a Professor of Development Studies at the Centre for Development Support at the University of the Free State (UFS). His research interests include housing policy, small cities and towns (mining and renewable towns and cities), and public health focusing on children.

Michael Marshall is a PhD candidate in Urban Studies and Planning at the University of Sheffield, UK. His research focuses on UK social housing policy, finance, and practice, as well as evaluating the impact of policy on housing needs.

Julie Tian Miao is an Australian Research Council Discovery Research Fellow and Associate Professor in Property and Economic Development at the University of Melbourne. She also holds honorary and visiting positions at Harvard, Shanghai Jiaotong, Hong Kong, and Adelaide Universities. Her research focuses on the 'intrapreneurial state' and 'innovation space' which explore the interfaces between planning, property, governance, innovation, and the knowledge economy.

Elelwani Mmbadi is a PhD candidate at the Centre for Development Support, University of the Free State. His research focuses on informal settlement upgrading and asset-building, with specific attention to the contribution of informal settlement upgrading to the economic inclusion of low-income households into the residential property market.

Ryan Powell is a Reader in Urban Studies in the Department of Urban Studies and Planning at the University of Sheffield, UK. His research focuses on urban marginality and housing inequality.

Alexander Puzanov is a General Director and an Acting Director of the Municipal Economic Development Unit at the Institute for Urban Economics, a Moscow-based think tank (NGO). He is also a Professor and Head of the Department of Urban Economics and Municipal Management at the National Research University 'Higher School of Economics'. His research focuses on housing policy issues with particular attention on rental housing development and coordination of housing and urban development policies.

Matilda Annemark Sandberg is a PhD student in Urban Studies at Malmö University. She is interested in matters concerning housing provision and housing inequalities. In her PhD project, Matilda studies housing quality in the Swedish context.

Jardar Sørvoll is a Research Professor at Norwegian Social Research (NOVA), Oslo Metropolitan University and Head of BOVEL, centre for housing and welfare research. His research covers a wide range of housing-related topics, including the history of housing policies in the Nordic countries, housing in the ageing society, and social housing for disadvantaged groups.

J. Rosie Tighe is an Associate Professor in the Department of Urban Studies in the Levin College of Urban Affairs. She was previously an Assistant Professor in the Department of Geography and Planning at Appalachian State University. She holds a PhD in Community and Regional Planning from the University of Texas at Austin and a Master's degree in Urban and Environmental Policy and Planning from Tufts University. Her research focuses on housing policy, race and ethnicity, and neighbourhood revitalization.

Piret Veeroja is a Research Fellow at the Centre for Urban Transitions, Swinburne University of Technology (Australia). Her current research focuses on housing conditions and aspirations as fundamental to just and healthy cities.

Steffen Wetzstein is an Adjunct Research Fellow in the School of Social Sciences, The University of Western Australia, and a Joint Working Group Coordinator 'Policy and Research' at the European Network for Housing Research (ENHR). His international research and leadership work focuses on urban strategic planning, political economy, housing, affordability and liveability, territorial go, and development.

1
INTRODUCTION

Martin Grander and Mark Stephens

Housing has emerged as one of the most pressing issues of societal importance in recent decades. The awareness of a housing crisis is seemingly significant: the UN, the World Economic Forum, the World Bank, and the IMF; national politicians; researchers and housing commentators all over the world – everyone seems to agree that there is a crisis across housing markets internationally. International growth regions are characterized by increasing housing shortages and housing segregation, growing homelessness, and other forms of exclusion in the housing market, as well as high and rising housing costs (World Economic Forum, 2022). The problems highlighted can be summarized as (1) a housing shortage in larger cities; (2) an increasing share of the population (not only the poorest) is being excluded from the formal housing market; and (3) that residential segregation is spreading.

The shortage of affordable housing is a global urban phenomenon. According to the Lincoln Institute of Land Policy (Kallergis et al., 2018), 90 per cent of 200 cities examined in a survey were unaffordable based on average house prices in relation to median income. Ultra-low interest rates in the period between the Global Financial Crisis (GFC) and the resurgence of inflation in the aftermath of the COVID-19 pandemic and the Russian invasion of Ukraine created conditions for rising house prices. These were exacerbated in some countries by low levels of housing construction, which contributed to shortages. While credit has been cheap, access to it has been limited by high house prices, often combined with restrictions on lending brought in after the GFC. Consequently, in many countries, access to homeownership has become more limited as households have been priced out, which has contributed to falling homeownership rates. Central banks responded to the rise in inflation by increasing interest rates, giving rise to fears that real estate bubbles would burst, with serious societal and economic consequences. This fear is especially acute in recently well-performing economies such as China, where the economy relies on a booming construction sector. A central tension in policy arises from the desire to expand homeownership among households with limited economic means and the dangers of a mortgage-driven housing bubble, a tension that was central to the US subprime crisis.

Combined with a shortage of affordable rented housing (Aalbers et al., 2017; Anacker, 2019; Fields & Uffer, 2016), the boom in the housing market has led to a situation where low-income households, young people, and immigrants are increasingly being excluded from the formal housing market, and many are being forced to live in formal or informal substandard housing. It is

 DOI: 10.1201/9781003212690-1

becoming increasingly difficult for young people to move away from their family homes and for newly arrived immigrants to access adequate housing (Murdie, 2010). Homelessness is increasing and changing in character. Private rental housing is often expensive, and the public and social housing sectors are often decreasing and being put under pressure. Consequently, many people are excluded from the rental market as well. Studies show that it is not always those who are worst off who experience precarious informal housing conditions. Grander (2023) identifies a "middle group" that cannot access the mainstream housing market due to low or irregular income but also does not qualify for housing through the safety nets of the welfare system. An increasing proportion of those who still have a home lacks real security of tenure as they sublet on uncertain terms. Reports of such informal rental markets can allow regulatory protections such as tenure security and the regulation of rents to be evaded, sometimes illegally. Scarcity also increases the potential for discrimination to be used as a rationing device. According to Dorling (2014), a "housing precariat" of people who are forced to move from one substandard and unsafe accommodation to another has emerged.

Residential segregation is widening and deepening. Most urbanized areas, in one way or another, reflect socioeconomic differences in their spatial distribution (Musterd & Ostendorf, 1998; Musterd et al., 2017). Segregation may be driven directly by market processes ("filtering") associated with gentrification, but more recently state-led forms of "gentrification" have developed, for example through the demolition of public housing estates as part of densification projects (Hochstenbach, 2017; Lees et al., 2016). Segregation can in turn cause stigmatization, labour market discrimination, and even social unrest (Musterd, 2020; Skifter Andersen et al., 2016).

The housing crisis was unfolding in these ways in the early 2020s in major cities across the world. Conversely, there are also regions with declining populations that have significant housing vacancies, which may threaten the finances of the local government as tax bases shrink. Clearly, in mainly market-based housing systems, a low or unstable income is a major barrier to being able to access secure and adequate housing at an affordable cost. However, as housing systems become more reliant on the market, and if the market increasingly fails to provide housing that is widely affordable, then housing becomes ever more detached from the notion of housing as a right, raising profound doubts about its role as part of a welfare state. Hence, a major problem in contemporary housing markets is the increasing differences in housing opportunities between different social groups – what could be described as increased housing inequality (Dorling et al., 2005; Grander, 2023; James et al., 2022).

The COVID-19 pandemic highlighted and exacerbated the relationship between housing inequalities and health outcomes. The home was central to the authorities' strategies for social distancing, meaning that access to and the characteristics of the home have become central to individuals' possibilities to isolate themselves. Members of overcrowded households in disadvantaged areas of cities had fewer opportunities to work from home because they were disproportionately employed in vital services such as health and social care and essential occupations such as food processing. This meant that they had to go to workplaces, mingle with colleagues, and thus were more exposed to the virus than others. Of course, their household members were exposed to the virus when they returned home. When counting the deaths from COVID-19, individuals in overcrowded households in disadvantaged city areas were clearly hit the worst (Kamis et al., 2021). Thus, the relevance of the welfare state in relation to housing seems pressing in the wake of the Covid-19 pandemic.

Since the pandemic receded, hopes that the greater societal cohesion that its containment required would lead to a reassertion of governmental willingness to intervene to reduce inequalities soon became lost in the rising tide of inflation, creating more societal tension and the loss of any post-pandemic optimism. With economies across the world under strain, the search for enduring answers to the questions of housing and welfare remains as acute as ever.

***

This handbook addresses the disputed interrelationship between housing and welfare. However, rather than discussing welfare, which is often understood as "governmental support" in the form of individual subsidies or benefits[1], we – like many housing researchers before us – take as our point of departure the concept of a "welfare regime", as proposed by Esping-Andersen in his seminal work *The Three Worlds of Welfare Capitalism* (Esping-Andersen, 1990). Here, the welfare regime concept encompasses not only arrangements for taxation and social security but also labour markets and how economies are organized. Thus, the concept provides for a deeper analysis in which distinctive housing systems can emerge in parallel to these wider welfare systems (Stephens, 2020a, 2020b).

Although housing scholars generally agree that housing should be regarded as part of such a wider welfare system, it has proven hard to pinpoint and operationalize its position within it. Similarly, according to Article 25 of the UN Declaration of Human Rights, everyone has a right to adequate housing, and many countries have consequently adopted constitutions or policies that propose the "right to housing". But what such a right implies in practice and who is responsible for fulfilling it are disputed. We subscribe to the argument advanced by Bengtsson (2001, p. 256), that the "right to housing" embraces all principles that constitute a national housing policy and that the right to housing is thus a political "marker of concern", highlighting housing as an area for welfare state policy.

The basis for the difficulties in discussing the relationship between housing and welfare is the different nature of housing in relation to other parts of welfare provision, for example, health and school education. While the latter two are often something that citizens are entitled to, free of charge or for a small cost, housing must be paid for. Housing is seldom free, and non-market allocation is reserved for forms of social rented housing. Consequently, most housing is distributed by market mechanisms, making analogies with other welfare sectors, where non-market allocation is normal, problematic. As discussed by Bengtsson (2001, p. 257), housing in all modern welfare societies "is *at the same time* defined as an individual market commodity and as a public good demanding state involvement". However, for the market to be able to provide housing for everyone, including the less affluent, the state has a crucial role, often described in terms of *correcting* the market (in the sense of market outcomes rather than in the sense of economists' concept of market failure). Such correctives might be present in the form of means- or need-tested public/social housing, rent regulations, housing allowances, or subsidized homeownership.

The duality of housing as a societal good (again, not a public good in the technical economic sense) and a market commodity is a tension most housing scholars seem to agree upon. Consequently, Torgersen (1987, p. 116) has labelled housing "the wobbly pillar under the welfare state", while Kemeny (2006, p. 9) refers to housing as "a welfare dimension where certain sector interests have an entrenched role". Grander (2023) discussed the complexity of connecting housing to welfare regarding the competing "use" and "exchange" values of housing. Thus, the discussion of housing must relate both to economic and financial dimensions as well as questions of its social and wider societal characteristics.

The arrangements and extent of market correctives in the housing-welfare nexus are crucial not only for the potential right to housing but also for understanding how housing systems relate to *inequality*. As the level of housing consumption (access to housing with diverse standards and locations) is almost always related to household income, the redistributive role of the welfare regime can have a pivotal impact on housing outcomes. Consequently, this book, through its focus on the interrelationship between housing and welfare, seeks to contribute to the large and growing body of research on housing inequality.

As we have noted, most researchers of housing and welfare are grounded in or inspired by Esping-Andersen's (1990) book, *Three Worlds of Welfare Capitalism,* which thus often serves as the baseline for analysis of how housing systems can emerge in parallel to wider welfare systems. Esping-Andersen himself, however, refrained from including housing in his well-known typology of welfare regimes. Instead, the work of Kemeny on housing regimes has become the natural point of departure for most research in housing and welfare in recent decades. In his early work, Kemeny (1981, 1995) linked the level of homeownership with the strength of the welfare state. He posited that "homeownership societies" tend to have weaker welfare states as voters with mortgage commitments experience the "front-end loading" of housing costs often at the same time as they have responsibilities towards children. Consequently, they prefer low taxation over social expenditure. In contrast, "cost rental societies", rely on a stronger welfare state, which contributes towards the creation of a housing system centred around a high degree of state-subsidized cost-rental housing. Thus, according to Kemeny, welfare systems and housing systems are causally linked.

However, the work of Kemeny and his understanding of the relationship between welfare and housing regimes have been contested more recently by both empirical and theoretical research (Grander, 2017, 2023; Stephens, 2016, 2020a). There is therefore a need to revisit the connection between housing and welfare, considering changes in both the wider welfare regimes and housing systems in different geographical contexts. This book, therefore, aims to provide an empirically grounded and theoretically updated perspective of how welfare regimes relate to housing systems and how changes in the housing system drive changes in the welfare system, thus looking into the causes rather than symptoms of contemporary housing problems (Lawson, 2006). Through this approach, the book seeks to answer the question of how housing can be understood as part of the wider welfare regime in the 2020s.

The book brings together 25 international housing researchers covering 15 countries worldwide. With such a global approach, the book aims to provide an updated empirical picture of different housing systems and their connection to the welfare regime in different national contexts. The book moves beyond the usual focus on affordable housing provision in the context of well-developed welfare regimes and includes countries from the global south, incorporating regions where it is debatable whether there are welfare systems present at all. Thus, the book aims to provide the reader with an insight into the large differences in housing provision in international contexts, as well as the large differences regarding how the welfare state is comprised. From these insights, we reflect on whether regime approaches continue to provide a suitable theoretical framework for understanding the relationship between housing and the wider welfare state.

Aim and structure of the book

The main research question of the book is framed as "How can the interrelationship between housing and welfare be understood in the global condition of housing provision in the 2020s?". By investigating the current situation and historical development of housing provision and welfare distribution in different contexts worldwide, the book aims to contribute to an expanded understanding of housing and welfare.

The book is organized into three main parts. The first part, *Understanding Housing and Welfare*, written by editors Martin Grander and Mark Stephens, provides a discussion of the prevailing ideas of how housing can be understood as part of the wider welfare regime. In the first chapter of this part, the basic concepts of housing and welfare are clarified. The notions of welfare states and welfare regimes and housing systems and housing regimes are defined and elaborated, discussing the distributional logic of welfare and housing in different international contexts. Taking Esping-Andersen's classic typology of welfare regimes as a point of departure, the chapter discusses the basis for Esping-Andersen's (1990) ideas as well as more recent developments, critiques, and alternative understandings of such a categorization. Regarding housing, we first introduce how comparative approaches that theorize housing have mirrored approaches more closely focused on welfare states. The "housing-welfare regime" framework, founded by Kemeny (1995, 2006) serves as the main point of departure. Here, we discuss Kemeny's theories on the evolution of dualist and unitary/integrated housing systems, which he argued developed differently in contrasting countries during the post-war economic recovery, depending on the relationship between state and market, and the relationship between labour and capital.

In the second chapter of this part, we discuss the current academic understanding of how housing interlocks with welfare. The chapter problematizes models of how housing and welfare can be analysed in tandem, building on the literature on decreasing housing affordability, more limited access to housing, and increasing housing inequality. Based on such a development and on recent critiques of the housing-welfare regime framework from housing scholars, we argue for an updated, empirically grounded understanding of the relationship. We present a framework (Stephens, 2020b) for an updated understanding of the interrelation, developing Kemeny's theories and the current paradigm of analysing housing systems from a comparative welfare state perspective. As we argue, such a developed understanding needs to be based on empirical data on housing and welfare from different contexts, which evidently have undergone important changes in recent decades.

The second and longest part of the book, *Housing and Welfare in a Global Context,* represents the main empirical contribution. Divided into 12 chapters covering 15 countries, this part of the book assembles 23 national housing experts from across the world, including established as well as early career researchers, to provide an updated state-of-the-art analysis of the relationship between housing and welfare and how this relationship has developed over time. Using a common template, the researchers have identified the basis for housing provision in diverse geographical areas, and how housing is regarded in relation to the welfare regime. These chapters are written from a country perspective; each one takes its point of departure in a national (or regional, as in the case of Catalonia) context, where the country could be a representative of a larger geographical area or group of countries while also highlighting differences within a region. Thus, even though the chapters are written from a country perspective, they relate to a larger geographical area. These chapters provide an updated overview of the housing regime in the wider welfare context, providing a basis for a synthesized analysis in the subsequent chapters (Table 1.1).

The third main part of the book, *Housing and Welfare in the 21st Century*, written by the editors, presents the main analysis of the book. In the first chapter of this part, conclusions on the changes in housing and welfare regimes are drawn based on the empirical findings from the national chapters. Here, the editors draw together the differences in housing and welfare regimes and the changes in the different contexts, contributing to the "bigger picture" of the gradual development and the current situation. The chapter shows that both housing and welfare systems, regardless of their earlier constitution, have changed due to processes such as marketization and income polarization.

The second chapter of this part, which addresses the embeddedness of housing in the welfare regime, takes the findings one step further and aims to develop the theoretical understanding of the embeddedness of housing in the wider welfare regime through an analytical framework

Table 1.1 Case study countries

Chapter	*Author(s)*	*Primary country/ countries covered*	*Other countries ("Regional outlook")*
Housing and welfare in England	Michael Marshall, Ryan Powell, and Stephen Hinks	England	Scotland, Wales, and Northern Ireland
Housing and welfare in the US	Rosie Tighe and Patrick Grogan-Myers	USA	
Housing and welfare in Australia	Kath Hulse, Piret Veeroja, and Zoe Goodall	Australia	New Zealand
Housing and welfare in Sweden, Norway and the wider Nordic region	Jardar Sørvoll, Carina Listerborn, and Matilda Annemark Sandberg	Sweden & Norway	Denmark, Finland, and Iceland
Housing and welfare in conservative central Europe: The cases of Germany, Austria and Switzerland	Steffen Wetzstein	Germany, Austria, and Switzerland	
Housing and welfare in Italy	Daniela Leonardi	Italy	
Housing and welfare in Catalonia, Spain	Melissa García-Lamarca	Spain (Catalonia)	
Housing and Welfare in Russia and the former Soviet Union	Alexander Kalyukin and Alexander Puzanov	Russia	Lithuania, Belarus, Armenia, Kazakhstan, Ukraine, Azerbaijan, Georgia, Tajikistan, and Kyrgyzstan
Housing and Welfare in East-Central Europe – The case of Hungary	József Hegedüs	Hungary	Estonia, Poland, Slovakia, Romania, and Slovenia
Housing and welfare in China	Julie Miao	China	Hong Kong, Singapore, South Korea, and Taiwan
Housing and welfare in the wider latin american context – the Chilean experience	Valentina Abufhele and Nicolás Angelcos	Chile	Costa Rica and Paraguay
Housing and welfare in South Africa	Elelwani Mmbadi, Lochner Marais, and Job Gbadegesin	South Africa	

constructed on the basis of earlier research and empirical findings. Here, we introduce a typology that aims to contribute to an updated understanding of how housing can be understood and analysed as part of the welfare state. In the final chapter of this part and of the book, we summarize the findings and reflect on the evolution of regime approaches as theoretical frameworks for understanding housing systems.

Note

1 For example, the Cambridge Dictionary defines welfare as "help given, especially by the state or an organization, to people who need it, especially because they do not have enough money".

References

Aalbers, M. B., Loon, J. V., & Fernandez, R. (2017). The Financialization of a Social Housing Provider. *International Journal of Urban and Regional Research, 41*(4), 572–587. https://doi.org/10.1111/1468-2427.12520

Anacker, K. B. (2019). Introduction: Housing Affordability and Affordable Housing. *International Journal of Housing Policy, 19*(1), 1–16. https://doi.org/10.1080/19491247.2018.1560544

Bengtsson, B. (2001). Housing as a Social Right: Implications for Welfare State Theory. *Scandinavian Political Studies, 24*(4), 255–275. https://doi.org/10.1111/1467-9477.00056

Dorling, D., Ford, J., Holmans, A., Sharp, C., Thomas, B., & Wilcox, S. (Eds.). (2005). *The Great Divide: An Analysis of Housing Inequality*. Shelter.

Dorling, D. (2014). *All That Is Solid: How the Great Housing Disaster Defines Our Times, and What We Can Do About It*. Penguin.

Esping-Andersen, G. (1990). *The Three Worlds of Welfare Capitalism*. Princeton University Press.

Fields, D., & Uffer, S. (2016). The Financialisation of Rental Housing: A Comparative Analysis of New York City and Berlin. *Urban Studies, 53*(7), 1486–1502. https://doi.org/10.1177/0042098014543704

Grander, M. (2017). New Public Housing: A Selective Model Disguised as Universal? – Implications of the Market Adaption of Swedish Public Housing. *International Journal of Housing Policy, 17*(3), 335–352. https://doi.org/10.1080/19491247.2016.1265266

Grander, M. (2023). The Inbetweeners of The Housing Markets – Young Adults Facing Housing Inequality in Malmö, Sweden. *Housing Studies, 38*(3), 505–522. https://doi.org: 10.1080/02673037.2021.1893278

Hochstenbach, C. (2017). State-led Gentrification and the Changing Geography of Market-Oriented Housing Policies. *Housing, Theory and Society, 34*(4), 399–419. https://doi.org/10.1080/14036096.2016.1271825

James, L., Daniel, L., Bentley, R., & Baker, E. (2022). Housing Inequality: A Systematic Scoping Review. *Housing Studies*. https://doi.org/10.1080/02673037.2022.2119211

Kallergis, A., Angel, S., Liu, Y., Blei, A. M., Galarza Sanchez, N., & Lamson-Hall, P. (2018). *Housing Affordability in a Global Perspective*. Working Paper WP18AK1. Lincoln Institute of Land Policy.

Kamis, C., Stolte, A., West, J. S., Fishman, S. H., Brown, T., Brown, T., & Farmer, H. R. (2021). Overcrowding and COVID-19 Mortality Across U.S. Counties: Are Disparities Growing Over Time? *SSM – Population Health, 15*, 100845. https://doi.org/10.1016/j.ssmph.2021.100845

Kemeny, J. (1981). *The Myth of Home Ownership: Private Versus Public Choices in Housing Tenure*. Routledge.

Kemeny, J. (1995). *From Public Housing to the Social Market: Rental Policy Strategies in Comparative Perspective*. Routledge.

Kemeny, J. (2006). Corporatism and Housing Regimes. *Housing, Theory and Society, 23*(1), 1–18. https://doi.org/10.1080/14036090500375423

Lawson, J. (2006). *Critical Realism and Housing Research*. Routledge.

Lees, L., Shin, H. B., & López-Morales, E. (2016). *Planetary Gentrification*. Polity.

Murdie, R. (2010). Precarious Beginnings: The Housing Situation of Canada's Refugees. *Canadian Issues*, Autumn 2010, 47–51. https://www.proquest.com/scholarly-journals/precarious-beginnings-housing-situation-canadas/docview/1009046004/se-2

Musterd, S. (Ed.) (2020). *Handbook of Urban Segregation*. Edward Elgar.

Musterd, S., Marcińczak, S., Van Ham, M., & Tammaru, T. (2017). Socioeconomic Segregation in European Capital Cities. Increasing Separation Between Poor and Rich. *Urban Geography, 38*(7), 1062–1083.

Musterd, S., & Ostendorf, W. (1998). *Urban Segregation and The Welfare State: Inequality and Exclusion in Western Cities*. Routledge.

Skifter Andersen, H., Andersson, R., Wessel, T., & Vilkama, K. (2016). The Impact of Housing Policies and Housing Markets on Ethnic Spatial Segregation: Comparing the Capital Cities of Four Nordic Welfare States. *International Journal of Housing Policy, 16*(1), 1–30. https://doi.org/10.1080/14616718.2015.1110375

Stephens, M. (2016). Using Esping-Andersen and Kemeny's Welfare and Housing Regimes in Comparative Housing Research. *Critical Housing Analysis, 3*(1), 19–29.

Stephens, M. (2020a). How Housing Systems are Changing and Why: A Critique of Kemeny's Theory of Housing Regimes. *Housing, Theory and Society, 37*(5), 521–547. https://doi.org/ 10.1080/14036096.2020.1814404

Stephens, M. (2020b). Towards a Multi-Layered Housing Regime Framework: Responses to Commentators. *Housing, Theory and Society*, *37*(5), 584–596. https://doi.org/10.1080/14036096.2020.1816573

Torgersen, U. (1987). Housing: The Wobbly Pillar of the Welfare State. In I. B. Turner, J. Kemeny, & L. Lundqvist (Eds.), *Between State and Market: Housing in the Post-Industrial Era* (pp. 116–121). Almqvist & Wiksell.

World Economic Forum. (2022). *What Has Caused the Global Housing Crisis - and How can We Fix It?* https://www.weforum.org/agenda/2022/06/how-to-fix-global-housing-crisis [Accessed: 30 August 2022].

PART I

Understanding housing and welfare

In this part of the book, we examine the concepts of welfare and housing. This first chapter examines welfare and housing separately, while the next one discusses the relationship between them. Together, they make the case for treating housing as being deeply intertwined with the wider welfare state, based not only on the theoretical literature on welfare and housing systems but also on empirical research on housing inequality. Thus, the two chapters establish the basis for the relationship between housing and welfare employed in the country chapters.

DOI: 10.1201/9781003212690-2

2
UNDERSTANDING HOUSING AND WELFARE

Martin Grander and Mark Stephens

Understanding welfare

Before Esping-Andersen's (1990) ground-breaking book, *The Three Worlds of Welfare Capitalism*, welfare state studies tended to be narrowly conceived, focusing on "the traditional terrain of social amelioration: income transfers and social service" (Esping-Andersen, 1990, p. 2). Esping-Andersen observed that empirically, analysis was often crude and one-dimensional, for example, based on the level of public expenditure on the welfare state. Moreover, theories of welfare states were similarly crude, for example, utilizing one-dimensional convergence theories such as the "logic of industrialism", whereby economic growth generated surpluses that could be diverted into social programmes (Wilensky & Lebeaux, 1958). Esping-Andersen's ambition was "to offer a reconceptualization and re-theorization on the basis of what we consider important about the welfare state" (Esping-Andersen, 1990, p. 2).

One of Esping-Andersen's innovations was to adopt a broad conception of the welfare state. An inquirer with such a broader view is thus interested in the state's role in managing and organizing the economy. He considers issues of "macro-economic steering", as well as meso- and micro-level questions regarding the labour market, employment, and wages, to be integral components of the welfare-state complex. Esping-Andersen calls this approach to studying the welfare state the "Keynesian welfare state" or "welfare capitalism", hence the title of his seminal book.

In the tradition of Esping-Andersen, we also subscribe to such a broad approach to welfare. Hence, we use the notion of *welfare regimes* to signal that we mean something more than the conventional social-amelioration policies traditionally associated with the welfare state. Welfare regimes, as discussed by Esping-Andersen (1990, 1999), are understood as the articulation of welfare programmes and institutions – including the state, markets, and households – insuring households against social risks and therefore promoting and protecting welfare. Such a definition also indicates that the relationship between the state and the market differs between contexts, not only in how traditional social welfare policies are constructed but also in terms of the labour market, education, and other aspects of society.

Esping-Andersen also offers a distinctive explanatory framework for welfare state regimes, differing from the "structural" approaches favoured by many of his predecessors. He suggests that power relations predicated on class and religion create distinctive clusters of welfare regimes – an

DOI: 10.1201/9781003212690-3

approach sometimes presented as a form of "class mobilization" theory. While he does consider the "class character of welfare states" (Esping-Andersen, 1990, p. 33) to be significant, he adopts a more subtle approach to explaining the development of welfare states and their evolution into distinctive types. The form that working-class mobilization takes is important, but so too is the nature of the coalitions that it forges with, for example, rural lobbies or the "new" middle classes to win elections and form governments. Other factors may shape the development of welfare states, but this advanced form of class mobilization, he maintains, explains the differences between them.

Welfare state regimes

But what did Esping-Andersen mean by "welfare state regime"? Although Esping-Andersen does not explicitly define the concept of a welfare regime (or a "welfare state regime", a term he also uses), he does describe it in terms of a relationship between the state and capital: "To talk of 'a regime' is to denote the fact that in the relation between state and economy a complex of legal and organizational features are systematically interwoven" (Esping-Andersen, 1990, p. 2).

In his study of 18 industrialized countries, Esping-Andersen identifies three different welfare state regimes: the liberal, the corporatist (or conservative) and the social democratic. His categorization of welfare state regimes is based on three criteria. First and foremost is the degree of *interplay between the state, market, and family*. By analysing this, Esping-Andersen provides an overview of "qualitatively different arrangements" (1990, p. 26), which in turn create distinctive systems that possess attributes relating to *decommodification* and *social stratification,* which are the second and third criteria.

He defines decommodification as "... the degree to which individuals, or families, can uphold a socially acceptable standard of living independently of market participation" (Esping-Andersen, 1990, p. 37). Stratification is more complex since it is "part and parcel" of welfare states, meaning that social policy may seek to reduce the impacts of stratification but may also be part of a stratifying process (p. 23). Therefore, Esping-Andersen deploys a historical comparative approach and relates social stratification to the core organizing principles of each identified welfare state regime.

Esping Andersen's first regime category is the *Liberal welfare states*, where the archetypical example is the US, but almost always includes the UK and often other Anglophone countries. Liberal regimes are dominated by a low degree of state regulation of labour markets and typically favour means-tested assistance, modest universal transfers, and parsimonious social insurance plans. Social benefits cater mainly to a clientele of low-income state dependents. Esping-Andersen identified this group as being "usually working-class". The progress of social reform in these countries has, according to Esping-Andersen's analysis, been circumscribed by traditional, liberal work-ethic norms, where limited welfare benefits would not provide a sufficient incentive to opt for welfare instead of work. This explains why entitlement rules are strict and often associated with stigma, and benefits are typically modest. In turn, the state encourages the market, either passively – by guaranteeing only a minimum – or actively – by subsidizing private welfare schemes. Governments thus usually ensure limited well-being for those with the lowest incomes, providing partial welfare services through means-testing. In sum, the liberal regime typically seeks to maximize the function of the market while minimizing the state's involvement. There is thus a low degree of de-commodification and high levels of stratification, as they, in their promotion of "residualism", means-testing, emphasize individual responsibility, and prefer the use of the private and voluntary sectors:

> The consequence is that this type of regime minimizes decommodification, effectively contains the realm of social rights, and erects an order of stratification that is a blend of a relative equality of poverty among state-welfare recipients, market-differentiated welfare among the majorities, and a class-political dualism between the two.
>
> (Esping-Andersen, 190, p. 27)

Castles (1994) made the case for Australia being treated separately as a "wage earners' welfare state", with welfare directed at the workforce while leaving those outside it dependent on rather sparce means-tested provision. Nonetheless, Deeming (2017, p. 416), while sceptical of the real-world accuracy of Esping-Andersen's characterization of liberal regimes in 1990, argues that "the liberal world that Esping-Andersen described in 1990 was, in fact, prescient for the one we find today in the English-speaking nations and their welfare states".

In the second category, the *Conservative welfare states*, exemplified by Austria, France, Germany, and Italy, the distribution of social rights is often based on class and status, consolidating divisions among wage earners. In these welfare states, which are characterized as being "corporatist" in their historic legacy (and eventually upgraded to cater to the new "post-industrial" class structure), the liberal principles of market efficiency and commodification never took hold. Consequently, the granting of social rights was hardly ever a seriously contested issue. Rather, what dominated these welfare states was the preservation of a view on rights attached to class and status. The state organization, with its corporatist principles, was thereby to displace the market as a provider of welfare. Private insurance and unemployment benefits, therefore, play a marginal role. However, as the state focuses on upholding status differentials, the redistributive impact of the state's measures is limited, according to Esping-Andersen. Instead, civil society plays a large role. Social policies characteristically maintain social differentiation with individual welfare conditions modified by non-state providers: faith-based communities, trade unions, kinship networks, etc. Corporatist regimes are often influenced by the Church and are hence strongly committed to the preservation of traditional familyhood. The state's family benefits encourage motherhood, while social insurance therefore often excludes non-working wives. Public or private family services such as kindergartens are often underdeveloped, as the state relies on the family's capacity to service its members. Only when such capacity is exhausted can the state interfere, according to the principle of "subsidiarity". Conservative regimes thus ensure both a measure of de-commodification and a high level of stratification, as they to a high degree differentiate between occupational groups.

Lastly, Esping-Andersen places the small group of Nordic countries (Sweden, Denmark, Norway, and Finland) in a specific type of welfare state regime, which he labels *social democratic welfare states*. (The Netherlands is characterized as being hybrid corporatist/social democratic.) The key principle of the social democratic welfare regime is that the state—rather than the market, family, or civil society – is mainly responsible for guaranteeing basic economic security through universal benefits without means-testing. The state dominates welfare provision through universal social rights, which are given to the entire population based on citizenship. The cost of raising a family is also socialized. Esping-Andersen discusses how social democratic regimes produce low levels of benefit differentiation through universalistic principles and a de-commodification of social rights, which were extended to the middle classes (in contrast to liberal regimes). In summary:

> Rather than tolerate a dualism between state and market, between working class and middle class, the social democrats pursued a welfare state that would promote an equality of the highest standards, not an equality of minimal needs as was pursued elsewhere. This

> implied, first, that services and benefits be upgraded to levels commensurate with even the most discriminating tastes of the new middle classes; and, second, that equality be furnished by guaranteeing workers full participation in the quality of rights enjoyed by the better-off.
>
> (Esping-Andersen, 1990, p. 27)

It is important to note that Esping-Andersen did see the welfare state regimes as ideal types and that he makes clear there is no single "pure case". For example, he states that even though the Scandinavian countries are predominantly social democratic, "they are not free of crucial liberal elements" (1990, p. 28). In the same manner, the American social security system can be regarded as redistributive and compulsory and far from actuarial, which is a characteristic of the social democratic regime.

Critique of the welfare state regime theory

Despite remaining the standard reference in welfare state research over the last three decades, the typology of Esping-Andersen has been criticized. One of the points of the critique has been regarding the scope of the study. Esping-Andersen's discussion is based on three ideal-typical welfare state regimes associated with states in North America and Western Europe. The empirical cases contain six English-speaking countries, including the US, the UK, and Australia; five countries in the middle corporatist group, including Germany, France, Finland, Italy (the only southern European country), and Japan (the only Asian country in the study); and six in the social democratic cluster (including Sweden, Norway, Denmark, and the hybrid case of the Netherlands). As Stephens (2016) notes, countries in Central and Eastern Europe, where communism collapsed the year before the book was published, were excluded, as were countries in Africa and Latin America. Except for Japan, there are no Asian countries covered in his analysis. As argued by Stephens (2016, p. 25), the triangle of state-market-family might be fully capable of analyses of other countries and contexts, but "the underlying political philosophy or political economy that is captured by terms such as 'liberalism' and 'social democracy' may be lacking". Therefore, the explanatory power of Esping-Andersen's classification is arguably limited.

The absence of certain regions and states has been a matter of concern for many welfare researchers following Esping-Andersen. Within Europe, a "Southern European" (sometimes referred to as "rudimentary" or "familial") regime has been developed. Here, a protected and highly regulated core labour market is accompanied by a large informal labour market. Access to the formal labour market may depend on clientelism. Poor coverage of social security (other than for pensions) leads to reliance on the family, including a greater proportion of multi-generational households (Allen, 2006). That said, there has been much variety in the welfare regime's development across the southern European countries.

Holliday (2000) argues that particularly southeast Asian countries exhibit "productivist" regimes where social protection is part of a system devoted to economic growth. Clapham (2018) elaborates on the nature of productivist regimes and describes them as occurring when

> Economic objectives largely define social policies, with the application of social security as a means to target economically important interest groups. Priority in policy formation is thus given to enhancing economic and social development, with state, family and market relationships aligned around these objectives.
>
> (Clapham, 2018, p. 38)

This model has become dominant in explanations of social policy and welfare system development in Asian countries, with South Korea as the exemplar, at least until the remarkable transformation of the Chinese economy.

Post-socialist countries in central and eastern Europe have also presented challenges for researchers. The socialist model based on the individual wage system and a high level of in-kind provision delivered through the workplace collapsed along with the Communist regimes. As new systems of social protection have developed, the sudden withdrawal of the state and the shift to market economies have led to a significant role for the family in delivering welfare. As in Southern European countries, in most post-socialist countries there is a high level of multi-generational households (Stephens et al., 2015).

Regime theory has been challenged not only by the expansion of the capitalist world eastward in Europe and in other world regions but also within the original industrialized nations, where welfare regimes have clearly changed since Esping-Andersen wrote his book.

Following the long period of economic growth that lasted for three decades after the Second World War (known in France as *les trente glorieuses*), western economic performance deteriorated. Inflation and sluggish growth following the oil crises in the 1970s gave rise to "stagflation", a loss of confidence in Keynesian economics, and the birth of the supply-side revolution in economics. This rejected the core Keynesian belief that demand management could ensure economic growth and instead posited that improving productivity on the supply side was the only route to rising prosperity. The focus shifted to many of the core institutions of the "broad" welfare state at a time when western economies were deindustrializing and shifting towards economies based more on services. Often, labour markets were reformed. Systems of centralized wage bargaining, associated with corporatist and social democratic regimes, were wound down and abandoned. Centralized wage bargaining tends to raise the floor at the bottom end of the labour market and lower the ceiling at the top. Labour markets were deregulated in favour of "flexibility", giving rise to insecurity and the growth of the "gig" economy. Sometimes increased conditionality was attached to out-of-work social security support. Examples of such welfare regime resets include the Reagan Revolution in the US during the 1980s, followed by welfare reform embraced by Democrat President Bill Clinton in the 1990s. He fulfilled his promise to "end welfare as we know it" by signing the legislation that replaced the main Great Society-era social assistance benefit (Aid for Families with Dependent Children) with a time-limited alternative (Temporary Assistance for Needy Families) (see DeParle, 2004). The Thatcher Revolution in the UK greatly weakened the position of trade unions and deregulated the labour market. In the 1990s, Germany's Social Democrat-led government introduced far-reaching labour market and social security ("Hartz") reforms in an effort to revitalize the undynamic economy with low levels of employment growth. Sweden, an example of the social democratic welfare regime, also abandoned its distinctive Rehn-Meidner economic framework and embraced labour market and social security reform following the economic crisis of the 1990s. In Esping-Andersen's terminology, western countries increased the commodification of labour, with the result that in most (but not all) of them poverty and inequality rose.

Moreover, demographics have changed drastically since 1990. Many of the Western and Nordic European countries that had not previously experienced post-colonial immigration have seen radically increased refugee immigration from Africa and the Middle East, changing their often ethnically homogenous demography. Sweden is perhaps the clearest example, where ethnic segregation has put a strain on the welfare system in terms of social stratification and, as we shall see, housing issues. Nationalist movements have seen increased support, and political majorities in some European countries are dependent on anti-immigration parties. For example, the Brothers of

Italy, whose lineage traces back to Mussolini, are expected to lead the next government in Italy, and Marine Le Pen's National Rally (previously known as the National Front) has been a political force for several decades in France. In Denmark, between 2015 and 2019, *Dansk Folkeparti* was the second-largest party in the parliament (before dropping radially to become the smallest party in the 2022 election). The Sweden Democrats, a party with neo-Nazi roots, have carefully laundered their brand and are today being extremely influential in Swedish national politics. They have yet to become members of the government, but, being the second largest party in the parliament, they have set several conditions for supporting the current right-wing government, many of them connected to immigration.

The principal parties of the centre-left and centre-right – Social Democrats and Christian Democrats in much of Western Europe – that provided ideological coherence to post-war welfare states have lost support in many countries with the rise of the far left and (particularly) far right, sometimes necessitating "grand" coalitions of formerly rival parties or coalitions with Green parties. Populism, with Donald Trump in the US and Viktor Orbán in Hungary being prominent examples, has challenged even the basic ground rules of liberal democracy, while parties representing independence movements have formed governments in Catalonia (Spain) and Scotland (UK). This is not without precedent. Large Communist parties in France and Italy required wide coalitions to exclude them from governments in the 1950s and 1960s. Moreover, path-dependent institutions and cultural attitudes perhaps mean that underlying ideological coherence should not yet be discounted.

The phenomenon of demographic ageing began in high-income countries but is now growing most rapidly in low- and middle-income countries. The World Health Organisation[1] notes that already there are more people aged over 60 than under five. The number of people aged over 60 will double to 2.1 billion by 2050, and those aged over 80 will triple by then. This has implications both for housing and for the operation of welfare regimes that are all linked in one way or another to labour market participation. So-called "dependency ratios" (the ratio of the non-working population to the working population) are set to rise.

Housing researcher Jim Kemeny (1995, 2006) criticized Esping-Andersen's categorization for relying too much on labour movement theory rather than corporatist theory in its analysis of the causes of welfare regimes. This results in it being too loose a shirt, notably because Esping-Andersen draws out two extremes (the social democrat and liberal regimes) and then places virtually all the nations that could not be allocated to these groups in the large middle group of corporatist regimes. We will shortly return to this argument in our discussion about the relationship between welfare and housing, but first we need to consider the meaning of housing and housing regimes.

Understanding housing

Just as with welfare, this book argues for a broad definition of housing. We argue that housing can be described neither as merely the material buildings in which we reside nor as a social construction, reduced to identity or the meaning of having a place called home. Rather, we suggest that housing is a complex object of social relations where material and discursive aspects are regarded as being equally important. Housing is perceived empirically as consisting of physical objects (houses, apartments, etc.), but it also refers to being "housed". Thus, while housing, both as a noun and as a verb (c.f., Ruonavaara, 2018), is a "basic foundation for life" (Clapham, 2018, p. 174), it is also the product of actions performed by housing institutions and individuals – actions that are caused by underlying and interacting social structures, powers, and mechanisms.

As noted in the introduction to this book, Esping-Andersen's study did not include a consideration of the role of housing, and many welfare researchers influenced by him have also refrained from doing so. However, Esping-Andersen does, tantalizingly if somewhat obscurely, note that even subscribers to a "narrow" definition of a welfare state might include "some token mention of the housing question" (Esping-Andersen, 1990, p. 1). Housing allowances, an important social security benefit in some regime types, especially in north-west Europe and Scandinavia, do not merit even such a "token mention" and are excluded in his calculations of the social welfare transfers. Yet the omission of housing has not diminished the attraction to many housing researchers of adopting and adapting the organizing principles of Esping-Andersen's welfare regimes as a basis for explaining housing systems (Stephens, 2016).

In an often-quoted article, Torgersen (1987) identified housing, alongside education, health care, and social security, as one of the four pillars of the welfare state. However, he describes housing as the "wobbly pillar", since housing, to a far greater degree than the other sectors, is a "market product" (Torgersen, 1987, p. 116f.). The equivocal position of housing between the state and market has thereby, according to him, led welfare researchers to omit housing from their studies of welfare states. Indeed, several housing researchers after him have argued that the lack of interest in housing among welfare researchers is due to housing having an ambiguous role within the wider welfare state (Lennartz, 2011; Malpass, 2008). Kemeny, who was a "housing sociologist" argued that, since the housing sector straddles both state and market, "it is probable that vested market interests are more prominent in housing than in other welfare sectors" (Kemeny, 2006, p. 1). Bengtsson, a political scientist, discussed housing as the market commodity within the welfare state and has argued – inspired by Kemeny – that housing [systems] should be defined as "the social, political, and economic organization of production, distribution, and consumption of housing" (Bengtsson, 2015, p. 25, our translation). In line with these scholars, we argue that studies of housing (systems) need an approach that, to a greater degree, takes into consideration the power balance between different interest groups.

Comparative housing research

Attempts to theorize housing from an international comparative perspective have mirrored those more closely focused on welfare states. Thus, David Donnison's (1967) pioneering book, *The Government of Housing,* falls within the convergence tradition of Wilensky by suggesting that housing systems are shaped by the wider economic development of a country (see Stephens & Hick, forthcoming). He identified, on the one hand, southern European countries as exhibiting "embryonic" housing systems with little state intervention and "comprehensive" ones with extensive state intervention on the other. Sweden and the Netherlands were identified as having comprehensive housing systems, whereas Anglophone countries (the US, the UK, and Australia) fell into an intermediate grouping where state intervention was targeted. Donnison envisaged that countries would gravitate towards the comprehensive model as economies developed, but by the time a revised edition of the book was published (Donnison & Ungerson, 1982), he was forced to conclude that the consensus in favour of improvement through state intervention no longer existed (Stephens & Hick, forthcoming). The convergence school continued after Donnison, with Harloe's (1995) book, *The People's Home*, linking the governments' willingness to support mass public housing programmes to the rise of Fordism and the winding down of such programmes to the fall of Fordism as advanced economies entered their post-industrial phase.

Since then, many accounts of the development of housing systems implicitly adopt a convergence framework, citing powerful structural forces, notably globalization (which, *inter alia,* weakens the position of labour, leads to the competitive erosion of corporate tax bases, and integrates financial markets) and neoliberalism (which provides the ideological justification for welfare state retrenchment). Clearly, these forces cannot be ignored as drivers of welfare regime or housing system change, but the continued diversity of welfare regimes and housing systems points to the importance of the distinctive national institutional structures through which they are mediated. For example, the GFC represented a worldwide exogenous shock, but the way it was mediated through the financial and housing systems differed greatly (see Forrest & Yip, 2011).

Kemeny's theory of housing regimes

Esping-Andersen's typology of welfare regimes profoundly affected the way in which comparative housing research was conducted. Kemeny and Lowe (1998) criticized the way in which much comparative housing research was overly descriptive and even fell short of being truly comparative (rather, it was "juxtapositional"), and comparative scholars became more aware of the need to adopt a more theoretical approach. Esping-Andersen's typologies have been frequently applied to housing, assuming that they can be "read across" (Stephens, 2016). There is some logic to this in the sense that one might expect the causes of welfare regimes to be much the same as those of housing regimes, and if the welfare regime sets a high store on decommodification, then one might expect the housing system to do the same. However, Stephens (2016) argues that this cannot be simply assumed, and the question as to whether the housing system replicates, reinforces, or counters the distributional tendencies of the welfare regime is ultimately an empirical and outcome-based one. Moreover, scholars should be wary about *assuming* distributional outcomes. Housing that is decommodified in the sphere of production, for example, may not be decommodified in the sphere of consumption. In the Soviet bloc, subsidized housing built by the state or state-owned enterprises was used to reward managers and party officials in lieu of relatively flat wage structures. Hence housing that was decommodified in the sphere of production was used to recommodify labour in the sphere of consumption (Stephens et al., 2015). Scholars therefore have good reason to be wary about simply "reading across" from Esping-Andersen, for they risk not just what Kemeny called "mindless classification" (2001, p. 61), but profoundly misunderstanding the housing system *and* its relationship with the welfare regime.

A parallel field of scholarship has grown up and become known as the "housing-welfare regime" framework, of which Kemeny was the founder and remains its "most influential" (Blackwell & Kohl, 2019, p. 1447). His thesis was developed in a series of five key publications (a book and four articles) published between 1995 and 2006. His starting point is sociological. He advocates a "middle-range" theory that synthesizes empiricism and theory to deduce how housing systems arise from cultural, ideological, or political dominance (Kemeny, 1995, p. 39). The empirical objects for the analysis consist of political and economic policies and processes. He discusses the housing market as part of the welfare state at large and relates different housing regimes to Esping-Andersen's (1990) welfare regimes. The housing market is considered by Kemeny (2006) to be a special dimension of the welfare state where a specific sector of stakeholders has a specific role, a so-called "sector regime".

In its initial iteration, Kemeny's thesis suggested that the design of what he called the cost rental sector could shape the rental sector as a whole. Later, he argued that it could shape the entire housing system through the relationship between rental and ownership sectors and by shaping the

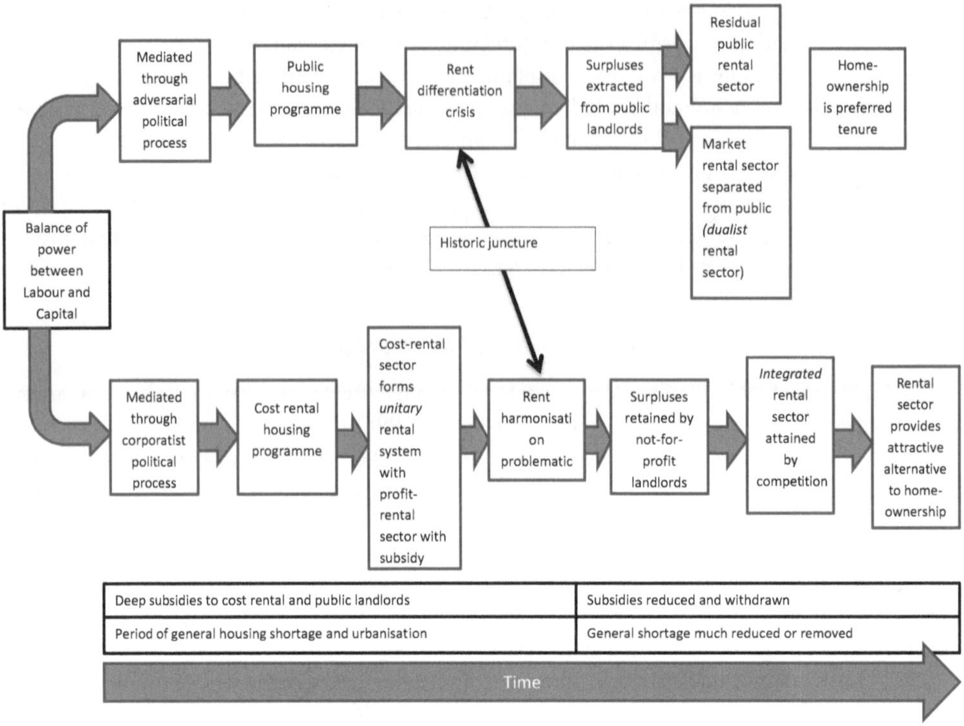

Figure 2.1 Evolution of dualist and unitary/integrated housing systems.
Source: Stephens (2020a)

wider political economy of housing. He links the rise of different rental markets (labelled unitary, integrated, and dualist rental markets; see below) to the development of different welfare states and how the balance of power between different interest groups in the various sectors of the welfare state has evolved over time.

The theory was developed from and applied to western advanced economies in the decades following the Second World War, when the housing shortage was evident in these countries (see Figure 2.1). In most countries, in response to the crisis, some form of state-subsidized housing provision programme was developed that addressed all or parts of the population. How these programmes developed and how extensive they became depended, according to Kemeny, on the political balance of power between capital and labour and, above all, on how this balance developed in corporatist/consensual (for example, in Germany) and adversarial political structures (for example, in the UK). Kemeny argues that during the post-war economic recovery, including the stabilization of the housing supply, different countries came to act in different ways in terms of housing supply, depending on the relationship between state and market and the relationship between labour and capital.

The rise of the unitary/integrated rental markets

The development of the corporatist welfare states, above all the German one, with origins in what are often called "social market economics", is assumed by Kemeny to be the basis for the balance between state and capital in the so-called *integrated rental market*. This type of market economy,

Kemeny suggested (1995, p. 11), attempts to construct markets by pointing to a balance between economic and social priorities and thereby mitigating the undesirable effects inherent in the market economy. In this way, the foundations can be laid for a rental market where public and private actors compete to achieve a balance between economic and social values.

Kemeny argued that unitary regimes, which he later evolves into integrated rental markets (see below), arise from a strategy based on "corporatist" political collaboration processes and lead to subsidized and cost-based rental market programmes, what Kemeny calls "cost rental housing programmes". (In most countries, cost-rental sectors would coincide with social rented sectors, but in Sweden's special case, the term "socially rented" is rejected because of its connotations of stigmatized public housing in the UK and US.) An interesting development in Kemeny's terminology emerges here. In his 1995 book, he uses the terms "unitary" and "integrated" synonymously. However, in later texts (especially Kemeny et al., 2005), he distinguishes between the two, depending on how rent-setting principles have developed in the relationship between the for-profit rental sector and the publicly controlled one.

During the initial years of subsidized housing programmes in corporatist states, according to Kemeny, a unitary rental market is achieved through a combination of, on the one hand, subsidies to the publicly regulated rental sector, which is governed by a cost principle, and, on the other hand, regulation of rents in the for-profit rental sector. In this formative phase, the cost-driven sector needs to be subsidized while the for-profit sector is regulated so that the former can compete with the latter and thus create what Kemeny terms a unitary rental market. (Kemeny assumed that established for-profit landlords would have lower costs than rental housing built with recently acquired debt.)

As housing shortages are met, the cost rental housebuilding programme is wound down, and cost rental landlords' debts begin to diminish, leading to rents generating surpluses – a process known as "maturation". According to Kemeny, this creates a problem with "harmonizing" the rents between the two sectors. The cost-rental sector exerts pressure to keep rents down in the market-driven sector, which leads to the subsequent abandonment of crude rent control. Here, Kemeny believes that a "historical juncture" occurs and the integrated rental market emerges, which would thus be a development from the unitary rental market. In the integrated rental market, rent control can be abandoned, although some other government interventions may remain. For example, a "second generation" rent regulation that limits rent increases can replace previous blunt forms of rent control (Kemeny et al., 2005, p. 859). According to Kemeny, without some form of rent regulation, there would be no conditions for competitive rents in the publicly controlled rental sector, and thus no competition between the two sectors. Kemeny further believed that strong tenure protection was a prerequisite for such competition.

An integrated rental market creates a form of social market that, with security of tenure, good-quality housing, and competitive rents, can compete with homeownership. The rental market thus attracts all income groups, and, in this way, a "cost-rental society" arises, which is characterized by competition between different forms of letting for all types of households, which is contrasted with a "home-ownership society", where owned accommodation is seen as more desirable. In this way, Kemeny (1995, p. 38) argues that the construction of the rental market has repercussions on the entire housing system, and this is why his theory of dualistic and integrated rental markets is also part of a broader theory of housing regimes.

Kemeny argued that the German principle of corporatism, which was thus based on "a system of cooperation and compromise between capital and labour, orchestrated by the state" (Kemeny, 1995, p. 65), spread to Germany's neighbouring countries, including Austria and the Netherlands, but also to northern Europe, not least Sweden and Denmark, and that integrated rental markets, therefore, came to develop in these countries. This is why (as mentioned above) Kemeny believes that the Nordic, social democratic countries should in fact be included in the family of "corporatist welfare regimes", rather than categorized as social democratic welfare regimes as Esping-Andersen (1990) does in his schematization. In later works, Kemeny (2006) identifies six countries: Sweden, the Netherlands, Germany, Switzerland, Austria, and Denmark, as countries with integrated rental markets. He also claims that France is a possible integrated market.

The rise of dualist rental markets

The rise the dualist rental markets is similarly linked by Kemeny to the development of the balance between capital and labour in what he terms adversarial political structures, with the UK as the typical example. The post-World War II subsidized housing programmes arose in the UK through adversarial rather than corporatist political relations. The consequence of such relationships was, according to Kemeny (1995), that the profit-driven rental market created acute problems for households with limited finances regarding access to housing, which in turn actualized the need for a public safety net in the form of a cost-based rental sector. This sector was kept separate from the profit-driven sector to "protect the market" (Kemeny, 1995, p. 9, 2006, p. 3). The rental sector was "socialized" and controlled through a set of bureaucratic rules, usually by means-testing and/or income ceilings, which would limit access to public rental housing for better-off households.

As the cost-rental sector develops and matures, its rents fall relative to those in the for-profit sector, resulting in what Kemeny calls a "rent differentiation crisis" (1995, p. 9). In a similar way as in the integrated rental sector, a historical juncture arises here. The state chooses not to allow rental surpluses to remain within the sector but instead finds ways to extract them, for example through the discounted sale of properties to tenants, boosting homeownership. The dualist rental market now arises because the state tightens control of the cost rental sector, which includes a forced increase in rents through regulation and a ban on reinvesting profits in rental housing. Kemeny likens such a system to the Soviet-style "command economy".

The result is that the cost-based rental sector becomes residual as the supply of housing in this sector becomes limited, of lower quality, and available only to low-income households. The room for manoeuvre for the landlords in this sector becomes small, and the rents are tightly regulated or controlled in relation to the rents in the profit-driven rental market. As access to the cost-controlled rental sector becomes limited and the for-profit rental sector becomes unattractive due to high rents, lower quality, and weak or non-existent security of tenure, there is a greater demand for owner-occupied housing.

This is where Kemeny takes the step from the rental market to the housing regime in general. Due to the problems in the rental market, the states find themselves forced to subsidize owner-occupied housing to enable access for lower-income groups. Tenants are offered the opportunity to buy their properties and homes at discounted prices, of which the UK is an example. Here, according to Kemeny, the foundation is thus laid for an ownership regime, a "home-ownership society", where households typically strive – and are guided by the government – towards owner-occupied housing.

Kemeny identifies Great Britain (especially England), Ireland, the US, Canada, Australia, and New Zealand as dualistic rental markets and home-owning societies. The Finnish model is also taken as an example and is seen by Kemeny as a clear contrast to the rental market that has developed in the neighbouring corporatist countries of Sweden and Denmark.

Therefore, Kemeny uses his model of the development of two-fold rental housing systems to question the utility of Esping-Andersen's categorization of welfare regimes (as Sweden and

Denmark have integrated rental markets, whereas Finland and Norway have dualist rental systems), arguing that "the Nordic countries do not comprise a uniform category as regards rental housing, contrary to what Esping-Andersen has claimed for the welfare state as a whole" (Kemeny, 2006, p. 6).

Conclusion

This chapter has introduced and summarized the two principal articulations of regime theory as applied to the welfare state and to housing, as well as the key studies that preceded them. It has also highlighted the political and economic context in which these regimes were created in the 20th century and how this has changed under the pressures of economic globalization and political fragmentation. This chapter also demonstrates how regime theory has bifurcated between housing and the wider welfare state, and this inevitably inhibits our ability to understand the relationship between them, particularly in a changing world. It is to this relationship that we turn in the next chapter.

Note

1 World Health Organization *Ageing and Health* https://www.who.int/news-room/fact-sheets/detail/ageing-and-health.

References

Allen, J. (2006). Welfare Regimes, Welfare Systems and Housing in Southern Europe. *European Journal of Housing Policy*, *6*(3), 251–277. https://doi.org/10.1080/14616710600973102

Bengtsson, B. (2015). Allmännyttan och Bostadspolitiken i går, i dag och i Morgon – Institutionella Förutsättningar i Förändring [Public Housing and Housing Policy Yesterday, Today and Tomorrow—Institutional Conditions in Change]. In T. Salonen (Ed.), *Nyttan med Allmännyttan* [The benefit of Public Housing] (pp. 25–48). Liber.

Blackwell, T., & Kohl, S. (2019). Historicizing Housing Typologies: Beyond Welfare State Regimes and Varieties of Residential Capitalism. *Housing Studies*, *34*(2), 298–318. https://doi.org/10.1080/02673037.2018.1487037

Castles, F. G. (1994). Comparing the Australian and Scandinavian Welfare States. *Scandinavian Political Studies*, *17*(1), 31–46.

Clapham, D. (2018). *Remaking Housing Policy: An International Study*. Routledge.

Deeming, C. (2017). The Lost and the New 'Liberal World' of Welfare Capitalism: A Critical Assessment of Gøsta Esping-Andersen's The Three Worlds of Welfare Capitalism a Quarter Century Later. *Social Policy and Society*, *16*(3), 405–422. https://doi.org/10.1017/S1474746415000676

DeParle, J. (2004). *American Dream. Three Women, Ten Kids and a Nation's Drive to End Welfare.* Viking Penguin.

Donnison, D. V. (1967). *The Government of Housing.* Penguin.

Donnison, D. V., & Ungerson, C. (1982). *Housing Policy.* Penguin.

Esping-Andersen, G. (1990). *The Three Worlds of Welfare Capitalism.* Princeton University Press.

Esping-Andersen, G. (1999). *Social Foundations of Postindustrial Economies.* Oxford University Press.

Forrest, R., & Yip, N.-Y. (Eds.) (2011). *Housing Markets and the Global Financial Crisis: The Uneven Impact on Households.* Edward Elgar.

Harloe, M. (1995). *The People's Home? Social Rented Housing in Europe and America.* Wiley.

Holliday, I. (2000). Productivist Welfare Capitalism: Social Policy in East Asia. *Political Studies*, *48*(4), 706–723. https://doi.org/10.1111/1467-9248.00279

Kemeny, J. (1995). *From Public Housing to the Social Market: Rental Policy Strategies in Comparative Perspective.* Routledge.

Kemeny, J. (2006). Corporatism and Housing Regimes. *Housing, Theory and Society*, *23*(1), 1–18. https://doi.org/10.1080/14036090500375423

Kemeny, J., Kersloot, J., & Thalmann, P. (2005). Non-Profit Housing Influencing, Leading and Dominating the Unitary Rental Market: Three Case Studies. *Housing Studies*, *20*(6), 855–872. https://doi.org/10.1080/02673030500290985

Kemeny, J., & Lowe, S. (1998). Schools of Comparative Housing Research: From Convergence to Divergence. *Housing Studies*, *13*(2), 161–176. https://doi.org/10.1080/02673039883380

Lennartz, C. (2011). Power Structures and Privatization Across Integrated Rental Markets: Exploring the Cleavage Between Typologies of Welfare Regimes and Housing Systems. *Housing, Theory and Society*, *28*(4), 342–359. https://doi.org/10.1080/14036096.2011.562626

Malpass, P. (2008). Housing and the New Welfare State: Wobbly Pillar or Cornerstone? *Housing Studies*, *23*(1), 1–19. https://doi.org/10.1080/02673030701731100

Ruonavaara, H. (2018). Theory of Housing, From Housing, About Housing. *Housing, Theory and Society*, *35*(2), 178–192. https://doi.org/10.1080/14036096.2017.1347103

Stephens, M. (2016). Using Esping-Andersen and Kemeny's Welfare and Housing Regimes in Comparative Housing Research. *Critical Housing Analysis*, *3*(1), 19–29. https://doi.org/10.13060/23362839.2016.3.1.250

Stephens, M., & Hick, R. (Forthcoming). Comparative Housing Research. In K. Jacobs, K. Flanagan, J. Verdouw, & J. de Vries (Eds.), *Research Handbook on Housing and Society.* Edward Elgar.

Stephens, M., Lux, M., & Sunega, P. (2015). Post-Socialist Housing Systems in Europe: Housing Welfare Regimes by Default? *Housing Studies*, *30*(8), 1210–1234. https://doi.org/10.1080/02673037.2015.1013090

Torgersen, U. (1987). Housing: The Wobbly Pillar of the Welfare State. In B. Turner, J. Kemeny, & L. Lundqvist (Eds.), *Between State and Market: Housing in the Post-Industrial Era* (pp. 116–121). Almqvist & Wiksell.

Wilensky, H. L., & Lebeaux, C. N. (1958). *Industrial Society and Social Welfare. The Impact of Industrialization on the Supply and Organization of Social Welfare Services in the United States.* Russell Sage Foundation.

3

UNDERSTANDING THE RELATIONSHIP BETWEEN HOUSING AND WELFARE

Martin Grander and Mark Stephens

The last chapter introduced the concepts of welfare and housing regimes, examined their evolution and summarized the most influential iterations of them. However, the study of housing regimes became somewhat disjointed from the study of wider welfare regimes, inhibiting understanding of the relationship between them. This chapter explores the distributional implications of welfare and housing regimes and the relationship between them. It also examines some of the critiques of Kemeny's theory of housing regimes and of regime theory in general. More recent variations on regime theory (the varieties of residential capitalism and growth regimes) as well as the phenomenon of financialization are also introduced. The chapter closes by outlining a framework for analysing housing systems within the context of wider welfare regimes and the macroeconomy.

Universality and selectivity in housing policy

Closing in on the relationship between housing and welfare, the central aspects of Kemeny's categories of housing regimes are the opposing principles of universalism and selectivity. Such principles – relating to whether (welfare) policies and measures are directed to the whole population (universal) or to those in greatest need (selective) – are directly transferable to the question of housing provision. There, the principles could be related both to the notions of the "right to housing" (Bengtsson, 2001) and stratification, i.e., how housing policy and housing systems differentiate between social groups, for example through means-testing, and thereby, in various degrees, promote residualism in housing.

A much-debated question among housing scholars and activists is the right to housing. Such a right is a universal and unconditional right and is included in several international treaties, notably the International Covenant on Economic, Social and Cultural Rights, under which there is a "right of everyone to … adequate … housing" (Article 11(1)). The right to housing also appears in many countries' constitutions (for example, Portugal, Spain and the Netherlands). However, it is seldom possible for an individual to use this "right" to secure housing for themselves because of the lack of any mechanism with which to enforce it. Instead, implicit in the view of housing as part of the wider welfare state is that the state intervenes in the market in order to ensure such a right, for example through a social or public rented housing sector that provides dwellings for economically disadvantaged households. Such a view of the role of housing in relation to the welfare state is characteristic

DOI: 10.1201/9781003212690-4

of a *selective* housing policy. In selective housing policies, which could be found in the dualist rental markets, the state is responsible for providing for households of lesser means alongside the general housing market. In contrast, a *universal* housing policy, characterizing countries with unitary and integrated rental markets, is dependent on the relationship between the state and the housing market actors in providing housing for all households. A universal housing policy is characterized by the state being expected to correct the general housing market so market actors can provide housing for all types of households, regardless of their economic situation (Bengtsson, 2001).

While a selective housing policy and a dualist rental market link to the liberal welfare regime in Esping-Andersen's typology, a universal housing policy and an integrated market could be clearly linked to the universal principles of the social democratic welfare regime. However, where there is a shortage of cost rental housing (as there invariably is), access to it must be rationed in some way. "Housing for all" means that everyone is in principle eligible, but it is who actually gets it that is relevant. The lines of inclusion and exclusion – what we might call stratification – must be drawn, and they are not always drawn in favour of those least able to access adequate housing through the market. As Esping-Andersen (1990, p. 23) observed, "it [the welfare state] is, in its own right, a system of stratification. It is an active force in the ordering of social relations." The same is true of cost rental housing.

Housing regimes' ability to counteract inequality

In 1990, Esping-Andersen wrote, "Equality has always been what welfare states were supposed to produce, yet the image of equality has always remained rather vague" (1990, p. 5). As discussed earlier, we here need to acknowledge that the welfare state both creates and dampens social stratification and that this is clearly connected to the degree of (de)commodification, i.e., how households can uphold a socially acceptable living standard independently of labour market participation. Some welfare regimes are built on the assumption that welfare benefits are the basis for diminishing inequalities, while others use them to maintain status differences. In other regimes, there has been a greater focus on the overall distribution of income in order to eradicate poverty.

To connect this to the housing issue, we need to examine, theoretically, how housing regimes replicate, exacerbate, or reduce inequalities produced by the wider welfare regime. In other words, does the housing system intensify or diminish inequalities arising from labour market income and redistributed by taxation and social security?

Theoretically, access to housing needs to be examined in relation to household income. If access to housing were directly related to household income, the level of housing consumption would depend on current income. Hence, the wider welfare regime would have a pivotal role in its distributional tendencies. If a housing system consisted solely of housing rented at market rents, inequality in current income would be directly replicated in housing consumption. However, housing distribution is a more complex issue, which is why the logic of market-based provision, welfare redistribution, and state correctives in different contexts needs to be empirically examined to establish whether housing regimes replicate, exacerbate or reduce inequalities produced by the wider welfare regime. Stephens and Van Steen (2011), for example, used the concept of "housing income" (which is similar to economists' notion of imputed rental income but also includes the value of tax breaks and housing allowances) to test the distributional impact of the housing system. They found that shortly before the GFC in England and the Netherlands, the housing system did reduce levels of poverty compared to cash income. However, it was equally clear that changes to the housing system – within the market as well as government policy – could produce different outcomes. Research from Sweden has shown that the rental market, with its combination of universal public housing and collectively negotiated rents (what Enström Öst et al. (2014) call "soft" rent control), has a redistributional character as low-income renters living in attractive areas of cities enjoy relatively low housing costs (thus counteracting socioeconomic segregation). On the other hand, such rent control in combination with universal public housing also has the capacity to subsidize well-off households, as public housing is allocated with regard to time spent in the housing queue, not according to means.

Our forthcoming analysis will therefore consider the features of housing systems that may reduce inequalities, which may include (a) social housing or other forms of below-market rents; (b) housing allowances; and (c) outright/unmortgaged ownership. (The latter case suggests that the market can become a source of decommodification since outright ownership can yield a high level of imputed rental income with little outlay from current income.) In the same vein, we will explore what features of housing systems may increase inequalities. Such features include (a) unequal access to ownership/mortgage credit; (b) inflationary housing markets (e.g., due to monetary policy); and (c) unequal access to "decommodifying" instruments.

Developing Kemeny's housing regimes

As noted, Kemeny developed the most influential iteration of the housing-welfare regime approach. Even though it has been almost three decades since it was first advanced, it retains this status. However, it should be, and to an extent has been, questioned, and it is important to consider whether the framework remains and indeed the housing-welfare regime approach remains useful, with or without refinement. In this section, we examine some of the critiques that have been applied to Kemeny's and the wider housing-welfare regime framework. Building on such a critique, we reflect on whether regime approaches continue to provide a suitable theoretical framework for understanding the relationship between housing and the wider welfare state and suggest a developed analysis using a multi-layered housing regime framework.

Interdependency with changing wider welfare regimes

Stephens' (2020a) Focus Article, published in *Housing, Theory and Society*, provided a critical analysis of Kemeny's thesis, using the case studies of Sweden, Germany and the UK to examine whether Kemeny's typology was ever valid and, if so, whether it remained so.

Stephens argued that Kemeny conflated two dynamics, namely whether the rental markets were dualist or unitary/integrated with their impact on the whole housing system. He argued that a unitary market had been created in Sweden, and in Germany, it had approached an integrated one in the 1990s. In both cases, these influenced the nature of the whole housing system. However, in the case of the UK, while a dualist rental market had been created, it was not the principal driver of the housing system, which he attributed to the polarization between homeownership and social (mainly public) rented housing, with the private rented sector marginalized to the point of becoming peripheral.

When examining whether these housing systems had changed, Stephens noted that the rise of mortgaged homeownership in Sweden indicated that the unitary housing market had lost its salience. While much of the institutional structure of the unitary housing market (such as the corporatist approach to rent setting, "soft" rent regulation and tenure security) had been maintained, marketization of the tenant-co-operative sector, privatization in the municipal housing sector, and the deregulation of the mortgage market created what Christophers (2013) labelled a "monstrous hybrid." Moreover, Grander (2017, 2023) uses the Swedish case to show how the categorization of rental markets into dualist and integrated rental markets is both misleading and outdated empirically. While the universal approach of Swedish public municipal housing has remained ideologically close to its origins in 1945, marketization, starting in the aftermath of the 1980s and 1990s financial crises and the eventual adoption of European legislation on competition in 2011, has drastically tilted the playing field for public housing. Public housing is still universal in principle, but in practice it is increasingly selective, albeit in a reversal of the traditional meaning. In search of minimizing financial risks, public housing companies increasingly turn to better-off households by tightening minimum income requirements, some even excluding applicants whose low incomes mean that they qualify for housing allowance. Thus, "reverse selectivity" occurs as municipal housing companies increasingly act in the same way as their private sector counterparts. In parallel, the public housing companies are cooperating with welfare institutions (municipal social services) that rent apartments from the public housing companies and sublet them to vulnerable households (the social service pays the rent). This, Grander argues, leaves a "middle group" of young adults and households with low and medium incomes excluded from all subsectors of the housing market.

Stephens (2020a) also noted that with markets tightening, Germany had retreated from an integrated rental market, with pressures on rents leading to a reversion to more onerous forms of rent regulation. Nonetheless, regulation maintained a unitary rental market that remained capable of competing with homeownership, as evidenced by the broad range of income groups within the rental sector, the subdued levels of home ownership and the apparent lack of appetite for mortgages.

Stephens (2020a) and Grander (2017) also examined *why* housing systems were changing. In both Sweden and Germany, a factor undermining the unitary/integrated rental markets is the weakening of the supply of cost rental housing. Both countries experienced the loss of cost rental housing due to different forms of privatization: in Germany, the sale of some municipal stocks to institutional investors; in Sweden, a limited form of sales to tenants to form ownership cooperatives; and, during the last decade, increasing financial demands that have made the municipal housing companies dependent on property sales (often to international investors) to finance new housing production. Privatization is, of course, a strategy associated with dualist rather than unitary systems. Moreover, inherent to the German cost-rental system is the provision for cost-rental housing provided by private landlords to "melt away" into the market sector once subsidized loans have been repaid, causing the sector to shrink. Kemeny assumed that surpluses generated by cost rental sectors in unitary rental markets would form the basis of replacement and expansion, even when subsidies had been much reduced or withdrawn. In the cases of Sweden and Germany, empirically, this did not occur, suggesting a degree of overoptimism in the original thesis.

Stephens (2020a) also points to qualitative changes in housing systems that arise from their relationship with what he calls the "wider welfare regime," that is, the labour market, tax and social security systems that have distinctive distributional tendencies. The UK, Sweden and Germany all reformed their wider welfare regimes in response to economic crises. In the UK, the emergency loan from the IMF in 1976 heralded a sea change in politics, leading to the deregulation of the labour market, tax cuts and a less solidaristic social security system, all contributing to a rise in poverty. In Sweden, the banking crisis of the early 1990s also led to a decisive shift away from the social democratic economic and social model, involving both labour market and social security reforms. Poverty and inequality in Sweden have risen, too, which we will come back to. In Germany, during the economic malaise of the 1990s, when employment growth lagged behind the US, the

Social Democrat-led government also introduced labour market and social security reforms (Hartz IV) that have accelerated the demise of the "male breadwinner" characteristic of the corporatist labour market as female employment has risen, along with so-called "mini-jobs."

Kemeny characterized the housing system as a distinct "sector regime," with his views on its dynamic connection with the wider welfare regime relating to the now rather dated debate about home ownership and social security. The argument was that high levels of homeownership lead young families to demand lower taxes due to the front-end-loaded costs of homeownership, leading to low levels of social security, which in turn reinforces the demand for homeownership in the hope that it can provide low-cost housing in retirement (assuming the mortgage has been repaid by then). This can be called an "old" model of asset-based welfare, which is now complemented by a "new" version whereby deregulated mortgage markets allow equity to be released as an income stream from unindebted properties.

Stephens (2020a) suggests a more subtle yet causally vital relationship between the wider welfare regime and the housing system. He argues that a wider welfare regime that produces a low level of poverty and inequality is a prerequisite for a unitary or rental market to operate inclusively. Low levels of poverty and inequality necessarily imply a small population with limited or no capacity to make market-like choices. Hence, they allow a "housing for all" approach to accessing cost rental housing (e.g., no income limits) *and* for it to operate as a social market, with tenants able to respond to price signals and exercise market-like choices. In contrast, high levels of poverty and inequality imply a larger pool of people with limited or no financial capacity to exercise market-like choices. It was the rise in poverty and inequality in the UK in the 1980s, combined with reductions in the size of the social rented sector, that led to increased targeting of social housing, the growth in homeownership for those who could afford it, and, of course, tenure polarization. Social tenants as a group became more dependent on social assistance/housing allowances to pay even below market rents and consequently could not exercise market-like choices, with Kemeny likening the system to a "command economy." In countries such as Sweden, faced with growing levels of poverty and inequality, the maintenance of a "housing for all" approach by public municipal housing companies is colliding with the European Union's demands on competition, leading to "reverse selectivity" (Grander, 2017) of the housing regime described above. The progressive outcomes of unitary systems assumed by Kemeny are consequently undermined, with lines of inclusion and exclusion stratifying access in such a way as to limit housing's decommodifying potential.

Within this account lie the drivers of mortgaged homeownership. Countries including the UK (from the dualist family) and Sweden (from the unitary family) deregulated their mortgage markets in the 1980s and subsequently experienced the growth of mortgaged home ownership. The power of cheap finance was intensified after the GFC when western monetary policy shifted towards ultra-low interest rates supported by Quantitative Easing (QE) – the acquisition of bonds by central banks "printing" electronic money. During the GFC, QE was aimed at purchasing bonds from banks to free up balance sheets to lend for the purchase of private sector assets (including housing). During the pandemic, attention shifted towards purchasing government bonds, which many would argue was a form of deficit financing that also exerted downward pressure on interest rates. Monetary policy supported house prices and eventually began to limit access to homeownership, creating additional pressures on rental markets. Yet in Germany, similar conditions were met by what Stephens (2020a) attributes to a cultural resistance to mortgage debt. This implies that convergent drivers of housing system change, such as monetary policy, are mediated through nationally distinct institutions and therefore can produce divergent outcomes.

Varieties of residential capitalism and financialization

The development depicted above speaks to much of the financialization literature, most closely associated with Manuel Aalbers (2016). Financialization may be attributed to the deregulation of national mortgage markets since the 1980s, the fall in interest rates attributable to the low levels of inflation and the release of excess savings derived from the integration of China into the world economy (what Aalbers (2016) refers to as the "global wall of money"), and more recently to "unconventional" monetary policies QE operated by central banks. Aalbers' (and others') studies demonstrate how financialization can take many forms, extending beyond the simple model of mortgaged homeownership to the varieties of institutional investment that became interested in German municipal rental housing (Wijburg & Aalbers, 2017), the importance of super-rich international individuals investing in properties in "world" cities (Paris, 2017) and more recently (based on the Netherlands) the importance of wealth as well as finance as a driver as older households become small-scale landlords (Ronald et al., 2015; Byrne & Norris, 2019; Hochstenbach & Aalbers, 2023).

Financialization is therefore highly variegated, and while it is easily understood intuitively, it lacks a common definition. Schwartz (2020, p. 503) provides one definition of financialization, which is "the transformation of more and more income streams into asset streams…" Moreover, it is an example of the "dependent variable problem" that is widely recognized in welfare state literature but neglected by housing studies (Hick & Stephens, 2023). This means it is difficult to measure its effects, which is one of the problems with one of the most significant contributions to the literature, namely Schwartz and Seabrooke's (2008) positioning of financialization within a regime and typology framework. This contrasts with the bulk of financialization literature, which examines it essentially as an independent variable so that it can be represented as being

> A form of 'soft' convergence in which an exogenous global force, or [Aalbers' (2016]) 'general mechanism' from outside the housing system (the 'global wall of money' [Aalbers 2016, p. 139]), is mediated through distinctive institutional structures which he [Aalbers] calls 'specific conditions... at work in different housing markets...' (ibid.).
>
> (Stephens, 2020a, p. 523)

Hence the importance of Schwartz and Seabrooke's conceptualization of financialization as a shaper of "varieties of residential capitalism" – so placing housing within the Varieties of Capitalism typology used to explain different economic models (rather than welfare states). They operationalize financialization by comparing the relative size of mortgage debt with the relative size of owner-occupied tenure in more than 25 (almost all advanced) economies. Countries with high levels of mortgage debt and high levels of owner-occupation were placed in a Liberal Market grouping; those with high levels of mortgage debt but low levels of owner-occupation were labelled Corporatist Market; those with low levels of mortgage debt and low levels of owner-occupation were State Developmentalist; and those with low levels of mortgage debt and high levels of homeownership were Familial.

Interestingly, this placed most of the countries regarded as having unitary rental markets in the corporatist market group (Germany, Sweden, the Netherlands, and Denmark). Similarly, the Liberal Market group is made up of only Anglophone countries with dualist rental markets (e.g., the US and UK), with the exception of Norway. The familial grouping contains both southern European countries (Spain, Italy and Portugal) and a number of formerly socialist countries from central and eastern Europe (e.g. Slovenia, Slovakia, Hungary and Russia). The smallest group, those countries combining low levels of mortgage debt with low levels of owner-occupation, is labelled state developmentalist and includes Austria, France, the Czech Republic and Japan. However, this operationalization raises some methodological concerns. The chosen benchmarks are the percentage point deviation from the OECD averages over the 1992–2002 period, which gives rise to concerns that include variation within the period, the suitability of a relative benchmark, and the countries included (and excluded) from the benchmark, which in any case do not match those to which it is applied.

Nonetheless, the authors succeeded in demonstrating that "…residential property markets must be included as both a source of knowledge on comparative political economy through studying different systems as well as within international political economy to trace the diffusion of policy and ideas" (Schwartz & Seabrooke, 2008, pp. 258–259). More recently, Reisenbichler (2021) linked the financialization of housing to a typology of "growth regimes" that have emerged as economies have transitioned away from Fordist models. He distinguishes between countries that have embraced financialization and those that have resisted it. He suggests that countries such as the US and the UK have actively promoted house price growth through mortgage credit, partly to compensate individual households for the reduced generosity of social welfare programmes but also as the basis for stimulating consumption as the basis of economic growth. These may be contrasted with price-sensitive export-oriented economies based on manufacturing, notably Germany, where housing finance is constrained. The highly financialized Nordic countries are placed in an "intermediate" category on the basis that "dynamic" housing markets neither reinforce nor contradict growth regimes that are based on high-tech industries.

However, much as it rings true, Resienbichler's account faces some tensions. He suggests that the financialized countries promote housing through instruments such as tax breaks and guarantees and contrasts this approach with Germany's decision to remove the main tax break for homeownership (*Eigenheimzulage*) and its replacement with a cheaper subsidized pension savings scheme (*Wohnriester*) in 2006. However, the introduction of a subsidy designed to promote home purchase among first-time buyers with children (*Baukindergeld*) is dismissed as being "hardly a growth strategy" (Reisenbichler, 2021, p. 340). Yet in the UK, what was once the single largest financial subsidy to housing (interest relief on homeowners' mortgages) was phased out by 2000. Nonetheless, Reisenbichler (2021, p. 330) argues that Germany "retained conservative housing finance structures while at the same time sacrificing moderate support for housing finance, such as tax breaks for homeowners, in the name of structural reform and fiscal consolidation." This could well be right, although Stephens (2020b) has argued that the question needs to be explored in more depth as it is entirely possible that, given the existence of the European Single Market, which allows banks to offer products across the EU, the relatively low levels of mortgage debt in Germany (and Austria) may be attributable to a lack of demand associated with a cultural aversion to debt.

Although it is derived from a regime framework that attempts to explain economic models, the Varieties of Residential Capitalism strand in the regime literature emphasizes the important relationship between housing and the welfare state. As described earlier, Kemeny (1981, 1995) posited that an inverse relationship existed between levels of homeownership and the generosity of the welfare state, arguing that the front-end loading of the costs of homeownership encouraged prime-age voters to prefer lower rates of taxation over higher welfare standards. Castles (1998) suggested that there is a narrower but "really big" trade-off between homeownership and the generosity of state pensions; so long as outright homeownership is attained by retirement age, it can compensate for relatively frugal state pensions. Kemeny (2006, p. 59) later argued that if countries with integrated rental markets experienced declines in welfare generosity, particularly among pensioners, "we can expect them to begin to transform into monotenural homeowning societies."

Schwartz and Seabrooke's (2008) contribution promoted the argument that the financialized growth in homeownership and house prices had helped to compensate households for stagnant or falling real wages. (It is often quoted alongside Crouch's (2009) "privatised Keynesianism" and Prasad's (2012) "mortgage Keynesianism.") Schwartz (2020, p. 503) elaborates on this argument in a more recent contribution:

> Political pressure for financialisation is strongest in societies with the weakest left-wing parties and welfare states, but those are precisely the societies where unequal and unstable incomes limit asset creation, and in which inadequate regulation is likely to trigger a crisis.

In a relatively neglected paper, Schwartz (2012) made a crucial distinction between what he calls the "macro-level" and "micro-level" welfare states. The micro-level welfare state refers to the household level – the functioning of labour markets, tax and social security systems, and the housing system. However, the "macro-level" welfare state refers to the systems of micro- and macro-prudential regulation surrounding finance, especially mortgage finance. As we have seen, these were relaxed in many countries from the 1980s onward, undermining this "form of social protection of capital" (Schwartz, 2012, p. 37). In the US, it facilitated the growth of the sub-prime and Alt-A mortgage markets, whose instability was at the root of the GFC. Putting the two arguments together,

> The contraction of the micro- and macro level welfare state respectively provided both necessary and sufficient conditions for a massive financial crisis, by allowing financial elites to gamble with households' single biggest liability, and thus banks' single biggest asset.
>
> (Schwartz, 2012, p. 37)

Yet, even here, we must be careful when specifying the role of the state. Schwartz argues that the fulfilment of financialization actually depends on state support. He argues its development depends on the ability of financial institutions to securitize mortgages, and, based on US experience, this requires some state backing. "Financialisation is," he argues, "in large part mortgage-isation, and mortgage-isation requires active state support" (Schwartz, 2020, p. 503–504). Some of the winding down of securitization in Europe since the GFC suggests a lack of financial support from Europe's authorities. Yet, as Stephens (2020b) observes, the emergence of unconventional monetary policy practiced by central banks since the GFC, primarily in the form of QE, helped to bring about a climate of ultra-low interest rates. The German constitutional court ruled against the European Central Bank's QE programme during the COVID-pandemic, in what the economist Adam Tooze called "an ordoliberal fantasy world" (quoted by Stephens, 2020b, p. 553). QE, Stephens argues, is part of what he calls "really big finance" that underpinned the pandemic and post-pandemic house price boom, halted only when inflationary spillovers exacerbated by the tardy recovery of post-pandemic supply chains Russia's invasion of Ukraine in February 2022 caused central banks internationally to hike interest rates and contemplate throwing QE into reverse (quantitative tightening).

Rejection or development?

The most stringent attack on the housing-welfare regime approach can be found in Ball (2020). Many of Ball's criticisms are, in fact, pragmatic and relate to poor execution. For example, he argues that housing regimes are "didactic devices [that] empirically foreclose analysis when it should be the beginning, because no country's institutional frameworks are the same as others and outcomes vary widely" (Ball, 2020, p. 37). However, this misunderstands the essential point about regimes: they *are* derived empirically. Certainly, if they are *assumed* without empirical foundation and employed as an independent variable, then the results may well be unsound. Similarly, Ball claims that regime theory is "heavily and often unreasonably value-laden" (2020, p. 38), for example, expressing a preference for state intervention and "housing for all" allocation policies in social rented housing. He rightly points to the context in which policies operate, but much of the literature on housing regimes debates and seeks to establish outcomes and the extent to which they are attributable to the design of housing systems (Mandic & Cirman, 2012; Hick et al., 2022). Institutional path dependence pre-dating the institutional developments associated with the growth of the welfare state (Blackwell & Kohl, 2019) and, perhaps more significantly, the general level of prosperity in a country are factors that can be (and often are) considered.

More serious is Ball's contention that regimes cannot reliably predict outcomes. Stephens (2016, p. 26) noted that Kemeny is "curiously silent" on housing outcomes, which might be attributable to the lack of available data when he developed his thesis. The development of reasonably comparable indicators at a general level (e.g., by the OECD) but more particularly at the household level (in the EU Statistics on Income and Living Conditions, EU-SILC) has opened up the field to analysis. Ball refers to an article by Dewilde (2017) in which she uses cluster analysis to examine the outcomes of housing regimes and concludes that

> The theoretical construct of ideologically driven power constellations defining housing regimes, and their translation into housing policies, can thus be confirmed by its outcomes.
> (Dewilde, 2017, p. 397)

Ball's (2020) interpretation is that there is substantial intra-regime variation and suggests that [a] reasonable conclusion would be that this result is to be expected as housing outcomes are affected by a wide range of factors beyond those defined as distinguishing Housing Regimes." (Ball, 2020, p. 40). Whether one emphasizes similarities or differences is a moot point. But more fundamentally, the available independent and dependent variables such as these are problematic (Stephens, 2016). In particular, the "independent variable" problem – whereby results depend heavily on the selection of the independent variable, or outcome indicator – has been highlighted by Hick and Stephens (2023).

Ball suggests a revived version of the "structures of housing provision" (which he originally proposed in 1986) as an alternative to regime approaches. The focus would be on "networks of relations between a variety of organisations existing at a point in time in particular types of housing provision and how they behave in the contexts of prevailing rules and practices, *which they themselves help to create*" (Ball, 2020, p. 41, emphasis in the original). The diagrammatic scheme presented refers, *inter alia*, to land, finance, housebuilding, users and stock, repairs and changes as circles and ovals, which, where they touch, exhibit direct relationships. Other factors (location and prices/ rents) are key differences between existing and new/mover households. It is essentially a bottom-up approach, which, as Ball argues, downplays policy. Practitioners of regime theory would argue that policy is but one element, and the idea of institutions shaping rules is something that is implied by, for example, Kemeny's focus on institutional culture in the form of consensual or adversarial systems. More fundamentally, Ball's scheme seems to lack organizational logic, which is precisely the strength of regimes.

Regimes represent organizing principles for both states and other institutions, which gives them coherence and explanatory power. The dependent (and to an extent independent) variable problems present challenges, but we would emphasize the congruence between different regime approaches: between Esping-Andersen's welfare regimes, Kemeny's housing regimes, Schwartz and Seabrooke's varieties of residential capitalism, and Resienbichler's growth regimes. Countries tend to fall into the same groupings, whichever framework is adopted.

Pragmatic objections relate to issues such as geographical coverage or poor execution. Poor execution may arise from the "mindless classification" that Kemeny (2001, p. 61) warned about and Aalbers' complaint that many comparative studies take a version of Esping-Andersen "for granted" (2016, p. 8). Poor execution may also arise from the inappropriate adoption of Kemeny's typology, which is limited temporally and geographically, as he conceded. It really applies to a relatively small number of advanced economies that developed subsidized rental programmes in the third quarter of the 20th century, so it is surprising that some authors have sought to apply it to rather incongruous settings. As Stephens (2016) observed, the welfare triangle of market-state-family derived from Esping-Andersen is more portable but may lack explanatory power, which is one of the challenges highlighted by Stephens et al. (2015) in their study of the former socialist countries in central and eastern Europe. A greater challenge arises when "there is no obvious underlying ideology (liberal, social democratic, etc.), then the core explanatory element is missing. This is the conundrum [that arises when] the former ideological coherence that underpinned distinctive housing systems in Western Europe is in decline" (Stephens, 2020b, p. 588).

Other essentially pragmatic objections relate to the focus on national housing systems. Matznetter (2020, p. 563), following Hoekstra (2020), argues that "regional to local housing markets and policies are becoming the appropriate scale for comparative research." While there is scope for variation in housing policy within nation states, Stephens (2011, 2020b) argues that the institutions of the wider welfare regime (i.e., macro-economic management, labour markets, tax and social security systems) are largely designed at the national level, and these set the "boundaries of possibilities" within which housing system variation is set. Even where federal governments have largely withdrawn from housing at the regional level (e.g., Germany), there is a strong path dependence arising from the era of large-scale federal-level intervention, as well as ideological/cultural factors that may be influential independently of legislative competence.

Conclusion

Regime approaches have become the dominant framework through which to characterize, understand and explain housing systems internationally. The review and discussion in this chapter have demonstrated how the regime approach seeks to capture organizing principles not merely to categorize but to explain how housing systems are formed, operate and change. A robust housing regime approach necessitates an understanding of the context in which it operates. The dynamic relationship between the housing system and the "micro welfare state" has been long discussed, while the "macro welfare state" in terms of macroeconomic policy and broader forces such as globalization need to be incorporated, acknowledging that the relationship is two-way.

We reproduce Stephens' (2020b) typology of such a multi-layered housing regime framework to illustrate our starting point (Figure 3.1). The core of the typology remains middle-range national-level sector regimes. Similar to other depictions of the housing sector (e.g., Burke, 2012), the housing system is divided into the three spheres of consumption, production and exchange, but these are set within the context of the "wider welfare regime" and macro-level drivers. The wider welfare regime accommodates labour market, tax and social security institutions that produce distinctive distributional outcomes and are akin to Esping-Andersen's welfare regimes. These are influential in establishing the "boundaries of possibilities" – the levels and distribution of

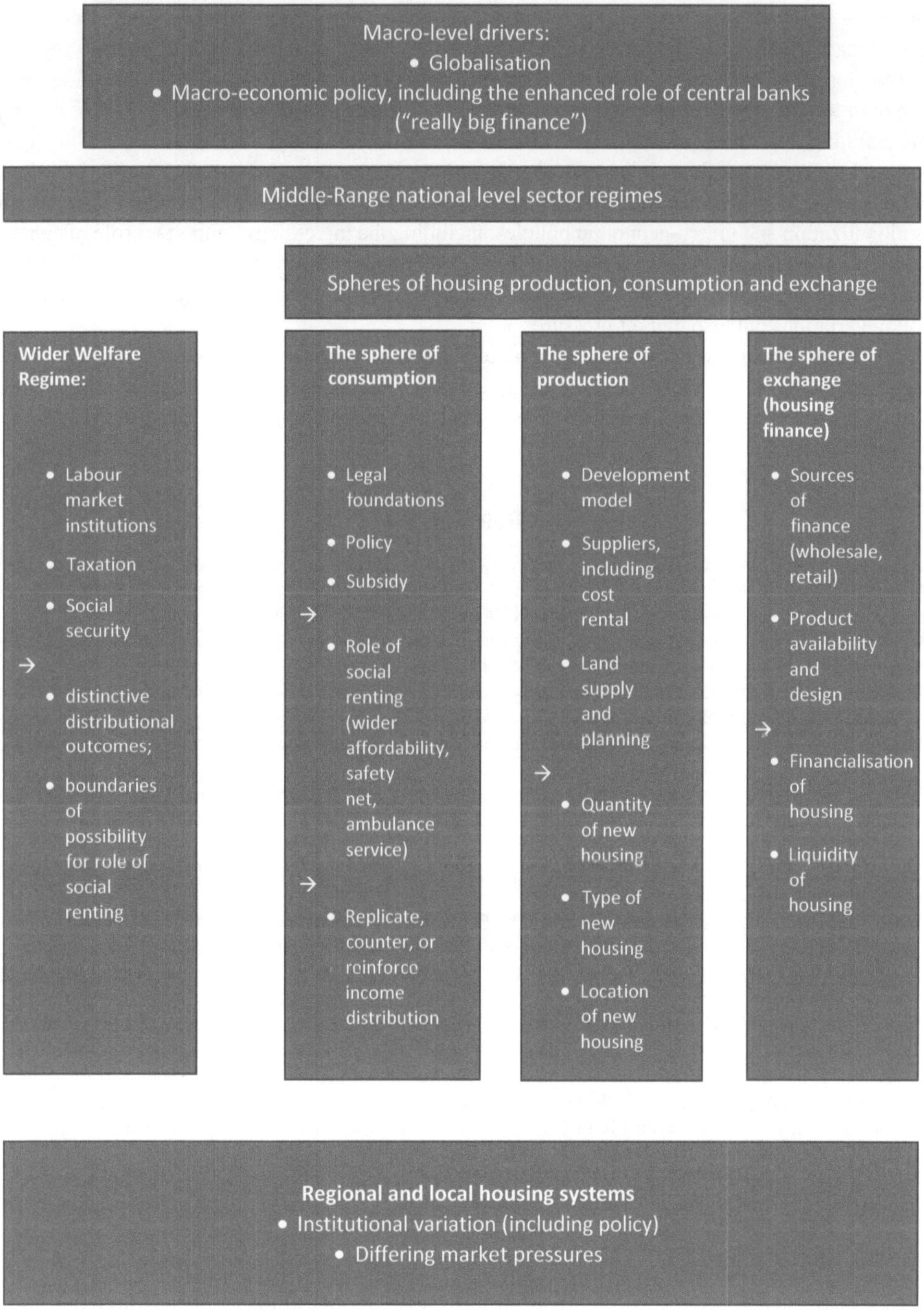

Figure 3.1 Typology of multi-layered housing regime framework.

Source: Stephens (2020b).

income – in which the housing system operates. The housing system itself is interpreted much more broadly than the organization of the rental sector, which is the driver in Kemeny's regime, and can replicate, counter or reinforce the pattern of income distribution created by the wider welfare regime. The sphere of production reflects the importance of supply, including the role of land and planning. The sphere of exchange is characterized by the housing finance system, which incorporates key aspects of financialization into the regime framework. Compared to earlier iterations of housing regimes, greater emphasis is placed on macro-level drivers, which include forces such as globalization and macro-economic policies, including the increasingly important role of central banks and their role in "really big finance." At the other end of the scale, beneath the middle-range national housing system lie regional and local housing systems, which may reflect institutional and policy variations but also market pressures.

The next part of the book contains chapters examining housing systems from the world's regions and will allow us to return to the questions of housing regimes and their relationship with welfare regimes. The framework presented here is then employed to analyse these countries in Part III.

References

Aalbers, M. B. (2016). *The Financialization of Housing: A Political Economy Approach*. Routledge.

Ball, M. (2020). Classifying Housing Regimes. Is it Worth Doing? What are the Alternatives? *Critical Housing Analysis*, *7*(1), 36–48. https://doi.org/10.13060/23362839.2020.7.1.502

Bengtsson, B. (2001). Housing as a Social Right: Implications for Welfare State Theory. *Scandinavian Political Studies*, *24*(4), 255–275. https://doi.org/10.1111/1467-9477.00056

Blackwell, T., & Kohl, S. (2019). Historicizing Housing Typologies: Beyond Welfare State Regimes and Varieties of Residential Capitalism. *Housing Studies*, *34*(2), 298–318.

Burke, T. (2012). The Australian Residential Housing Market: Institutions and Actors. In R. Tomlinson (Ed.), *Australia's Unintended Cities* (pp. 35–50). CSIRO.

Byrne, M., & Norris, M. (2019). Housing Market Financialization, Neoliberalism and Everyday Retrenchment of Social Housing. *Environment and Planning A: Economy and Space*, *54*(1), 182–198. https://doi.org/10.1177/0308518X1983261

Castles, F. G. (1998). The Really Big Trade-Off: Home Ownership and the Welfare State in the New World and the Old. *Acta Politica*, *33*(1), 5–19.

Christophers, B. (2013). A Monstrous Hybrid: The Political Economy of Housing in Early Twenty-first Century Sweden. *New Political Economy*, *18*(6), 885–911. https://doi.org/10.1080/13563467.2012.753521

Crouch, C. (2009). Privatised Keynesianism: An Unacknowledged Policy Regime. *The British Journal of Politics and International Relations*, *11*(3), 382–399. https://doi.org/10.1111/j.1467-856X.2009.00377.x

Dewilde, C. (2017). Do Housing Regimes Matter? Assessing the Concept of Housing Regimes Through Configurations of Housing Outcomes. *International Journal of Social Welfare*, *26*, 384–404. https://doi.org/10.1111/ijsw.12261

Enström Öst, C., Söderberg, B., & Wilhelmsson, M. (2014). Household Allocation and Spatial Distribution in a Market Under ("Soft") Rent Control. *Journal of Policy Modeling*, *36*(2), 353–372. https://doi.org/10.1016/j.jpolmod.2014.01.002

Esping-Andersen, G. (1990). *The Three Worlds of Welfare Capitalism*. Princeton University Press.

Grander, M. (2017). New Public Housing: A Selective Model Disguised as Universal? – Implications of the Market Adaption of Swedish Public Housing. *International Journal of Housing Policy*, *17*(3), 335–352. https://doi.org/10.1080/19491247.2016.1265266

Grander, M. (2023). The Inbetweeners of the Housing Markets – Young Adults Facing Housing Inequality in Malmö, Sweden. *Housing Studies*, *38*(3), 505–522. https://doi.org/10.1080/02673037.2021.1893278

Hick, R., Pomati, M., & Stephens, M. (2022). *Housing and Poverty in Europe: Examining the Interconnections in the Face of Rising House Prices*. Cardiff University.

Hick, R., & Stephens, M. (2023) Housing, the Welfare State and Poverty: On the Financialization of Housing and the Dependent Variable Problem. *Housing, Theory and Society*, *40*(1), 78–95. https://doi.org/10.1080/14036096.2022.2095438

Hochstenbach, C., & Aalbers, M. B. (2023). The Uncoupling of House Prices and Mortgage Debt: Towards Wealth-Driven Housing Market Dynamics. *International Journal of Housing Policy*. https://doi.org/10.1080/19491247.2023.2170542

Hoekstra, J. (2020). Comparing Local Instead of National Housing Regimes? Towards International Comparative Housing Research 2.0. *Critical Housing Analysis*, *7*(1), 74–85. https://doi.org/10.13060/23362839.2020.7.1.505

Kemeny, J. (1981). *The Myth of Home Ownership: Private versus Public Choices in Housing Tenure*. Routledge.

Kemeny, J. (1995). *From Public Housing to the Social Market: Rental Policy Strategies in Comparative Perspective*. Routledge.

Kemeny, J. 2001. Comparative housing and welfare: Theorising the Relationship. *Journal of Housing and the Built Environment, 16*(1), 53–70. https://doi.org/10.1023/A:1011526416064

Kemeny, J. (2006). Corporatism and Housing Regimes. *Housing, Theory and Society*, *23*(1), 1–18. https://doi.org/10.1080/14036090500375423

Mandic, S., & Cirman, A. (2012). Housing Conditions and Their Structural Determinants: Comparisons within the Enlarged EU. *Urban Studies*, *49*(4), 777–793. https://doi.org/10.1177/0042098011405688

Matznetter, W. (2020). How and Where Non-profit Rental Markets Survive – a Reply to Stephens. *Housing, Theory & Society*, *37*(5), 562–566. https://doi.org/10.1080/14036096.2020.1816570

Paris, C. (2017). The Super-Rich and Transnational Housing Markets: Asians Buying Australian Housing. In R. Forrest, S. Koh, & B. Wissink (Eds.), *Cities and the Super-Rich. The Contemporary City* (pp. 63–83). Palgrave Macmillan. https://doi.org/10.1057/978-1-137-54834-4_4

Prasad, M. (2012). *The Land of Too Much: American Abundance and the Paradox of Poverty*. Harvard University Press.

Reisenbichler, A. (2021). Housing Finance between Social Welfare and Growth Strategies. In A. Hassel & B. Palier (Eds.), *Growth and Welfare in Advanced Capitalist Economies* (pp. 320–347). Oxford University Press.

Ronald, R., Kadi, J., & Lennartz, C. (2015). Homeownership-Based Welfare in Transition. *Critical Housing Analysis*, *2*(1), 52–64. https://doi.org/10.13060/23362839.2015.2.1.176

Schwartz, H. M. (2012). Housing, the Welfare State, and the Global Financial Crisis: What is the Connection? *Politics & Society*, *40*(1), 35–58. https://doi.org/10.1177/0032329211434689

Schwartz, H. M. (2020). Covering the Private Parts: The (Re-) Nationalisation of Housing Finance. *West European Politics*, *43*(2), 485–508. https://doi.org/10.1080/01402382.2019.1582254

Schwartz, H., & Seabrooke. L. (2008). Varieties of Residential Capitalism in the International Political Economy: Old Welfare States and the New Politics of Housing. *Comparative European Politics*, *6*, 237–261. https://doi.org/10.1057/cep.2008.10

Stephens, M. (2011). Comparative Housing Research: A 'System-Embedded' Approach. *International Journal of Housing Policy*, *11*(4), 337–355. https://doi.org/10.1080/14616718.2011.626598

Stephens, M. (2016). Using Esping-Andersen and Kemeny's Welfare and Housing Regimes in Comparative Housing Research. *Critical Housing Analysis*, *3*(1), 19–29.

Stephens, M. (2020a). How Housing Systems are Changing and Why: A Critique of Kemeny's Theory of Housing Regimes, *Housing, Theory and Society*, *37*(5), 521–547. https://doi.org/10.1080/14036096.2020.1814404

Stephens, M. (2020b). Towards a Multi-layered Housing Regime Framework: Responses to Commentators. *Housing, Theory and Society*, *37*(5), 584–596. https://doi.org/10.1080/14036096.2020.1816573

Stephens, M., & Van Steen, G. (2011). 'Housing Poverty' and Income Poverty in England and the Netherlands. *Housing Studies*, *26*(7–8), 1035–1057.

Stephens, M., M. Lux, P. Sunega 2015. Post-socialist Housing Systems in Europe: Housing Welfare Regimes by Default? *Housing Studies, 30*(8): 1210–1234. https://doi.org/10.1080/02673037.2015.1013090.

Wijburg, G., & Aalbers, M. B. (2017). The Alternative Financialization of the German Housing Market. *Housing Studies*, *32*(7), 968–989. https://doi.org/10.1080/02673037.2017.1291917

PART II

Housing and welfare in a global context

In this part of the book, 23 authors contribute up-to-date overviews of how housing and welfare are connected in different international contexts, more specifically in 15 countries. We start with three chapters covering the Anglo-sphere countries (the US, England, and Australia), then proceed with two Western European chapters, one covering Germany, Austria, and Switzerland, and the other covering the Nordic countries (Sweden, Norway, and Finland). Thereafter, we turn to southern Europe, with one chapter on Italy and another on Spain, with a particular focus on the region of Catalonia. After that, three chapters present the post-socialist countries of Russia, Hungary, and China, respectively. Finally, this part of the book rounds off with chapters covering two countries from the global south, Chile and South Africa, whose housing systems are still shaped by their "radical" regime pasts.

The chapters have a common framework, pinpointing the basis for housing provision in the diverse geographical areas and how housing is regarded in relation to the welfare regime. Apart from Catalonia, the subchapters are written from a country perspective but relate to the larger geographical area as a whole. They provide a current overview of the housing and welfare systems but also highlight changes of importance to the wider context, providing the basis for the synthesized analysis in Part III of the book.

DOI: 10.1201/9781003212690-5

4
HOUSING AND WELFARE IN ENGLAND

Michael Marshall, Stephen Hincks and Ryan Powell

The current state of housing in the region

Overview

The consensus view is that England is in the midst of a "housing crisis". Discourse is dominated by an economistic diagnosis of the so-called crisis revolving around the need to increase housing supply to improve affordability and expand homeownership. Significant attention has centred on the thwarted housing aspirations of younger adults giving rise to the portmanteau term "Generation Rent" (Hoolachan et al., 2017). Successive governments have committed to targets for delivering new homes in England (between 250,000 and 300,000 per year), but as with previous governments, have consistently fallen short. Not only is the economistic diagnosis of the issue simplistic but it is also characterized by a "housing system amnesia", where a present-centred fixation on "housing supply" obfuscates longer-term shifts in the housing system (Powell & Robinson, 2019). Taking a longer-term view reveals that the only period during which housing targets were met was when there was substantial public sector delivery in the 1960s (Cole & Furbey, 1994). Yet social housing has been marginalized from English policy responses to the "housing crisis", driven by the ideology of homeownership since the 1980s. In contrast, the devolved administrations of Scotland, Wales and Northern Ireland have been more supportive of social housing in recent years.

In this chapter, we show that the English "housing crisis" is much wider than a lack of supply, stunted homeownership aspirations and Generation Rent. We locate the origins of today's housing challenges in the commodification and financialization of the housing system. Commodification foregrounds the privileging of housing's *exchange value* over its *use value*, whereby the function of bricks-and-mortar housing is subordinated to its value as a commodity asset to be exchanged (Madden & Marcuse, 2016). Housing financialization reflects the usage of housing as collateral in the production and circulation of financial assets, and the treatment of housing as an asset that derives returns via future income streams (Aalbers, 2016). In short, finance has come to dominate the UK housing system, enabled by economic deregulation, long-run shifts in housing delivery and the promotion of private home ownership as *the* aspiration. This, in turn,

DOI: 10.1201/9781003212690-6

has had profound impacts on the housing orientations and attitudes of individuals as the ideology of homeownership, and housing debt shape and frame household locational and financial decision-making while also inform the stigmatization of social housing and private renting (see Slater, 2018; Watt, 2021).

These trends have contributed to deepening housing polarization. The spread of pervasive housing insecurity at the bottom end of the housing system, and "spectacular speculation" and extraction at the other end (McKee et al., 2020; McKenzie & Atkinson, 2019). Such insecurity is typified by declining affordability across several regions (especially in London), persistently high levels of homelessness, increasing numbers of evictions from a loosely regulated private rented sector (PRS), and rising rent prices in the social rented sector (Fitzpatrick et al., 2021; Preece et al., 2020; Watt, 2018; Whitehead & Meen, 2020). Moreover, the dominant focus within the government on the quantity of housing supply obscures issues of geography and housing quality. The latter is rife in the UK, including overcrowding, a squeeze on space standards, fuel poverty and inadequate fire safety standards viscerally illustrated by the Grenfell Tower tragedy. On the morning of 14 June 2017, Grenfell Tower – a public housing tower block in North Kensington, London – was engulfed by a devastating fire that claimed the lives of 72 people (Apps, 2022; Clair, 2021; Hodkinson, 2019). For Hodkinson (2018), the tragedy of Grenfell was rooted in clear policy failure and negligence. Yet he is also clear that Grenfell symbolizes the neoliberal underpinnings of (social) housing policy that are exemplified by privatization, stock transfer and the sharpening focus on classed and racialized dynamics that underpin contemporary housing inequality in England.

Turning specifically to social housing, the sector is proportionately larger in the UK than in other Anglophone nations, and a case has been made that its relative size acts as a "saving grace" for living standards that mitigates some of the consequences of an otherwise residual UK welfare regime (Bradshaw et al., 2008). However, commodification and financialization have contributed to the social housing sector declining in size (especially in England), and rental price increases associated with new intermediate tenures and the growth of the PRS. One consequence has been the partial erosion of housing's "saving grace" role. Low-income households are increasingly subject to comparatively poor PRS housing conditions (Arundel, 2017; Tunstall, 2022) and well-documented precarity owing to its expense and tenure insecurity (Lombard, 2021; McKee et al., 2020). Housing commodification and financialization have interacted with a post-2010 austerity project to fragment what was previously understood as a dualist housing system, creating complex vehicles of public-private-third sector delivery (Hodkinson, 2019; Watt, 2021) and new barriers to secure and affordable housing (Preece et al., 2020).

Notes on tenures and terminology

The remainder of the chapter is divided into four further sections. Sections "A brief overview of the historical development of the housing regime: back to precarity and the ascendancy of finance, The welfare regime and its development and The interrelation between housing and welfare" focus on housing policy and the welfare regime in England, which is overseen by the Westminster Parliament (Table 4.1). We avoid conflating England with the UK, as this obscures the nuances that emerge from the contrasting approaches to housing and social security in Scotland, Wales and Northern Ireland (McKee et al., 2017). However, in Section "Regional outlook" we reflect on housing and welfare within the devolved administrations of Scotland, Wales and Northern Ireland relative to the English context.

Table 4.1 Notes on tenures and terminology (England)

Term	*Description*
Social housing	An umbrella term for either rented or homeownership housing at below market prices, including social rent, affordable rent and shared ownership. Provided by either councils (also known as local authorities) or non-profit housing associations.
Social rent	Non-market housing, with rents well below market rates, is nominally equivalent to Kemeny's "cost renting" category.
Affordable rent	Non-market housing with rents up to 80% of market rates. Typically more expensive than social rent, although they house largely the same populace.
Shared ownership	An intermediate tenure allows households to own a partial stake in their home alongside their landlord, to whom they pay the remaining rent.[1]
Private rented sector (PRS)	For-profit rental housing. Equivalent to Kemeny's "profit renting" category.
Homeownership	Owner-occupied housing, which may be owned either with or without a mortgage.

A brief overview of the historical development of the housing regime: Back to precarity and the ascendancy of finance

1945–1975

New housing delivery reached its peak during the post-war welfare consensus. The public sector took on a major role in housing delivery and local authorities became the primary providers of social housing. Social renting grew to be comparatively much larger than in other Anglophone nations (Kemeny, 1995, p. 106). The majority of state expenditure on social housing was "bricks and mortar" subsidies, namely supply-side subsidies to local authorities that kept rents low, or capital expenditure for building new housing (Hills et al., 1990). However, because social insurance benefits did not account for housing costs, many low-income households were reliant upon means-tested social assistance to meet rental costs (Glennerster, 2004, p. 82; see below).

Despite the expansion of council housing, housing in England failed to reach the status of universal social rights that rivalled private provision. From the 1950s onwards, social renting grew largely as an instrumental means of expediting slum clearance and meeting an acute post-war housing shortage. By the end of the 1950s, there was a political consensus that homeownership was the preferred tenure, which indeed became the largest tenure, and that social renting would be supplementary (Cole & Furbey, 1994, pp. 68–69). By contrast, the PRS went into decline as much of it was demolished through slum clearance, and was constrained by rent controls (Stephens, 2020, p. 540).

In the immediate post-war period, the social rented sector was largely an aspirational, mixed-income tenure (Ravetz, 2001). However, the quality of newly constructed social housing began to decline as policy overwhelmingly focused on the number of homes, compromising on issues such as space standards (see Cole and Furbey, 1994, p. 70). Although many residents had a strong sense of attachment to council homes and communities, stereotypes emerged that council housing – most commonly sub-standard high-rise housing – was synonymous with poverty and crime and that local authorities were ineffective landlords (Ibid, p. 174). This provided a conducive environment not only for the stigmatization of social renting and disinvestment in housing estates but also for the emergence of a distinctive "property machine" and lobby that would come to drive new urban development and housing delivery and displace notions of the pro-social (Atkinson & Jacobs, 2020; Slater, 2018; Watt, 2021).

1976–2009

The post-war consensus ended with the economic crises of the mid-1970s that necessitated reductions in state expenditure and wage restraint. Rates of new council house buildings fell sharply from the mid-1970s. Over time local authorities increasingly targeted allocations of social housing more explicitly to households in acute need. These changes in the social housing system combined with the Right to Buy (RTB) and geographies of deindustrialization initiated a process of *residualization* as social housing became increasingly a tenure for low-income and workless households (Jacobs, 2019, p. 80).

A political break occurred in 1979 with the election of a neoliberal Conservative government led by Margaret Thatcher. Housing was centrally implicated in the reimagining of citizen-state-market relations, with Council housing in particular framed as symbolic of state failure (Jacobs, 2019; McKee, 2012). Privatization, homeownership and deregulation were the answers, illustrating the central role of the state in driving commodification and financialization (Aalbers, 2016). A range of policy measures sought to privatize council housing and create a "property-owning democracy". The flagship policy underpinning this move was RTB, a policy that allowed tenants to purchase their council home at a discount well below market rates. Central government restrictions on the use of sales receipts also meant RTB could not be used to build or renovate council homes. And central government had already imposed restrictions upon local authorities' borrowing capacity in the 1970s. The combined effect was to end the role of councils as major providers of new housing (Gibb, 2021, p. 223). Legislation passed in 1989 prevented councils from using non-housing revenue to subsidize council housing rents, causing rents to rise, further incentivizing the exercise of RTB. Bricks and mortar subsidies fell throughout the 1980s. And as rents increased, demand-side housing assistance became the predominant source of government social housing expenditure (Hills et al., 1990, pp. 152–154).

The Labour government elected in 1997 embraced many aspects of this policy legacy, but differed in some notable respects. Labour explicitly cast themselves as a government of homeownership, but similarly saw the expansion of private housing supply as *the* solution to affordable homeownership, setting a target of 200,000 new homes annually (Ronald, 2008, p. 131). Delivery of new social housing continued to lag behind private supply, and Labour refused to reignite the development of new council housing, instead delivering new social housing via housing associations. A large amount of council stock was also transferred to HAs, often controversially so (Daly et al., 2005) and against residents' wishes. The 1988 Housing Act allowed for this transfer of stock as investment in the sector did not count as public sector debt (Atkinson & Jacobs, 2020, p. 21). However, the rate of stock transfers accelerated under Labour due to integration with the Decent Homes programme, which established a minimum standard for the quality of social housing and carried out major works to meet those standards. However, accessing government funding to meet the standard was conditional upon stock transfer to a housing association, transferring stock management to a quasi-public arms-length management organization, or engaging in a consortium with private finance (National Audit Office, 2010).

A key factor facilitating the expansion of homeownership was the deep and widespread financialization of housing, but the promise of homeownership for all failed to materialize. Financial deregulation in the early to mid-1980s liberalized the mortgage markets, and following innovations in financial markets in the late 1990s, cheap credit (leveraged against homeownership) became abundantly available (Wainwright, 2009). Two important consequences of financialization were the increased importance of house price appreciation for "economic growth" via debt-fuelled consumption (see Section "The interrelation between housing and welfare"), and an increasingly unequal distribution of housing wealth according to age and income (Arundel & Ronald, 2020). The 2000s were a period of house-price boom, but the overall homeownership rate started to decline in 2004 reflecting the pricing out of younger households. The PRS was the fastest growing tenure in the 2000s, benefitting from the relaxation of rent controls in 1989, and the expansion of buy-to-let mortgages (Forrest & Hirayama, 2015; Stephens, 2020, pp. 540–541). In London, by 2014, 36 per cent of all properties sold by London councils were let by private landlords. Perversely, this inflated the Housing Benefit (HB) bill and drove the massive cuts to HB in the PRS implemented in 2012 (Beatty et al., 2014; Powell, 2015).

2010-present

The global financial crisis (GFC) exposed the instability of financialized homeownership, with risky mortgage lending centrally implicated in the crash, as well as the failures of past policies and the erosion of a pro-social approach to housing. Despite the need for change in improving and widening access to shelter, post-GFC housing policy is marked mostly by the continuity of the previous epoch, with some oscillation in the approach to rented housing contingent upon political events. The primary focus of housing policy has been to boost supply and support first-time buyers into owner-occupation. The Coalition government (2010–2015) introduced the Help to Buy scheme to assist first-time buyers through governmental mortgage guarantees, equity loans and Individual Savings Accounts (ISA) linked to a cash bonus upon purchase (Tunstall, 2016, p. 128). The stamp duty tax imposed on house purchases has also been waived for first-time buyers (HMRC, 2017). As these reforms were premised on stimulating housing demand, one of their consequences was to support asset prices, contributing to house prices rising above their pre-GFC levels.

Supporting first-time buyers has not meant reversing the financialization of housing. Rather, investor strategies have shifted in focus towards safer returns; London has become a focal point for the storage of wealth in housing, being described as a "safe deposit box" for high-wealth individuals (Fernandez et al., 2016); and more recently Manchester (Goulding et al., 2023). Private finance has become more integral to the provision of social housing. The Coalition implemented an austerity programme that included dramatic funding cuts for local authorities and social housing. Housing associations have filled the funding gap through a combination of private finance, the development of homeownership for private sale, and the provision of Affordable Rents, which were introduced by the government in 2011 (Smyth, 2019) (see Section "The interrelation between housing and welfare").

The PRS grew to be larger than the social sector by 2012/2013, and the insufficient supply of social housing has contributed to the PRS increasingly accommodating low-income households reliant upon housing assistance to meet rental costs (Clair, 2021; Powell & Robinson, 2019). The PRS is relatively deregulated and predominantly comprised of small-scale amateur (or sideline) landlords. PRS housing is more likely to be low quality and insecure (Marsh & Gibb, 2019). PRS evictions and ending of tenancies have also driven homelessness (Table 4.2). The number of homeless households grew from 2010 to 2017 in England, with the ending or non-renewal of a short-term PRS tenancy as the primary factor for the first time since records began, a consequence of tenure insecurity and cuts to housing assistance (Fitzpatrick et al., 2017, pp. 61–64; see Section "The welfare regime and its development").

Table 4.2 Tenures as a percentage of total households in England (1953–2021)

	England				
	1953	*1981*	*1997*	*2010*	*2021*
Homeownership (own outright)				31.68	34.74
Homeownership (with mortgage)				35.71	30.12
Homeownership (total)	32	57.24	68.57	67.39	64.86
PRS	50	11.06	10.46	15.56	18.51
Social housing (council housing)				8.10	6.55
Social housing (housing association)				8.95	10.08
Social housing (total)	18	31.70	20.98	17.05	16.63

Source: English Housing Survey Table FT1101 (S101).

The welfare regime and its development

1945–1975

The foundation of the post-war social security system was largely based on the 1942 Beveridge Report and implemented by the 1945 Labour government. This applied across the UK, not just to England. The UK already had a patchwork of social insurance schemes, which Beveridge set out to consolidate and extend. The post-war social insurance model provided benefits during periods of need – e.g., sickness, retirement, and temporary unemployment – to those who had made the necessary contributions. Benefits were to be set at levels above the minimum needed for subsistence. And social insurance was supplemented by a family allowance, and a means-tested social assistance safety net (Glennerster, 2004, pp. 79–80). The British social insurance model had a number of distinctive features that set it apart from many of its European equivalents. Firstly, benefits were flat rate as opposed to earnings-related. Secondly, they were funded through a flat rate national insurance tax, and due to budgetary constraints benefits were comparatively less generous than in corporatist or Social Democratic countries (ibid., pp. 82–83). Thirdly, housing costs were not catered for by social insurance, meaning that low-income households were often reliant upon social assistance for their rental costs (Alcock, 2006, p. 213). A contributory state pension was also established, but it was set below subsistence levels, and many pensions relied on social assistance benefits (while others did not claim them).

The combination of full employment with social insurance meant poverty levels in the immediate post-war period declined. However, issues in the design of social security would plague the post-war settlement, and the welfare regime became more dependent upon means-tested social assistance. Low social insurance benefit rates meant many retired workers supplemented state benefits with occupational pensions, whereas people with interrupted working histories struggled to make the necessary contributions (Glennerster, 2004). This contributed to the so-called "rediscovery of poverty" in the 1960s. Abel-Smith and Townsend (1965) found that 18 per cent of households lived below the poverty line, which they defined as the sum of social assistance, rent allowance and some common extra allowances. In response, new means-tested benefits were introduced to support particular groups, including Supplementary Benefit (SB) in 1966 – initially for pensioners not qualifying for state pensions, but later for the unemployed who did not qualify for Unemployment Benefit – and Family Income Supplement (FIS) in 1970.

1976–2009

The most significant change in the welfare regime in the early years of Conservative rule was extensive liberalization of the labour market, reductions in the power of trade unions due to the desocialization of wage labour, deindustrialization, and a resultant increase in income inequality and poverty (Lund, 2008). In 1981 social security benefits were made less generous by linking increases to prices, breaking the link with earnings growth for pensions (Hills, 2004, p. 93). Following a review of social assistance, SB became Income Support, FIS became Family Credit (FC), and housing assistance was titled HB. These changes made social assistance a more central component of the welfare regime. They also extended the principle of conditionality in social security by making benefit receipt for jobseekers conditional upon active job search rather than passive declaration of availability (Jessop, 2003, p. 11). The Major administration (1990–1997) further increased the financial penalties for non-compliance with work search requirements under its Jobseekers' Allowance (JSA) scheme (Lund, 2008). The long-term reduction in supply-side subsidies for social housing, combined with the reliance upon social assistance to cover rental costs, meant that housing assistance became the dominant form of state housing expenditure (Hills, 2007, p. 64).

New Labour sought to promote social inclusion via the labour market through various carrots and sticks. The former were measures designed to "make work pay", including the establishment of a national minimum wage, and means-tested, in-work benefits such as the Child Tax Credit and Working Tax Credit, which were more generous replacements for previous schemes such as FC (Lund, 2008). The sticks in Labour's approach were part of the New Deal programme that embraced the conditionality principle. The New Deal provided packages of job search support, work experience, and training, and was compulsory for the young and long-term unemployed – with the threat of benefit withdrawal for non-compliance – and voluntary for lone parents and the disabled (Dean, 2006, p. 13). Income inequality remained higher than in corporatist or social democratic countries. Labour maintained the neoliberal legacy of labour market flexibility and low taxation, and economic growth became increasingly reliant upon consumption vis-à-vis housing-fuelled debt (Reisenbichler, 2021, see Section "The interrelation between housing and welfare").

2010-present

The post-2010 austerity agenda involved a welfare reform programme entailing social security cuts and enhanced conditionality that fell disproportionately on lower-income households. The effect of welfare reform has been to widen the gaps in the social security safety net and exacerbate housing stress.

Key changes to social security include the benefit cap, bedroom tax, and local housing allowance (LHA) cuts. The benefit cap imposed an overall limit on the amount a working-age person can claim. The bedroom tax reduced the housing assistance entitlement of working-age social housing residents with spare bedrooms, bringing the sector more in line with the PRS. LHA is used to determine housing assistance entitlement in the PRS and is calculated as a proportion of local private rents. The Coalition reduced that proportion from the 50th percentile to the 30th percentile, and uprated LHA rates by general inflation rather than actual rent price growth (Powell, 2015). The Coalition's flagship social security reform was Universal Credit (UC), which consolidates six benefits into a single payment, including HB. Notable changes made under UC include the imposition of a "two-child limit", which prevents UC claimants from receiving additional money for dependent children past their second child. The default payment of the housing element of UC to the claimant, whereas HB was (usually) paid directly to landlords. UC rollout also combined with a peak in the frequency and severity of sanctions for non-compliance with claimant conditions and work search requirements in the mid-2010s (Watts et al., 2014).

Perhaps surprisingly, the headline rates of poverty and income inequality have not changed significantly post-2010. Yet, in part, the relative poverty rate has remained stable because median incomes initially fell in the post-recession era and have seen limited real growth since (Corlett, 2020; Hills & Stewart, 2016, p. 253). Moreover, headline figures mask significant nuances, group and spatial variation in terms of growing inequalities. Child poverty rates have risen recently as the two-child limit and the benefit cap have taken effect (JRF, 2022). Similarly, Fitzpatrick et al. (2016) identified young single men and migrants as particularly at risk of destitution – defined as being unable to afford two or more of six essential items (e.g., shelter, food, heating) – due to commonly falling through the social security safety net and being frequently subject to sanctions. Housing assistance reform has also had profound impacts on housing stress. LHA reforms increased the gap between entitlements and rents and evidence suggests it was associated with increased overcrowding (Clair, 2021). Likewise, both controlled trials and observational data of UC recipients suggest rental arrears have increased due to making direct payment of the housing element of UC the default (Hickman et al., 2017; Hunter, 2020).

The interrelation between housing and welfare

To classify and understand change in the contemporary English housing system two factors are significant. The first is the increasingly residual welfare regime characterized by ongoing cuts to housing payments and the privatization of social housing. The second is the financialization of housing, which has driven house price inflation, facilitated the residualization of the welfare regime, and contributed to the segmentation of social housing.

In Kemeny's (1995) framework, England was classified as a *dualist* rental system, defined by a strict division between the cost and market renting sectors. Kemeny argued that the maturation of the social renting sector would result in a *rent differential crisis* as social renting would become competitive with private housing (1995, pp. 45–48). The policy response in dualist systems would manage any threat to private housing by containing the social renting sector by reducing demand and supply (ibid., pp. 51–53). As Section "A brief overview of the historical development of the housing regime: back to precarity and the ascendancy of finance" details, Kemeny's framework provides a compelling description of social housing policy from the latter part of the post-war period to the mid-1990s. However, Stephens (2020, p. 540) argues that the English system was not defined by a social renting vs. market renting dynamic, rather the primary division was between social renting and homeownership. Instead of being sheltered from competition, the PRS was a marginal tenure subject to relatively strict regulation until 1989 (ibid.).

Stephens goes further to suggest that Kemeny mistook the relationship between housing and the welfare regime (ibid., p. 542). The ability to develop an *integrated* rental market in England is necessarily constrained by the residual nature of the welfare regime. Integrated rental markets are complementary to relatively *de*-commodified welfare regimes, as lower levels of poverty help households exercise choice between social and market-rented options (ibid.). By contrast, comparatively high levels of poverty contribute to levels of housing need beyond that which can be accommodated in the social sector, incentivizing the rationing of social housing and the funnelling of marginalized households into the PRS.

Yet, the interrelation between housing and welfare is complex, as both are centrally implicated in a wider, financialized growth regime that acts as a powerful source of inertia. In the post-war golden age, economic growth was primarily driven by productivity and wage growth.

In the post-industrial era, however, a new growth regime has been embedded. "Growth" is driven primarily by financial services and debt-fuelled consumption. This debt is often leveraged against housing, especially where real wage growth is low or negative, making continued house price inflation foundational to the economy (Hassel & Palier, 2021; Reisenbichler, 2021). Households could theoretically use the wealth stored in property during retirement, for instance by downsizing, or could support consumption by borrowing against the equity in their home or deriving incomes from private landlordism (Ronald et al., 2015). Thus, the welfare regime does not just constrain the housing system, but housing can facilitate the reduction of state support by providing a form of "asset-based welfare". However, in practice, the potential for asset-based welfare has been limited by the wealth inequities produced by housing financialization (ibid.).

While the concept of the growth regime captures the interrelation between housing and economic activity at a macro level, an important phenomenon for understanding housing and welfare post-2010 is the extent to which financialization has driven change *within* the social sector. Financialization is manifest in changes in social housing provision, actors, intermediaries, practices, ownership and regeneration logics (Hodkinson, 2019; Watt, 2021). As such, the strict binary between social and market provision implied by Kemeny's *dualist* category may no longer apply.

Housing associations became increasingly distant from social renting in the 2010s with questions over how far the social element had been displaced by financial imperatives and new governance logics. As discussed, the provision of affordable rents, combined with private sales and increased borrowing, has been used by housing associations to meet the funding gap initiated by the austerity agenda (Smyth, 2019). In this context, housing associations have adopted new practices and objectives. A small minority of housing associations stated that the restrictions and uncertainty associated with this funding regime meant they could only develop new housing for intermediate homeownership tenures such as shared ownership (Morrison, 2017). The financial and regulatory environment for social housing has encouraged housing associations to generate capital by treating their homes as a tradable asset, selling social homes to either other housing associations or to the private market (Smyth, 2019). Not all housing associations have unquestioningly embraced this new practice, but some have been proactive and adopted a policy of earmarking high-value homes for disposal when they become vacant (Morrison, 2017). An active approach to selling housing assets has had consequences for the ownership of social housing, with an increasing number of sales being made to for-profit equity investors. These changes in landlord provision, practice and ownership can have material consequences for access to affordable housing. Marsh (2018) raises concerns that sales of social housing may contribute to socio-spatial segregation. The dual pressures of higher rents and unprecedented welfare cuts since 2010 have contributed to social landlords increasingly using affordability assessments in screening and excluding prospective tenants (Preece et al., 2020).

It is important not to overstate the extent to which these changes have transformed English social housing. Equity investment in social housing is a small, but growing, phenomenon. Most social housing remains social renting. And the housing assistance programme means for many households the difference between social and affordable rent does not affect their disposable income. Nonetheless, they point to an extension of housing financialization into the social sector in which changes in funding and ownership are associated with changing landlord practices and new mechanisms of exclusion. Blessing (2016, p. 167) argues that the "borders between social and commercial activities, once strictly patrolled, have been reconfigured as trading frontiers". Thus, the *dualist* binary may no longer reflect a more porous boundary between market and social housing.

Regional outlook

The preceding discussion has primarily been concerned with housing and welfare policy in England. However, since devolution in the late 1990s, there has been increased divergence between the four nations of the UK with approaches often rooted in contrasting views on the purpose of housing (McKee et al., 2017). The divergence between the nations demonstrates the relevance of party politics for UK housing outcomes while also the constraining role of the Westminster Parliament in setting the "boundaries of possibility" for devolved administrations. Particularly, in relation to the labour market and social security systems that define the wider welfare regime (Stephens, 2019). England emerges as an outlier among the UK nations in terms of a relative lack of housing policies aimed at alleviating poverty and housing marginalization, a more overt denigration of social housing, and its light-touch approach to regulating the PRS. In short, tackling housing inequalities and power imbalances has been more of a priority in the devolved nations than in England, with those challenges accentuated by unprecedented social security cuts emanating from Westminster.

The Scottish Parliament has responsibility for housing policy, powers to vary the rates and thresholds of income tax (but not the tax-free personal allowance), and some limited control over social security, including scope to vary the "housing cost element" within UC. Since 2007, the Scottish National Party (SNP) has run mostly minority but for a time a majority administration and has campaigned on a broadly anti-austerity platform. The SNP has sought to extend the safety net provided by social housing by abolishing the RTB in Scotland, extending the duty of local authorities to secure accommodation for homeless people to childless single people and couples, and varying the housing cost element in UC to mitigate fully the effects of the bedroom tax (ibid., p. 53–55). Scotland retains a larger social housing sector than the other countries of the UK (see Table 4.3) and has committed to ambitious affordable housing targets and built a consensus across the sector (Dunning et al., 2020; Powell et al., 2015). In contrast to the English social housing delivery programme, which has focused predominantly on affordable rent and shared ownership, the Scottish programme has delivered proportionately more social renting and has been funded with greater levels of governmental capital subsidy (Gibb, 2021). Scotland also leads the way in terms of PRS reforms aimed at addressing the imbalance in the landlord-tenant relationship (Gibb et al., 2019). Greatly enhanced tenure security was introduced in 2017 and the current administration reached an agreement with the Scottish Greens to introduce a national system of rent control. Further, legislation passed in 2022 has ended all mandatory grounds for eviction by introducing a test of reasonableness, and the government also introduced a time-limited rent freeze in response to the cost-of-living crisis in 2022–2023.

Table 4.3 Tenure as a percentage of total dwellings in UK countries in 2018

	England[1]	*Scotland*[1]	*Wales*[1]	*Northern Ireland*[2]
Homeownership	63.34	62.15	69.71	69
PRS	19.75	14.93	14.31	14
Social housing (council housing)	6.59	12.05	6.10	12
Social housing (housing association)	10.14	10.83	9.82	4

[1] *Source:* ONS Dwelling Stock, 201.

[2] *Source:* Department for Communities, Northern Ireland Housing Statistics, Table T1.3.

The Welsh Parliament has responsibility for housing and limited powers to vary income tax rates, but parts of social security policy remain under the Westminster remit. The social housing sector is slightly smaller in Wales than in England, but the Welsh Parliament has sought to shore it up by ending RTB (Stephens, 2019, p. 51). In 2014, the Welsh Parliament extended the duties placed upon local authorities to provide meaningful preventative assistance to households at risk of becoming homeless, with many of these measures being adopted into Westminster law in 2017. Wales has also legislated to improve the quality of provision and security of tenure in the PRS. In 2014 private landlord registration was made mandatory, and in 2016 courts were given the discretion to refuse eviction orders if the landlord was seeking an eviction to evade their repair responsibilities (Smith & Mackie, 2021, pp. 14–18; Wilson et al., 2019, p. 24).

The Northern Irish Assembly has legislative power over housing and social security. However, it has been suspended from time to time and this has hampered the development of policy. The social housing sector is predominantly public sector, with voluntary sector housing associations providing a smaller proportion than in England. The public sector is likely to decline in size, as all new social housing has been provided by housing associations since the early 2000s, and the RTB remains in place (Frey, 2018). The Northern Irish approach to the PRS is closer to England's in that it is relatively deregulated, although landlord registration is mandatory (Wilson et al., 2019, p. 13). Historically Northern Ireland's social security policy has been based on the "parity principle", which asserts Northern Irish citizens should be entitled to the same social security provision as people in Great Britain, meaning it has adopted Westminster policy in practice. However, social policy diverges in some respects in Northern Ireland, with the Assembly legislating to continue to pay UC directly to landlords and securing additional funding to protect the incomes of households affected by the bedroom tax for a time-limited period (Frey, 2018).

Concluding discussion

Housing has long been described as the "wobbly pillar" in the UK welfare regime, failing to reach the level of universal social rights (Malpass, 2003). By the 1960s the housing system was already dualist, with a clear distinction between social renting and private homeownership. Nevertheless, the post-war period was one in which the social sector expanded and provided good quality, affordable housing for a range of income groups. The post-war welfare regime, despite being less generous than its corporatist and social democratic comparators, produced lower levels of poverty and income inequality than have endured since the 1980s.

Following the end of the post-war consensus, two long-term processes have been central to the putative *housing crisis*, commodification and financialization. Not only have they transformed UK housing and orientations towards it, but they have also contributed to a transformation of the state (Jacobs, 2019). The commodification of housing was facilitated by the transfer of ownership that occurred via RTB and then compounded by the ending of municipal housing delivery. The dominance and availability of deregulated finance within the housing system and increased investment in housing contributed to unprecedented house price inflation from the late 1990s, playing a central role in the GFC. Financialization has continued apace post-GFC but in a different guise. The strategies of investors have shifted towards what is seen as the safe returns provided by urban real estate and social housing.

Public discourse has focused largely on the decline of homeownership, but this obscures housing issues occurring across the income distribution. Financialization and deregulation initiated the re-emergence of the diverse PRS and thus accentuated housing insecurity for low-income households. The social housing funding and regulatory environment has actively sought to replace public subsidy with private investment, often with disastrous consequences (see Hodkinson, 2019; Watt, 2021). Commercial practices have expanded and new forms of exclusion emerged in social housing (e.g., more expensive intermediate tenures, sales of high-value housing, and affordability assessments). As such, the public-private distinction set out by Kemeny, while always being blurred, is by now unable to capture the fluid contemporary reality of a housing system driven by international investment and financial logic.

Increased housing supply has been the purported solution to housing issues from the mid-2000s onwards. Increased supply is undoubtedly necessary – especially in regions of high housing demand and growth in household formation such as London (Mulheirn, 2019) – but there is important nuance in terms of what forms of new housing are provided, and how it is supplemented with wider reform to decommodify the housing system. Common across all UK regions is a shortage of social rented housing, which could take the form of new supply, and schemes such as the London Mayor's right-to-buy-back that allows councils to repurchase properties lost via RTB that are now in the PRS. New supply also has limited impact on issues of housing quality, which requires stronger regulation of rental housing, and implementation of an updated and ambitious Decent Homes Programme. However, it is worth reiterating that the financialization of social housing is a critical constraint on the ability to meet these (and other) social objectives. In the absence of sufficient state subsidy, the costs of borrowing make it prohibitive for social landlords to rapidly increase social rented provision while expending significant resources to improve the quality of existing stock (Twomey, 2022).

Moreover, change is necessary beyond the housing system, most notably in the welfare and growth regimes with which it interacts. Social security has become more residual, conditional and punitive over the long term. Gaps in the safety net have been introduced and the transition to UC is associated with higher levels of housing stress. High-income inequality interacts with residual social support to produce levels of housing need and poverty that incentivize the rationing of social housing. Moreover, the reliance upon rising house prices for "economic growth" creates a vested interest in the continuation of rising house prices.

Yet the contrasting approaches in the devolved administrations of the UK demonstrate that the existing trajectory is amenable to *some* change. The proportionately higher levels of social rented housing delivery in Scotland, and the tighter regulation of the PRS in both Scotland and Wales, illustrate the continued importance of policy in affecting housing outcomes. Furthermore, devolution has facilitated policy innovations that have later been transferred into English law. Nonetheless, the autonomy of devolved administrations is constrained by Westminster, where the overall direction of travel remains centred on homeownership, debt and financial imperatives, to the detriment of increasingly acute social and environmental urgencies.

Note

1 Confusingly, shared ownership is classified as social housing for funding purposes, but the occupiers are classified as homeowners (including in Table 4.2 in this chapter). *Social housing* in Table 4.2 is the sum of social rent and affordable rent.

References

Aalbers, M. B. (2016). *The Financialization of Housing: A Political Economy Approach*. Routledge.
Abel-Smith, B., & Townsend, P. (1965). *The Poor and the Poorest*. G. Bell & Sons.
Alcock, P. (2006). *Understanding Poverty*. 3rd ed. Palgrave Macmillan.
Apps, P. (2022). *Show Me the Bodies: How We Let Grenfell Happen*. Oneworld.

Arundel, R. (2017). Equity Inequity: Housing Wealth Inequality, Inter and Intra-generational Divergences, and the Rise of Private Landlordism. *Housing, Theory and Society*, *34*(2), 176–200. https://doi.org/10.1080/14036096.2017.1284154

Arundel, R., & Ronald, R. (2020). The False Promise of Homeownership: Homeowner Societies in an Era of Declining Access and Rising Inequality. *Urban Studies*, *58*(6), 1120–1140. https://doi.org/10.1177/0042098019895227

Atkinson, R., & Jacobs, K. (2020). *What Do We Know and What Should We Do About Housing?* Sage.

Beatty, C., Cole, I., Powell, R., & Sanderson, E. (2014). *Monitoring the Impact of Recent Measures Affecting Housing Benefit and Local Housing Allowances in the Private Rented Sector: The Response of Landlords* (Research Report No. 870). Department for Work and Pensions.

Blessing, A. (2016). Repackaging the poor? Conceptualising Neoliberal Reforms of Social Rental Housing. *Housing Studies*, *31*(2), 149–172. https://doi.org/10.1080/02673037.2015.1070799

Bradshaw, J., Chzhen, Y., & Stephens, M. (2008). Housing: The Saving Grace in The British Welfare State? In S. Fitzpatrick & M. Stephens (Eds.), *The Future of Social Housing* (pp. 7–25). Shelter.

Clair, A. (2021). The Effect of Local Housing Allowance Reductions on Overcrowding in the Private Rented Sector in England. *International Journal of Housing Policy*, *22*(1), 119–137. https://doi.org/10.1080/19491247.2021.1964253

Cole, I., & Furbey, R. (1994). *The Eclipse of Council Housing*. Routledge.

Corlett, A. (2020). Charting The UK's Lost Decade of Income Growth. *Resolution Foundation*. https://www.resolutionfoundation.org/comment/charting-the-uks-lost-decade-of-income-growth/ [Accessed 30 September 2022].

Daly, G., Mooney G., Poole, L., & Davis, H. (2005) Housing Stock Transfer in Birmingham and Glasgow: The Contrasting Experiences of Two UK Cities. *European Journal of Housing Policy*, *5*(3), 327–341. https://doi.org/10.1080/14616710500342234

Dean, H. (2006). *Activation Policies and the Changing Ethical Foundations of Welfare*. In ASPEN/ETUI Conference: Activation Policies in the EU, 20–21 October 2006, Brussels.

Dunning, R., Moore, T., Ferrari, E., Hoolachan, J., Keskin, B., Powell, R., & O'Brien, P. (2020). *Affordable Housing Need in Scotland Post 2021*. Shelter Scotland. https://assets.ctfassets.net/6sqqfrl11sfj/2GyBuBrrm22xlfDhisIS2Q/7a59fe65eb062884230a3a072cc854fe/Affordable_Housing_Need_in_Scotland_Post-21_final_report.pdf [Accessed: 14 October 2022].

Fernandez, R., Hofman, A., & Aalbers, M. B. (2016). London and New York as a Safe Deposit Box for the Transnational Wealth Elite. *Environment and Planning A*, *48*(12), 2443–2461.

Fitzpatrick, S., Bramley, G., Sosenko, F., Blenkinsopp, J., Johnsen, S., Littlewood, M., Netto, G., & Watts, B. (2016). Destitution in the UK. *Joseph Rowntree Foundation (JRF)*.

Fitzpatrick. S., Pawson, H., Bramley, G., Wilcox, S., & Watts, B. (2017). *The Homelessness Monitor: England 2017*. Crisis.

Fitzpatrick, S., Pawson, H., Bramley, G., Wood, J., Watts, B., Stephens, M. & Blenkinsopp, J. (2021). *The Homelessness Monitor: England 2021*. London: Crisis. https://www.crisis.org.uk/media/244702/crisis-england-monitor-2021.pdf

Forrest, R., & Hirayama, Y. (2015). The Financialisation of the Social Project: Embedded Liberalism, Neoliberalism and Home Ownership. *Urban Studies*, *52*(2), 233–244. https://doi.org/10.1177/0042098014528394

Frey, J. (2018). Social Housing in Northern Ireland: Challenges and Policy Options. *UK Collaborate Centre for Housing Evidence*. https://housingevidence.ac.uk/wp-content/uploads/2018/10/R2018_SHPWG_02_Social-housing-in-Northern-Ireland_webready.pdf [Accessed: 18 February 2018].

Gibb, K. (2021). Divergent Approaches to Affordable Housing Supply in a Devolved Policy System: Scotland and England After 2010. *International Journal of Urban Sciences*, 25(sup1), 218–240. https://doi.org/10.1080/12265934.2020.1730935

Gibb, K., Livingston, M., & Berry, K. A. (2019). *Overview of Private Rented Housing Reforms in Scotland*. Scottish Parliamentary Corporate Body.

Glennerster, H. (2004). Poverty Policy from 1900 to the 1970s. In H. Glennerster, J. Hills, D. Piachaud, & J. Webb (Eds.), *One Hundred Years of Poverty and Policy* (pp. 63–91). Joseph Rowntree Foundation (JRF).

Goulding, R., Leaver, A., & Silver, J. (2023). From Homes to Assets: Transcalar Territorial Networks and the Financialization of Build to Rent in Greater Manchester. *Environment and Planning. A*, *55*(4), 828–849. https://doi.org/10.1177/0308518X221138104

Hassel, A., & Palier, B. (2021). *Growth and Welfare in Advanced Capitalist Economies: How Have Growth Regimes Evolved?* Oxford University Press.

Hickman, P., Kemp, P. A., Reeve, K., & Wilson, I. (2017). The Impact of the Direct Payment of Housing Benefit: Evidence from Great Britain. *Housing Studies, 32*(8), 1105–1126. https://doi.org/10.1080/02673037.2017.1301401

Hills, J. (2004). The Last Quarter Century: From New Right to New Labour. In H. Glennerster, J. Hills, D. Piachaud, & J. Webb (Eds.), *One Hundred Years of Poverty and Policy* (pp. 92–131). Joseph Rowntree Foundation (JRF).

Hills, J. (2007). *Ends and Means: The Future Roles of Social Housing in England. Centre for Analysis of Social Exclusion*. CASE report 34. https://sticerd.lse.ac.uk/dps/case/cr/CASEreport34.pdf [Accessed: 18 February 2022].

Hills, J., Hubert, F., Tomann, H., & Whitehead, C. (1990). Shifting Subsidy from Bricks and Mortar to People: Experiences in Britain and West Germany. *Housing Studies, 5*(3), 147–167. https://doi.org/10.1080/02673039008720681

Hills, J., & Stewart, K. (2016). Socioeconomic Inequalities. In R. Lupton, T. Burchardt, J. Hills, K. Stewart, & P. Vizard (Eds.), *Social Policy in a Cold Climate: Policies and Their Consequences Since the Crisis* (pp. 245–265). Policy Press.

HMRC. (2017). *Stamp Duty Land Tax: Relief for First Time Buyers*. https://www.gov.uk/government/publications/stamp-duty-land-tax-relief-for-first-time-buyers/stamp-duty-land-tax-relief-for-first-time-buyers [Accessed: 18 February 2022].

Hodkinson, S. (2018). Grenfell Foretold: A Very Neoliberal Tragedy. In C. Needham, E. Heins, & J. Rees (Eds.), *Social Policy Review* 30 (pp. 5–26). Policy Press.

Hodkinson, S. (2019). *Safe as Houses*. MUP.

Hoolachan, J., McKee, K., Moore, T., & Soaita, A. M. (2017). 'Generation Rent' and the Ability To 'Settle Down': Economic and Geographical Variation in Young People's Housing Transitions. *Journal of Youth Studies, 20*(1), 63–78. https://doi.org/10.1080/13676261.2016.1184241

Hunter, P. (2020). *Falling Behind: The Impact of Universal Credit on Rent Arrears for Council Tenants in London*. The Smith Institute. https://www.smith-institute.org.uk/book/falling-behind-the-impact-of-universal-credit-on-rent-arrears-for-council-tenants-in-london/ [Accessed: 18 February 2022].

Jacobs, K. (2019). *Neoliberal Housing Policy: An International Perspective*. Routledge.

Jessop, B. (2003). *From Thatcherism to New Labour: Neo-Liberalism, Workfarism, and Labour Market Regulation*. Department of Sociology, Lancaster University. https://www.lancaster.ac.uk/fass/resources/sociology-online-papers/papers/jessop-from-thatcherism-to-new-labour.pdf [Accessed: 18 February 2022].

Joseph Rowntree Foundation (JRF). (2022). *UK Poverty 2022: The Essential Guide to Understanding Poverty in the UK*. Joseph Rowntree Foundation (JRF). https://www.jrf.org.uk/report/uk-poverty-2022) [Accessed: 30 September 2022].

Kemeny, J. (1995). *From Public Housing to the Social Market: Rental Policy Strategies in Comparative Perspective*. Routledge.

Lombard, M. (2021). The Experience of Precarity: Low-Paid Economic Migrants' Housing in Manchester. *Housing Studies, 38*(2), 307–326. https://doi.org/10.1080/02673037.2021.1882663

Lund, B. (2008). Major, Blair and the Third Way in Social Policy. *Social Policy & Administration, 42*(1), 43–58. https://doi.org/10.1080/13676261.2016.118424110.1111/j.1467-9515.2007.00591.x

Madden, D., & Marcuse, P. (2016). *In Defense of Housing*. Verso.

Malpass, P. (2003). The Wobbly Pillar? Housing and the British Postwar Welfare State. *Journal of Social Policy, 32*(4), 589–606. https://doi.org/10.1017/S0047279403007177

Marsh, A. (2018). *Social Housing Governance: An Overview of the Issues and the Evidence*. Collaborative Centre for Housing Evidence.

Marsh, A., & Gibb, K. (2019). *The Private Rented Sector in The UK: An Overview of the Policy and Regulatory Landscape*. UK Collaborative Centre for Housing Evidence.

McKee, K. (2012). Young People, Homeownership and Future Welfare. *Housing Studies, 27*(6), 853–862. https://doi.org/10.1080/02673037.2012.714463

McKee, K., Muir, J., & Moore, T. (2017). Housing Policy in the UK: The Importance of Spatial Nuance. *Housing Studies, 32*(1), 60–72. https://doi.org/10.1080/02673037.2016.1181722

McKee, K., Soaita, A. M., & Hoolachan, J. (2020). 'Generation Rent' and the Emotions of Private Renting: Self-Worth, Status and Insecurity Amongst Low-Income Renters. *Housing Studies, 35*(8), 1468–1487. https://doi.org/10.1080/02673037.2019.1676400

McKenzie, R., & Atkinson, R. (2019). Anchoring Capital in Place: The Grounded Impact of International Wealth Chains on Housing Markets in London. *Urban Studies, 57*(1), 21–38. https://doi.org/10.1177/0042098019839875

MHCLG. (2020). The Charter for Social Housing Residents: Social Housing White Paper. *Ministry of Housing, Communities and Local Government.* https://assets.publishing.service.gov.uk/government/uploads/system/uploads/attachment_data/file/936098/The_charter_for_social_housing_residents_-_social_housing_white_paper.pdf [Accessed: 18 February 2022].

Morrison, N. (2017). Selling The Family Silver? Institutional Entrepreneurship and Asset Disposal in the English Housing Association Sector. *Urban Studies, 54*(12), 2856–2873. https://doi.org/10.1177/0042098016653874

Mulheirn, I. (2019). *Tackling the UK Housing Crisis: Is Supply the Answer*? UK Collaborative Centre for Housing Evidence.

National Audit Office (NAO). (2010). *The Decent Homes Programme. National Audit Office* [online]. https://www.nao.org.uk/wp-content/uploads/2010/01/0910212.pdf [Accessed 12 July 2022].

Powell, R. (2015). Housing Benefit Reform and the Private Rented Sector in the UK: On the Deleterious Effects of Short-term, Ideological "Knowledge". *Housing, Theory and Society, 32*(3), 320–345. https://doi.org/10.1080/14036096.2015.1027830

Powell, R., Dunning, R., Ferrari, E. T., & McKee, K. (2015). *Affordable Housing Need in Scotland.* Shelter Scotland/CIH Scotland/SFHA.

Powell, R., & Robinson, D. (2019). Housing, Ethnicity and Advanced Marginality in England. In J. Flint & R. Powell (Eds.), *Class, Ethnicity and State in the Polarized Metropolis: Putting Wacquant to Work* (pp. 187–212). Palgrave Macmillan.

Preece, J., Hickman, P., & Pattison, B. (2020). The Affordability of "Affordable" Housing in England: Conditionality and Exclusion in a Context of Welfare Reform. *Housing Studies, 35*(7), 1214–1238. https://doi.org/10.1080/02673037.2019.1653448

Ravetz, A. (2001). *Council Housing and Culture: The History of a Social Experiment.* Routledge.

Reisenbichler, A. (2021). Housing Finance between Social Welfare and Growth Strategies. In A. Hassel & B. Palier (Eds.), *Growth and Welfare in Advanced Capitalist Economies* (pp. 320–347). Oxford University Press.

Ronald, R. (2008). *The Ideology of Home Ownership: Homeowner Societies and the Role of Housing*. Palgrave Macmillan.

Ronald, R., Lennartz, C., & Kadi, J. (2015). *Whatever Happened to Asset-based Welfare? Shifting Approaches to Housing Wealth and Welfare Security*. (HOUWEL Working Paper Series; No. 7). Centre for Urban Studies, University of Amsterdam.

Slater, T. (2018). The Invention of the 'Sink Estate': Consequential Categorisation and the UK Housing Crisis. *The Sociological Review, 66*(4), 877–897. https://doi.org/10.1177/0038026118777451

Smith, B., & Mackie, P. (2021). Housing in Wales: Evidence for Welsh Government 2021–26. *UK Collaborative Centre for Housing Evidence*. https://housingevidence.ac.uk/wp-content/uploads/2021/04/Welsh-paper.pdf [Accessed: 18 February 2022].

Smyth, S. (2019). Embedding Financialization: A Policy Review of the English Affordable Homes Programme. *Housing Studies, 34*(1), 142–161. https://doi.org/10.1080/02673037.2018.1442561

Stephens, M. (2019). Social Rented Housing in the (DIS)United Kingdom: Can Different Social Housing Regime Types Exist within the Same Nation State? *Urban Research & Practice, 12*(1), 38–60. https://doi.org/10.1080/17535069.2017.1381760

Stephens, M. (2020). How Housing Systems are Changing and Why: A Critique of Kemeny's Theory of Housing Regimes. *Housing, Theory and Society, 37*(5), 521–547. https://doi.org/10.1080/14036096.2020.1814404

Tunstall, R. (2016). Housing. In R. Lupton, T. Burchardt, J. Hills, K. Stewart, & P. Vizard (Eds.), *Social Policy in a Cold Climate: Policies and their Consequences Since the Crisis* (pp. 125–146). Policy Press.

Tunstall, R. (2022). *Is Housing The 'Saving Grace' In Britain's Otherwise 'Ungenerous' Welfare State?* Housing Studies Association Conference 2022. https://www.researchgate.net/publication/360684415_Is_housing_the_'saving_grace'_in_Britain's_otherwise_'ungenerous'_welfare_state [Accessed: 12 October 2022].

Twomey, J. (2022). Moody's: HAs to Scale Back Development and Lever Up Amid £20bn Safety and Decarb Costs. *Social Housing*. https://www.socialhousing.co.uk/news/news/moodys-has-to-scale-back-development-and-lever-up-amid-20bn-safety-and-decarb-costs-74718 [Accessed: 19 October 2022].

Wainwright, T. (2009). Laying the Foundations for a Crisis: Mapping the Historico-Geographical Construction of Residential Mortgage Backed Securitization in the UK. *International Journal of Urban and Regional Research, 33*(2), 372–388. https://doi.org/10.1111/j.1468-2427.2009.00876.x

Watt, P. (2018). "This Pain of Moving, Moving, Moving": Evictions, Displacement and Logics of Expulsion in London. *L'Année sociologique, 68*, 67–100. https://doi.org/10.3917/anso.181.0067

Watt, P. (2021) *Estate Regeneration and its Discontents*. Policy Press.

Watts, B., Fitzpatrick, S., Bramley, G., & Watkins, D. (2014). Welfare Sanctions and Conditionality in the UK. *Joseph Rowntree Foundation*. https://www.jrf.org.uk/sites/default/files/jrf/migrated/files/Welfare-conditionality-UK-Summary.pdf [Accessed: 18 February 2022].

Whitehead, C., & Meen, G. (2020). Housing Affordability Has Worsened for Many as Incomes Stagnate. *LSE Blogs*. https://blogs.lse.ac.uk/lselondon/housing-affordability-has-worsened-for-many-as-incomes-stagnate/ [Accessed: 18 February 2022].

Wilson, W., Baxter, J., Berry, K., & Murphy, E. (2019). Comparing Private Rented Sector Policies in England, Scotland, Wales and Northern Ireland. *House of Commons Library*. https://researchbriefings.files.parliament.uk/documents/CBP-7624/CBP-7624.pdf [Accessed: 19 October 2022].

5

HOUSING AND WELFARE IN THE UNITED STATES

J. Rosie Tighe and Patrick Grogan-Myers

The current state of housing in the region

Despite being one of the wealthiest countries in the world, the United States provides very little assistance to the poor. Social welfare spending in the United States is well below the average of all OECD countries (OECD Social Expenditure Report). Furthermore, recent studies have shown that social spending in the United States is allocated primarily to the elderly, rather than to working families. (Tikkanen & Schneider, 2020) This is particularly true in the realm of housing, where only one-quarter of families who qualify for housing assistance receive it. The reasons for this failure are numerous – from a conception of housing as an investment or asset rather than home or shelter to a particularly American ideology that values the protestant work ethic and individualism over egalitarianism (Erikson & Tedin, 2003; Hays, 1985). The United States is more politically conservative than most other OECD countries. The average American is much more likely to view government as a cause of, rather than a solution to, social problems. Thus, *limiting* government intervention in the housing sphere is viewed by most Americans as the most effective mechanism to alleviate housing problems.

Another aspect of this value system is a belief that poverty in a "land of opportunity" like the United States is largely the fault or failure of the poor person (Bobo, 1991; Jencks, 1992; Massey & Denton, 1990; Wilson, 1987). This combination of ideology and perceptions of the poorly shaped housing and welfare policies from tenement reform in the early 1900s to welfare reform in the 1990s to 21st century calls for poverty deconcentration (Gilens, 1996; Goetz, 2000; Imbroscio, 2008). Furthermore, the protection of private property ownership and investment remains one of the most staunchly defended roles of the government in the United States (Alesina, Glaeser & Sacerdote, 2001; Hartman, 1998; Iglesias, 2007). Because of this emphasis on private property rights, government policies designed to aid the poor through housing have had to balance the need for such housing with a desire to protect private property and investment. Therefore, while Americans widely support aid for those who cannot help themselves, the methods, means, and extent of such aid are fiercely contested (Berinsky, 2002; Erikson & Tedin, 2003; Schneider & Ingram, 1993). Ideology and perceptions of the poor impeded the development of a consistent and sound welfare policy in the United States.

DOI: 10.1201/9781003212690-7

Over time and in spite of these challenges, health and safety regulations have gradually improved housing quality for poor people. According to the latest (2020) census, there are only about 6 million "substandard" housing units. The narrow definition of substandard[1] in the United States, however, belies the millions of families housed in environmentally hazardous conditions. A recent study found that up to 45 per cent of metropolitan households live in conditions hazardous to their health (National Center for Healthy Housing, 2022). Such poor quality housing is implicated in a vast array of health detriments due to pest infestation, heavy metal poisoning, lack of maintenance and proximity environmental hazards (Bennett, 2004; Currie & Sorensen, 2019; Evans et al., 2000; Suglia et al., 2011). Furthermore, in the United States, there are currently half a million unhoused people living in temporary shelters and camps. Nearly a third of these are families with children. The unhoused have a life expectancy of only 50 years – roughly one-third shorter than average.

In the housing policy realm, government subsidies have consistently supported middle and upper-class households rather than the poor. Only one in four eligible low-income households receives housing assistance through the primary subsidized housing vehicles: vouchers; tax credits; and public housing. Currently, the average time to wait for a voucher is 2.5 years, with many waiting as long as eight years to receive affordable housing (CBPP 2022). Furthermore, finding and paying for housing is challenging even for those lucky enough to receive benefits, because landlords are not required to accept tenants on housing vouchers (Tighe et al., 2017). While some cities and even a few entire states have passed laws eliminating this form of discrimination, studies show that there is little enforcement, and voucher holders continue to encounter discrimination in their housing searches (City of Chicago Commission on Human Relations, n.d.; Flagg, 2011; Healy & Lepley, 2016; Tighe et al., 2017).

The lack of tenant protections in most areas of the United States compounds an uneven and unfair market where landlords hold nearly all the power (Hatch, 2017). Rent control measures are nearly unheard of in the United States, so landlords can raise rents to what the market will bear at almost any time. Standard lease terms in the United States are comparatively short – one year or less – meaning landlords have ample opportunity to raise rents and force out tenants. The Covid-19 crisis highlighted this (Goodman & Magder, 2020; Liu & Su, 2021). Even with a national eviction ban, thousands of families ended up on the street in the middle of a pandemic due to inconsistent application of the Centers for Disease Control eviction moratorium or landlords challenging the use of the moratorium (Nova, 2020). As rents rise, landlords are eager to capitalize; evicting or failing to renew the leases of long-term tenants in favour of those who can pay more. So far, governmental policies have failed to address the housing needs of low-income Americans either through direct subsidy or regulation of the private market. Meanwhile, homeowners enjoy long-term (typically 30-year) fixed-rate mortgages and are able to deduct the interest from their income taxes, ensuring relative stability for those households.

A brief overview of the historical development of the housing regime

The United States has traditionally eschewed direct public housing development and management and systematically dismantled its 20th-century public housing program through poor policy design and deliberate mismanagement (Goetz, 2013). Since 1970, housing policy decision-making has been devolved to states that devise their own policies using federal funding. Overwhelmingly, those policies have relied on public-private partnerships to facilitate the development, management, and finance of subsidized housing in the United States (Bauman et al., 2000; Schwartz, 2021).

Housing policy: Finance

The first major Depression-era housing action was the creation of the Federal Housing Administration (FHA) in 1934. The primary purpose of the FHA was to stem foreclosures and spur home construction by insuring loans (Radford, 1996; Schwartz, 2014). The creation of the FHA and associated government-sponsored enterprises, however, failed to directly aid low-income households, which poses a significant ongoing problem for the US housing market as the financing of single-family housing has remained largely unchanged since the 1930s.

The benefits of the financial institutions and instruments created between the Great Depression and the end of the Second World War dramatically reduced the risk of lending and consequently made homeownership available to a much larger segment of American households. Furthermore, these changes "stabilized the flow of funds to housing during the Depression and set the stage for strong growth in the immediate postwar years" (Lea, 1996, p. 163). FHA mortgage lending, in conjunction with other policies that promoted and enabled suburban homeownership (including the GI Bill and the National Highway Act), increased the homeownership rate from 47 per cent before the FHA to 62 per cent in 1960 (HUD, 2004).

This system enabled an entirely new class – the middle class – to become homeowners by establishing, encouraging, and insuring mortgage loans. However, it did little to address the needs of the most disadvantaged populations. Despite the dramatic expansion in the availability of mortgage credit, discriminatory practices codified in FHA regulations allowed only white residents to utilize the new system (Freund, 2010; Rothstein, 2017). There were two housing markets – one white, with cheap money and ample supply, and one Black, with expensive money and very little supply. (Lake, 2017; Taylor, 2012, 2019) As a result, many Black people were paying far more for housing than it was actually worth, and gaining no equity or wealth from it (Taylor, 2012). Furthermore, the investment potential of their property was limited because if they wanted to sell, and capitalize on that investment, they could only find it in the form of other Black buyers, who lacked access to capital. Well after racial discrimination in lending was banned, Black people continued to pay considerably more for their loans than their white counterparts (Taylor, 2019). In 2020, high-income Black homeowners paid higher interest rates than low-income white homeowners (Hanifa, 2021).

In the case of FHA mortgages, which broadly made homeownership a reality for the middle class for the first time, not only were many minority borrowers excluded from obtaining home loans, but entire neighbourhoods were off-limits to potential investors (Immergluck, 2009). In order to obtain a mortgage, lending rules required that the home be located in an area occupied with "substantial, law-abiding, sober-acting, sane-thinking people of acceptable ethnic standards" (Vale, 2000, p. 169). These regulations, commonly known as "redlining," utilized a hierarchical rating system to rank neighbourhood suitability for mortgage investment (Briggs, 2005; Jackson, 1985; Massey & Denton, 1990). The designers asserted that proximity to Black areas and households would make a loan more likely to be foreclosed on because the neighbourhood might become "unstable." Thus, the government's assessment of risk directly corresponded to proximity to whiteness (Bond & Williams, 2007). All-white neighbourhoods surrounded by all-white neighbourhoods typically received the highest rating, "A," or "green" indicating minimal investment risk (Michney, 2021; Nelson et al., 2022).

Unlike their ethnic white counterparts, upwardly mobile Black families couldn't move out to the suburbs due to nearly universal deed restrictions in new suburban developments which barred them from ownership. As a result, such households became trapped in neighbourhoods that were, "cut off from sources of new investment[,] their homes and communities deteriorated and lost

Table 5.1 Market shares of housing tenures in the United States 1945–2020[2]

	1940	*1950*	*1960*	*1970*	*1980*	*1990*	*2000*	*2010*	*2020*
Total Housing Units	37.4M	46.2M	58.3M	68.7M	88.4M	102.3M	116KM	131.7M	140.5M
Total Occupied Units[3]	N/A	N/A	N/A	N/A	N/A	N/A	N/A	N/A	N/A
Owner Occupied	43.6%	55.0%	61.9%	62.9%	64.4%	64.2%	66.2%	65.1%	65.3%
Renter Occupied	56.4%	45.0%	38.1%	37.1%	35.6%	35.8%	33.8%	34.9%	34.7%
Public Housing Units[4]		0.65%	0.73%	1.21%	1.35%	1.37%	1.11%	0.89%	0.68%

value in comparison to those homes and communities that FHA appraisers deemed desirable." (Oliver & Shapiro, 2013). Meanwhile, "African Americans who desired and were able to afford homeownership found themselves consigned to central-city communities where their investments were affected by the "self-fulfilling prophecies" of the FHA appraisers." (Oliver & Shapiro, 2013) (Table 5.1).

Housing for low-income households in the United States is increasingly scarce, and little is provided directly by the federal government. The US Constitution provides the framework for the sharing and execution of powers, with state governments and the people exercising those powers not specifically delegated to the federal government or prohibited for the State governments to exercise (US Const. amend. X). The result is that the 50 states that make up the United States have significant control over their own housing policy. As such, there are significant differences across states and cities in terms of the extent and effectiveness of state and local housing regulations and subsidies.

Housing policy: Regulation

By the late 19th Century, industrialization had spurred intense overcrowding in America's tenement districts. New York City passed the first Tenement Reform Act in 1867, serving as a model for other cities around the country (Krieger & Higgins, 2002; Marcuse, 1978). Such minimum housing standard regulations (for example, requiring windows and fire escapes), however, did little to alleviate conditions in tenements, since the urban populations continued to grow. As a result, organizers and advocates continued to lobby for federal action to improve tenement conditions (Bauman et al., 2000; Marcuse, 1978). As the perception of the tenements' increasingly negative impact on society spread, calls for reform grew louder. Policymakers were driven to action by the "fear of what an excluded, alienated, disgruntled, badly housed population, to a large extent immigrant and not imbued with 'American' ideals and patterns of acceptable behaviour, might produce, particularly if crowded together in congested neighbourhoods with only their like" (Marcuse, 1978, p. 73). Thus, the emphasis of tenement reform involved not only improving the physical structure of the buildings but also changing the tenants' behaviour and their familial and social structures.

Zoning emerged in the nineteen teens as a mechanism to protect the health, safety, and welfare of the population, and quickly became the primary tool to shape local land use (Hirt, 2015; Wolf, 2008). Zoning has become, "a shorthand of the unstated rules governing what are widely regarded as correct social categories and relationships – that is, not only how land uses should be arranged, but how land users, as social categories, are to be related to one another (Perin, 1977, p. 3). Roughly three-quarters of all metropolitan land in the United States is zoned for single-family residential use. This hyper-protection of the single-family home has driven housing costs higher across time and space and has been maintained even as other forms of residential exclusion have been dismantled.

Cities and towns widely employ zoning to ensure that residential areas stay separate from industrial and commercial developments in order to avoid undue exposure to the potentially harmful health, safety, and environmental externalities associated with many commercial or industrial uses (Fischel, 2004; Pogodzinski, 1991). However, restrictive zoning measures can effectively exclude lower-income residents from moving into a community by limiting residential development at a scale affordable to them (Cowan, 2006; Ihlanfeldt, 2004). Some estimates show that 75 per cent or more of housing in US cities is located in single-family zones, meaning that only single-family homes can be built there (Badger & Bui, 2019). No duplexes; no multi-family housing; no corner stores; no cafes. Single-family zoning is a reason that so many artists, scholars, and writers have decried the "soullessness" of American suburbia (for example, Duany et al., 2001; Dunham-Jones & Williamson, 2011; Hayden, 2003; Miller, 2018). Such regulatory mechanisms historically deepened segregation by both race and class, and continue to do so, albeit in a less obvious and explicit manner than the overt policies of the pre-Civil Rights Act era (Orfield, 2006; Seitles, 1998).

Housing policy: Present-day policy and tools

Public housing – subsidized units managed by the state – has been steadily dismantled since Nixon's election in 1968. Filling the gap are three mechanisms used to provide subsidized housing to moderate and low-income households in the United States – vouchers, block grants, and tax credits. Each of these tools centres on property rights, the private market, and individualism.

Housing Choice Vouchers (formerly known as Section 8) provide recipients with a coupon they can use to pay up to 75 per cent of their housing costs. Vouchers take advantage of private-sector housing, not requiring direct construction or management of housing by the government (Turner, 2003). Vouchers appeal to many policymakers because they "involve less active interference in the production of housing by the private market than any other type of public subsidy" (Hays, 1985, 140). From a free-market, limited government standpoint, vouchers provided a good alternative to the costly and intrusive government-sponsored production programs of the 1960s (Burchell & Listoken, 1995; Krumholz, 2004). Housing advocates value their flexibility, as well as the intention of de-concentrating the poor by allowing them to move to any neighbourhood (Hays, 1985; Marcuse & Keating, 2006). However, they have yet to succeed in providing mobility to most of their holders, limiting the positive benefits they might have in the realm of poverty deconcentration and access to opportunity (Goetz, 2013; Imbroscio, 2008; McClure et al., 2015; Wang, 2018).

While theoretically, voucher holders can settle anywhere, research shows that recipients are no more likely than unsubsidized households to penetrate discriminatory market barriers and find rental accommodations in integrated living environments (McClure et al., 2015; Wang, 2018). This largely stems from the fact that the law does not require landlords (in most states and cities) to accept vouchers (Tighe et al., 2017). Because housing policy is diffused to state and local governments in the United States, enactment and enforcement of source of income protections—protecting those who derive their housing payments from vouchers—is patchwork, despite households with vouchers being able to move and take their voucher with them (McClure et al., 2015; Tighe et al., 2017). Furthermore, the rent ceiling imposed on the program prevents most recipients from moving into non-poor neighbourhoods (Imbroscio, 2008; McClure et al., 2015).

Second, states and municipalities receive Block Grants to use, for the most part, at their discretion (Orlebeke, 2000). While certain parameters exist regarding income targeting and some preferences for location, states and cities maintain significant control over how they choose to utilize their block grant funding (Burchell & Listoken, 1995). Two grant programs specifically address housing affordability: the Community Development Block Grant and the HOME program. The passage of the Housing and Community Development Act, which established CDBG, in 1974 represents one of the earliest steps in the retrenchment process, eliminating much of the federal oversight and guidelines for housing development and instead distributing funds using specific formulae to states and cities (Orlebeke, 2000). However, in practice, CDBG did little to stimulate affordable housing production (Goetz, 1995).

Working to address this issue, the 1990 Housing Act created the first instalment of the HOME program, which strove to remove the bureaucratic entrenchment of HUD and reinstate local authority over housing (Burchell & Listoken, 1995). Today, HOME is, "the largest federal block grant to state and local governments designed exclusively to create affordable housing for low-income households" (HUD, 2004). Rental housing programs comprise more than half of the nearly $2 billion in annual funding, which is split equally between production and preservation programs (Arigoni, 2001).

Third, with voucher programs in place to aid low-income families without production and block grants supplying flexible spending to local governments, the federal government eventually used tax incentives to increase opportunities for private development. Since its inception in 1986, the Low Income Housing Tax Credit (LIHTC) evolved into the principal mechanism for the production of affordable housing in the United States (Burchell & Listoken, 1995; Orlebeke, 2000), codifying the desire to "bring the efficiency and discipline of the private market to the building of affordable rental housing" (Cummings & DiPasquale 1999, p. 252). Qualifying LIHTC properties must contain at least 20 per cent units affordable to households earning 50 per cent or less of area median income (AMI) or 40 per cent to households earning less than 60 per cent of AMI (Cummings & DiPasquale, 1999; McClure, 2000). Unit affordability must be maintained for at least 15 years, but in some instances, the term of affordability is longer (Wallace, 1995).

In practice, however, developments using LIHTC funding are nearly always fully income restricted (McClure, 2008). The LIHTC also contains a preference for development in "qualified census tracts" – an area where at least half of the households earn less than 60 per cent of the AMI (Cummings & DiPasquale, 1999; Mueller & Schwartz, 2008). This requirement ostensibly promotes the development of affordable housing in communities where it is most needed, yet it runs counter to the goal of dispersing housing and promoting its construction in non-poor areas to avoid concentrating on the poor. As a result of such program guidelines as well as community pressures, development occurs primarily in low-income, metropolitan neighbourhoods, compounding segregation and inequality (Cashin, 2021; Trounstine, 2018).

The welfare regime and its development

Over time, housing in the United States has been viewed primarily as a commodity; asset; or investment vehicle rather than as shelter or home. As such, Americans do not view housing as a human right, despite an international movement calling for such (Kluegel & Smith, 1986). Since the early 1970s, the US government has gradually withdrawn from the provision of necessary goods and services to the poor. As a result, income and wealth inequality is higher in the United States than in almost any other developed country. Only four OECD countries have a higher GINI index than the United States, but the truly startling comparative measures are those that assess wealth. The United States is a massive outlier in wealth distribution, where the wealthiest 10 per cent of households own close to 80 per cent of total wealth. The country with the second highest concentration of wealth is the Netherlands, at 56 per cent. The United States provides significantly fewer housing and welfare benefits than its counterparts, and most of that spending is directed at older Americans in the form of Medicare, social security, and other retirement benefits than at working-age Americans and children in the form of housing and childcare (Fischer & Sard, 2017; Tikkanen & Schneider, 2020). In 2022, the United States virtually abandoned the idea of direct state provision of housing, relying almost completely on private developers and landlords.

The Great Depression of the 1930s created a housing crisis that transcended the tenements and immigrant ghettos, thereby compelling government intervention. Despite this acknowledgement, philosophical and political disagreements continued to hinder government action (Vale, 2000). On one side, public housing advocates sought to reform and aid the poor by creating a living environment, "antithetical to the urban slum" with proper light, heat, and plumbing (von Hoffman, 1996, 426). On the other, the real estate lobbies vehemently opposed government involvement in housing production. Most policymakers shared the views of the latter group, repeatedly blocking the establishment of a public housing program (Bratt, 1989).

While redlining deprived Black communities of investment, it allowed white communities to thrive. When Black people tried to integrate white communities, the opposition was organized, immediate, and often violent (Abrams, 1955; Meyer, 2000). Together, institutional discrimination, redlining, and communal racism excluded minority households from the opportunities afforded whites. Instead, upwardly mobile Black families, trapped in increasingly poor neighbourhoods, received no access to capital for improving their homes or establishing businesses (Satter, 2009; Taylor, 2019).

Federally sponsored housing was one of the few options left for those who could not qualify for homeownership. Ultimately, in spite of ideological opposition and economic concerns regarding public housing's impact on the private housing market, the federal government established a limited program in 1937. The goal of public housing was to provide shelter for the "submerged middle class" until they could purchase their own home (Bloom et al., 2015; Vale, 2000). The creation of this program represented a huge step forward for government involvement in housing, and by 1960, it had facilitated the construction of 400,000 public housing units to accommodate the poor (Bratt, 1989; Burchell & Listoken, 1995). However, these projects tended to be located on marginal tracts of land on the outskirts of town, in undesirable areas (Bickford & Massey, 1991; Vale, 2002).

Due to a combination of redlining, suburbanization, and white flight, America's inner cities became increasingly associated with blight and crime. In response, the public, members of the business community, and policymakers began to take an interest in the revitalization of urban centres (Anderson, 1964; Blessett, 2020). As a result, urban policy shifted its focus to economic development and growth in America's cities, seeking to reverse the deterioration of urban neighbourhoods and job centres (Jackson, 1985). These new priorities manifested in the Housing Act of 1949, designed to improve the public perception of American cities and to combat the fact that publicly built housing failed to offer relief to those in the greatest need (Lang & Sohmer, 2000; von Hoffman, 1996). The act is primarily remembered for the laudable goal of providing "a decent home and a suitable living environment for every American family" (Lang & Sohmer 2000, p. 291). The means to accomplish this were threefold: slum clearance, increased authorization of FHA loans, and the development of more than eight hundred thousand public housing units (Lang & Sohmer, 2000).

The majority of the funding stemming from the 1949 Housing Act went into the slum clearance portion of the program (generally known as "Urban Renewal") (Dreier, 2000; Krumholz, 2004).

In place of the slums and disinvestment that blighted American neighbourhoods, the Housing Act authorized the construction of thousands of brand-new housing units. However, the implementation of the urban renewal program landed in the hands of private developers, with very little federal oversight. Consequently, these measures seldom resulted in the construction of as many units as they condemned (Bennett et al., 2006; Krumholz, 2004; Lane, 1995). Furthermore, the housing constructed in place of those demolished was generally reserved for middle- and upper-class households, while housing for low-income families was placed in isolated areas far from established residential and job centres, as well as their previous neighbours and community ties (Bartelt, 1993; Gans, 1962). By the mid-1960s, 1 million poor and working-class residents of urban neighbourhoods had been displaced by Urban Renewal (Marcuse, 1978).

As units built in the 1940s aged and the wealth of the tenants decreased, less money was available for maintenance precisely when the buildings required significant repair (Bennett et al., 2006; Goetz, 2013; Vale, 2002). By the 1960s, the crisis escalated to such a point that HUD had two options that could salvage public housing: continue escalating rents in order to cover costs, leaving the neediest families outside of the benefit structure; or re-work the program to increase coverage of operating costs through federal subsidies (Bratt et al., 2009). It did neither.

By this time the public housing program was receiving criticism from all sides. Housing authorities across the country simply did not have the money to maintain the projects because the federal government had failed to provide it to them. This forced them to rely on rents, and the rents were too low to sustain maintenance. Housing Authorities across the country were forced to raise rents while tenants lived in deplorable conditions (Freidrichs et al., 2011). While housing officials, "showed some resentment at being treated like slumlords… [they also contended] that pure economics forced their hand." (Adams, 1968). As conditions in the projects became national news, Nixon took office and almost immediately withdrew support for low-income housing policies. Following the demolition of Pruitt-Igoe in 1971 – a mere 17 years after the first residents moved in – Nixon declared a moratorium on all new housing projects. Since the 1972 moratorium, federal dollars for public housing dried up. Few dollars went towards maintenance and, by 1980, many projects were nearing 50 years old, with very little upkeep (Goetz, 2013). They were falling apart, and the only people left were the most marginalized, most desperate people who had no place else to go.

The interrelation between housing and welfare

All social policies are subject to arguments about the nature of poverty and the poor (Abramovitz, 2001; Banfield, 1968; Gilens, 1999; Jencks, 1992; Katz, 1996). Public support for such policies often hinges on the perception of whether the poor are perceived to be responsible for their situation, or whether they are instead victims of societal or economic failures (Bobo, 1991; Wilson, 1996; Zucker & Weiner, 1993). As a result, public policies aimed at, "improving 'human capital,' correcting 'community pathology,' breaking the 'culture of poverty,' healing the 'broken family' all tended to restrict the problem to a 'disadvantaged' population outside what was considered a basically sound 'mainstream'" (Jackson, 1993, p. 416). Race continues to be a key factor in how a primarily white, middle class "mainstream" continues to be differentiated from a primarily non-white "underclass" in both academic and policy circles. This perception – widely held in the United States – has limited the establishment of a robust welfare state similar to its European counterparts. Over the past 50 years, while many EU countries have augmented their provision of services to the poor, the United States has steadily reduced welfare benefits for the non-elderly poor (Tikkanen & Schneider, 2020). This has had a disproportionate effect on people of colour who were shut out of the wealth-building opportunities of previous generations.

By the late 1960s, racial discrimination and redlining were illegal, but decades of institutionalized racism had created segregated metropolitan areas across the country (Cashin 2004; Massey & Denton, 1990; Seitles, 1998; Wilson, 1987). Residential integration was one of the primary goals of the civil rights movement, yet the obstacles seemed daunting, and the tools were limited (Anderson, 2002; Cashin, 2004). Local attempts at compliance faced vehement opposition from residents of neighbourhoods forced to deal with integration policies such as mandatory busing to integrate school districts (McConahay, 1982; Lukas, 1985). Such vehement community opposition to forced desegregation hindered the realization of the principles embodied in the Civil Rights Act of 1964. The housing element of that Act took four more years to pass because of the racism so deeply entrenched in white America when faced with sharing neighbourhoods with people of colour (Bobo, 1991; Bobo et al., 2012; Cashin, 2004; Schuman & Bobo, 1988; Tighe, 2012).

Most American communities remained segregated by both race and income and racial discrimination in lending continued to occur for decades after the Fair Housing Act supposedly eliminated it. Lending data showed that huge gaps remained between white and nonwhite Americans in terms of access to homeownership opportunities – particularly for those living outside of urban areas. Meanwhile, American homes so regularly increased in value, they were eventually turned into a global investment vehicle. As the 1970s experienced rising inflation and interest rates, savings and loan institutions, which can be thought of as smaller banks focused on pooling together deposits to make loans to purchase homes, struggled to return profits and remain competitive due to federal regulations and the changing economic conditions (Grabowski et al., 1994). Deregulation broadened products that could be offered by nonbanking institutions, introducing more competition for banks and, it was argued, allowing for more efficient market operations (Grabowski et al., 1994).

The potent combination of economic factors and deregulation, which encouraged riskier behaviour, resulted in the collapse of the savings and loan market. By the end of 1986, 47 per cent of the savings and loan assets were held by savings and loans that were insolvent or near insolvency; the total cost to the taxpayers for closing the insolvent savings and loans between 1986 and 1995 was estimated to be $152.9 billion (Curry & Shibut, 2000). In spite of the savings and loan collapse, deregulation continued during the first decade of the 21st century, allowing banks to make riskier loans to less creditworthy customers ("subprime") and introducing even more risk into what were once considered stable investments. Risky subprime mortgages were packaged into global investments with the assistance of credit default swaps, which act as insurance against default; the growth in usage of credit default swaps was seen as the market went from approximately $180 billion in 1998 to $57 trillion in June 2007 (Stulz, 2010). Riskier lending backed by the perceived safety of credit default swaps flooded the markets until it eventually spurred the 2007 housing crisis, which had worldwide implications.

Homelessness rose by 61 per cent in the years following the foreclosure crisis. The Great Recession witnessed millions of foreclosures, largely on low-income households and communities. This was particularly devastating to communities of colour, who were just starting to see some closing of the racial homeownership gap (Taylor, 2019). The policy responses to the crisis often compounded the problems of widespread foreclosure by focusing on shoring up financial institutions rather than supporting borrowers, and providing enormous amounts of funding for the demolition of properties in low-income neighbourhoods (Immergluck, 2011; Mallach, 2009).

Regional outlook

Because the policies that make up the United States' housing and welfare regime are patchwork, developed at the federal, state, and local levels, and overly reliant on private development to meet the supply of both affordable and market-rate housing, housing in the United States is subject to the fluctuations in the economy. Housing supply is not static across the country. While New York City and San Francisco residents are desperate for rental housing and paying upwards of $5 per square foot for the privilege, thousands of units sit vacant and abandoned in cities across the Rust Belt (e.g., Detroit & Cleveland) as manufacturing and population decline coupled with white flight spurred a decades long decline. While such cities enjoy relatively[5] low housing costs, their job markets are weak and wages are low.

Since the advent of public housing in 1937, real estate interests have steadily influenced the elimination of programs to support housing for low-income families and communities. Meanwhile, the United States subsidizes middle- and upper-class homeowners through the mortgage interest tax deduction – the cost of which dwarfs the money spent on low-income housing (Schwartz, 2014). Government provision of housing for the poor has steadily been abandoned in favour of subsidizing the private sector to take on that role – particularly when it comes to building and managing low-income housing. Yet the private sector has not done so for a complex set of reasons. Some advocates for housing affordability focus on how local land use policies such as zoning, health and safety requirements, and parking regulations increase the cost of housing.

Zoning in particular has become a popular target for those seeking greater housing affordability in high-cost areas. Since Minneapolis became the first major city to eliminate SF-only zoning (also known as "upzoning") in 2019 cities across the country have considered following suit. While these measures will certainly make it easier for developers to build multi-family housing, increasing the supply of market-rate housing is not sufficient to ensure affordability for low-income households. In fact, studies suggest that upzoning zoning alone does little to improve affordability, and could spur the displacement of low-income populations (Davis, 2021; Freemark, 2020; Rodríguez-Pose & Storper, 2020).

And affordability across the United States is worsening. The Case-Shiller Home Price Index, one measure of home sales price trends, recorded over 29 per cent increase between March 2020 and December 2021 and an almost 19 per cent increase in 2021 alone (S&P Dow Jones Indices LLC, 2022). Numerous factors can impact this, including low interest rates, low inventory for sale, and the ability for workers to work remotely (Joint Center for Housing Studies of Harvard University, 2021; Tully, 2022). The ability for workers to live in one place and work for a company that is located in a far-flung city or state is not new; however, the widespread ability for people to use this live-work situation due to the COVID-19 pandemic has dramatically impacted the urban core. Early analysis of the COVID-19 pandemic's impact suggests that people are moving away from the denser urban core to less dense suburban areas (Liu & Su, 2021), leading to declining rent and purchase prices in the urban core and increasing rent and purchase prices in suburban and exurban areas of a metropolitan region (Gupta et al., 2021).

With the dramatic decrease in housing prices during the Great Recession, particularly in Midwestern and legacy cities, the American housing supply – traditionally held by individual households – is now increasingly held by corporate interests and investors. Investors purchased one-quarter of all homes on the market in 2021 – an increase of nearly 100 per cent from a decade earlier (Henderson, 2022). Such owners have little incentive to keep the house occupied by long-term tenants when they can make more money renting through Airbnb or even keeping it empty because the tax implications are so positive. These trends constrain supply that is already limited by zoning and other land use regulations that prevent new housing (especially new multi-family and/or subsidized housing) from being built in nearly every community in the country. In the past 20 years, new home construction (supply) has outpaced household formation (demand), yet home prices continue to rise. In New York City alone, there are over 350,000 vacant units, up from 248,000 vacant units in 2017 (New York City Department of Housing Preservation and Development, 2022). Despite this, rents increased by 24 per cent with the median rent at $3,700 per month in 2022.

Concluding discussion

Whether it is housing, food insecurity, jobs and workforce development, or education, all social policies are subject to arguments about the nature of poverty and the poor (Abramovitz, 2001; Banfield, 1970; Gilens, 1995; Jencks, 1992; M. B. Katz, 1996). Public support for such policies often hinges on the perception of whether the poor are perceived to be responsible for their situation, or whether they are instead victims of societal or economic failures (Bobo, 1991; Wilson, 1996; Zucker & Weiner, 1993). In either case, policymakers and the public commonly view the poor as different or distinct from the remainder of society (Bartelt, 1993; Orfield, 2006). As a result, public policies aimed at, "improving 'human capital,' correcting 'community pathology,' breaking the 'culture of poverty,' healing the 'broken family' all tended to restrict the problem to a 'disadvantaged' population outside what was considered a basically sound 'mainstream'" (Jackson, 1993, p. 416). The characterization of the poor as a group culpable for, rather than victims of, their station in life serves to perpetuate both their social and spatial isolation (Briggs, 2003; Jargowsky, 2003; B. Katz, 2006).

Where there are policies that seek to protect people from predatory behaviour, such as consumer protections for borrowers, they have also been under attack. Whereas the mortgage crisis spurred some policy action, notably the creation of Consumer Financial Protection Bureau and the establishment of the Volcker Rule due to the passage of the Dodd-Frank Act, many of these protections were administratively removed or limited under the Trump administration and have yet to be reinstated. For example, the US Department of Housing and Urban Development moved to end a 2013 rule that allowed banks and lending institutions from creating apparently race-neutral policies and procedures but ultimately served to discriminate against minority borrowers (Arnold, 2019; Jan, 2020).

This rule change proved controversial and almost immediately faced challenges in court as housing discrimination continued. The Fair Housing Act bans discrimination in home purchasing and leasing based on a variety of characteristics – race; presence of children; disability; gender; and others – but enforcement is lax, and discrimination continues. While housing and tenant advocates rightly fight to expand those protections to other discriminated-against groups, the laws are irrelevant if they are not enforced.

Support for programs, such as the Low-Income Housing Tax Credit, gives reason for hope. As the country's largest affordable housing program with broad-based support, LIHTC demonstrates that there is both a demand for and interest in building more affordable housing. However, one program in the tax code is not enough; one program in the tax code does not constitute a coherent housing policy.

The history of housing development in the United States illustrates how policy intervention came as the result of wartime readiness, desires to exclude groups of people from opportunity, court cases, tax code changes, and civil rights legislation. Without a fundamental belief in the right to housing, let alone a constitutional amendment, American housing policy will continue on this fractured path. The United States is an outlier among its peer nations when it comes to housing and welfare policy. Its failure to provide basic protections for tenants; protect against predatory lending; or tackle homelessness has had a severe impact on poor Americans. As home prices continue to rise, these challenges are now being faced by more and more Americans. Questions remain as to how and whether the United States will address this crisis, yet doing so would require a fundamentally different approach to housing and welfare.

Notes

1 Lacking sufficient kitchen and/or bathroom facilities.
2 Data gathered from (Devaney, 1994; Olsen, 2001; U.S. Census Bureau, 2000, 2012, 2013, 2021; U.S. Federal Housing Administration, 1941; U.S. Housing and Home Finance Agency, 1951).
3 Data on vacancy rates is difficult to calculate due to a lack of data regarding occupied housing units. Current vacancy estimates in the United States are roughly .5 per cent for homeownership and 6 per cent for rental housing.
4 Public housing units are calculated as a percentage of total housing units, not occupied units (see previous footnote). Raw numbers: 1940 – 33,204; 1950 – 302,146; 1960 – 425,481; 1970 – 830,454; 1980 – 1,192,000; 1990 – 1,404,870; 2000 – 1,281,833; 2010 – 1,168,503; 2020 – 957,971.
5 Nowhere in the United States is it "affordable" – not even in weak market cities.

References

Abramovitz, M. (2001). Everyone is Still on Welfare: The Role of Redistribution in Social Policy. *Social Work*, *46*(4), 297–308. https://doi.org/10.1093/sw/46.4.297

Abrams, C. (1955). *Forbidden Neighbors: A Study of Prejudice in Housing*. Harper.

Adams, R. (1968, December 31). Rent Strike, Last-Resort of Slum Tenant, May Get Big Test Here. *St. Louis Post-Dispatch*, 1; 7.

Alesina, A., Glaeser, E., & Sacerdote, B. (2001, September 7). *Why Doesn't The Us Have A European-Style Welfare State?* The Brookings Panel on Economic Activity.

Anderson, E. S. (2002). Integration, Affirmative Action, and Strict Scrutiny. *New York University Law Review*, *77*(13:46), 1195–1271.

Anderson, M. (1964). *The Federal Bulldozer. A Critical Analysis of Urban Renewal, 1949-1962*. M.I.T Press.

Arigoni, D. (2001). *Affordable Housing and Smart Growth: Making the Connection* (95) [Report]. Smart Growth Network.

Arnold, C. (2019, July 13). A New Trump Rule Could Weaken A Civil Rights Era Housing Discrimination Law. *National Public Radio*. https://www.npr.org/2019/07/31/747006108/a-new-trump-rule-could-weaken-a-civil-rights-era-housing-discrimination-law

Badger, E., & Bui, Q. (2019, June 18). Cities Start to Question an American Ideal: A House With a Yard on Every Lot. *The New York Times*. https://www.nytimes.com/interactive/2019/06/18/upshot/cities-across-america-question-single-family-zoning.html, https://www.nytimes.com/interactive/2019/06/18/upshot/cities-across-america-question-single-family-zoning.html

Banfield, E. (1970). *The Unheavenly City*. Little Brown Press.

Bartelt, D. W. (1993). Housing the Underclass. In Michael B. Katz (Ed.), *The "Underclass" Debate* (pp. 119–157). Princeton University Press.

Bauman, J. F., Biles, R., Szylvian, K. M., Bauman, J. F., Biles, R., & Szylvian, K. M. (2000). *From Tenements to The Taylor Homes: In Search of an Urban Housing Policy in Twentieth-Century America*. Penn State University Press.

Bennett, L., Smith, J. L., & Wright, P. A. (Eds.) (2006). *Where are Poor People to Live? Transforming Public Housing Communities*. M.E. Sharpe.

Bennett, M. (2004). Cities in the New Millennium: Environmental Justice, The Spatialization of Race, and Combating Anti-Urbanism. *Journal of African American Studies*, *8*(1), 126–141. https://www.jstor.org/stable/41819049

Berinsky, A. J. (2002). Silent Voices: Social Welfare Policy Opinions and Political Equality in America. *American Journal of Political Science*, *46*(2), 276–287.

Bickford, A., & Massey, D. S. (1991). Segregation in the Second Ghetto: Racial and Ethnic Segregation in American Public Housing, 1977. *Social Forces*, *69*(4), 1011–1036. https://doi.org/10.2307/2579300

Blessett, B. (2020). Urban Renewal and "Ghetto" Development in Baltimore: Two Sides of the Same Coin. *The American Review of Public Administration*, *50*(8), 838–850. https://doi.org/10.1177/0275074020930358

Bloom, N. D., Umbach, F., & Vale, L. J. (2015). *Public Housing Myths: Perception, Reality, and Social Policy*. Cornell University Press.

Bobo, L. (1991). Social Responsibility, Individualism, And Redistributive Policies. *Sociological Forum (Historical Archive)*, *6*, 71–92.

Bobo, L. D., Charles, C. Z., Krysan, M., Simmons, A. D., & Fredrickson, G. M. (2012). The Real Record on Racial Attitudes. In Peter V. Madsen (Ed.), *Social Trends in American Life: Findings from the General Social Survey Since 1972*, (pp. 38–83). Princeton University Press.

Bond, C., & Williams, R. (2007). Residential segregation and the transformation of home mortgage lending. *Social Forces*, *86*(2), 671–698.

Bratt, R. G. (Ed.). (1989). *Rebuilding a Low-Income Housing Policy*. Temple University Press.

Bratt, R., Hartman, C., & Stone, M. (2009). *A Right to Housing: Foundation for a New Social Agenda*. Temple University Press.

Briggs, X. de S. (2003). Housing Opportunity, Desegregation Strategy, and Policy Research. *Journal of Policy Analysis and Management*, *22*(2), 201–206. https://www.jstor.org/stable/3325821

Briggs, X. de S. (Ed.) (2005). *The Geography of Opportunity*. Brookings Institution Press.

Burchell, R., & Listoken, D. (1995). Influences on United States Housing Policy. *Housing Policy Debate*, *6*(3), 559–617.

Cashin, S. (2004). *The Failures of Integration: How Race and Class are Undermining the American Dream*. Public Affairs.

Cashin, S. (2021). *White Space, Black Hood: Opportunity Hoarding and Segregation in the Age of Inequality*. Beacon Press.

City of Chicago Commission on Human Relations. (n.d.). *Section 8 Vouchers and Source of Income Discrimination* (p. 2).

Cowan, S. M. (7). Anti-Snob Land Use Laws, Suburban Exclusion, and Housing Opportunity. *Journal of Urban Affairs*, *28*(3), 295–313.

Cummings, J., & DiPasquale, D. (1999). The Low-Income Housing Tax Credit: An Analysis of the First Ten Years. *Housing Policy Debate*, *10*(2), 251–307.

Currie, M. A., & Sorensen, J. (2019). Repackaged "Urban Renewal": Issues of Spatial Equity and Environmental Justice in New Construction, Suburban Neighborhoods, and Urban Islands of Infill. *Journal of Urban Affairs*, *41*(4), 464–485. https://doi.org/10.1080/07352166.2018.1474081

Curry, T., & Shibut, L. (2000). The Cost of the Savings and Loan Crisis: Truth and Consequences. *FDIC Banking Review*, *13*, 26–35.

Davis, J. (2021). How Do Upzonings Impact Neighborhood Demographic Change? Examining the Link Between Land Use Policy and Gentrification in New York City. *Land Use Policy*, *103*, 105347. https://doi.org/10.1016/j.landusepol.2021.105347

Devaney, F. J. (1994). *Tracking the American Dream: 50 Years of Housing History from the Census Bureau: 1940-1990* (H121/94-1; Current Housing Reports, p. 74). Bureau of the Census.

Dreier, P. (2000). Labor's Lost Love? Rebuilding Unions' Involvement in Federal Housing Policy. *Housing Policy Debate*, *11*(2), 327–392. https://doi.org/10.1080/10511482.2000.9521371

Duany, A., Plater-Zyberk, E., & Speck, J. (2001). *Suburban Nation: The Rise of Sprawl and the Decline of the American Dream*. North Point Press.

Dunham-Jones, E., & Williamson, J. (2011). *Retrofitting Suburbia: Urban Design Solutions for Redesigning Suburbs*. John Wiley & Sons.

Erikson, R. S., & Tedin, K. L. (2003). *American Public Opinion: Its Origins Content and Impact* (6th ed.). Addison Wesley Longman.

Evans, G. W., Chan, H.-Y. E., Wells, N. M., & Saltzman, H. (2000). Housing Quality and Mental Health. *Journal of Consulting and Clinical Psychology*, *68*(3), 526–530. https://doi.org/10.1037//0022-006x.68.3.526

Fischel, W.A. (2004). An Economic History of Zoning and a Cure for its Exclusionary Effects. *Urban Studies*, *41*(2), 317–340.

Fischer, W., & Sard, B. (2017). Chart Book: Federal Housing Spending Is Poorly Matched to Need. *Center on Budget and Policy Priorities, March 8.*

Flagg, K. (2011). Mending The Safety Net Through Source of Income Protections: The Nexus Between Antidiscrimination and Social Welfare Law. *Columbia Journal of Gender and Law*, 20(1). https://doi.org/10.7916/cjgl.v20i1.2616

Freemark, Y. (2020). Upzoning Chicago: Impacts of a Zoning Reform on Property Values and Housing Construction. *Urban Affairs Review*, *56*(3), 758–789. https://doi.org/10.1177/1078087418824672

Freidrichs, C., Woodman, B., Freidrichs, J., & Henry, J. (2011). *The Pruitt-Igoe Myth*.

Freund, D. M. (2010). *Colored Property: State Policy and White Racial Politics in Suburban America*. University of Chicago Press.

Gans, H. (1962). *The Urban Villagers*. The Free Press.

Gilens, M. (1995). Racial Attitudes and Opposition to Welfare. *The Journal of Politics*, *57*(4), 994–1014. https://doi.org/10.2307/2960399

Gilens, M. (1996). Race and poverty in America: Public misperceptions and the American News Media. *Public Opinion Quarterly*, *60*(4), 515–541. https://psycnet.apa.org/doi/10.1086/297771

Gilens, M. (1999). *Why Americans Hate Welfare: Race, Media, and the Politics of Antipoverty Policy*. University of Chicago Press.

Goetz, E. G. (1995). Political Effects of Federal Policy Devolution on Local Housing Expenditures. *Publius*, *25*(3), 99–116.

Goetz, E. G. (2000). The Politics of Poverty Deconcentration and Housing Demolition. *Journal of Urban Affairs*, *22*(2), 157–173. https://doi.org/10.1111/0735-2166.00048

Goetz, E. G. (2013). *New Deal Ruins*. Cornell University Press.

Goodman, L., & Magder, D. (2020). *Avoiding a COVID-19 Disaster for Renters and the Housing Market*. Urban Institute.

Grabowski, R., Rangan, N., & Rezvanian, R. (1994). The Effect of Deregulation on the Efficiency of U.S. Banking Firms. *Journal of Economics and Business*, *46*(1), 39–54. https://doi.org/10.1016/0148-6195(94)90020-5

Gupta, A., Mittal, V., Peeters, J., & Van Nieuwerburgh, S. (2021). Flattening the Curve: Pandemic-Induced Revaluation of Urban Real Estate. *Journal of Financial Economics,* 146(2), 594–636. https://doi.org/10.1016/j.jfineco.2021.10.008

Hartman, C. (1998). A Case for a Right to Housing. *Housing Policy Debate*, *9*(2), 223–246.

Hanifa, R. (2021, February 16). High-Income Black Homeowners Receive Higher Interest Rates Than Low-Income White Homeowners. *Housing Perspectives*. https://www.jchs.harvard.edu/blog/high-income-black-homeowners-receive-higher-interest-rates-low-income-white-homeowners

Hatch, M. E. (2017). Statutory Protection for Renters: Classification of State Landlord–Tenant Policy Approaches. *Housing Policy Debate*, *27*(1), 98–119. https://doi.org/10.1080/10511482.2016.1155073

Hayden, D. (2003). *Building Suburbia: Green Fields and Urban Growth, 1820–2000*. Pantheon.

Hays, R. A. (1985). *The Federal Government and Urban Housing: Ideology and Change in Public Policy*. State University of New York Press.

Healy, L., & Lepley, M. (2016). *Housing Voucher Mobility in Cuyahoga County*. Housing Research & Advocacy Center.

Hirt, S. A. (2015). *Zoned in the USA: The Origins and Implications of American Land-Use Regulation*. Cornell University Press.

Iglesias, T. (2007). A Place to Call Home? Our Pluralist Housing Ethics and the Struggle for Affordability. *Wake Forest Law Review*, *42*, 511–593.

Ihlanfeldt, K. R. (2004). Exclusionary Land-use Regulations within Suburban Communities: A Review of the Evidence and Policy Prescriptions. *Urban Studies*, *41*(2), 261–283.

Imbroscio, D. (2008). "United and Actuated by Some Common Impulse of Passion": Challenging the Dispersal Consensus in American Housing Policy Research. *Journal of Urban Affairs*, *30*(2), 111–130. https://doi.org/10.1111/j.1467-9906.2008.00381.x

Immergluck, D. (2009). *Foreclosed: High-Risk Lending, Deregulation, and the Undermining of America's Mortgage Market*. Cornell University Press.

Immergluck, D. (2011). Too Little, Too Late and Too Timid: The Federal Response to the Foreclosure Crisis at the 5-Year Mark (September 19, 2011). Available at SSRN: https://doi.org/10.2139/ssrn.1930686

Jackson, K. T. (1985). *Crabgrass Frontier: The Suburbanization of the United States*. Oxford University Press.

Jackson, T. F. (1993). The State, The Movement, and the Urban Poor. The War on Poverty and Political Mobilization in the 1960s. In Michael B. Katz (Ed.), *The "Underclass" Debate: Views from History* (pp. 403–409). Princeton University Press.

Jan, T. (2020, October 22). New Federal Rule Will Make It Harder to Challenge Discrimination in The Housing Industry, Lawsuits Allege. *Washington Post*. https://www.washingtonpost.com/business/2020/10/22/housing-discrimination-lawsuit-hud/

Jargowsky, P. (2003). Stunning Progress, Hidden Problems: The Dramatic Decline of Concentrated Poverty in the 1990s. Redefining Urban and Suburban America: Evidence from Census 2000, Vol 2.

Jencks, C. (1992). *Rethinking Social Policy: Race, Poverty and the Underclass*. Harvard University Press.

Joint Center for Housing Studies of Harvard University. (2021). *The State of the Nation's Housing*. Harvard College.

Katz, M. B. (1996). *In the Shadow of the Poorhouse: A Social History of Welfare in America*. Basic Books.

Katz, M. B. (2006). *Racial Division and Concentrated Poverty*. Brookings Institution.Kluegel, J. R., & Smith, E. R. (1986). *Beliefs About Inequality: Americans' Views of what Is and What Ought to Be*. Aldine de Gruyter.

Krieger, J., & Higgins, D. (2002). Housing and Health: Time Again for Public Health Action. *American Journal of Public Health*, *92*(5), 758–768.

Krumholz, N. 2004. The Reluctant Hand: Privatization of Public Housing in the U.S. Paper presented to the CITY FUTURES conference, Chicago.

Lake, R. W. (2017). *The New Suburbanites. Race and Housing in the Suburbs*. Routledge.

Lang, R., & Sohmer, R. R. (2000). Legacy of the Housing Act of 1949: The Past, Present, and Future of Federal Housing and Urban Policy (Editors' Introduction). *Housing Policy Debate*, *11*(2), 291–298.

Lane, V. (1995). Best Management Practices in U.S. Public Housing. *Housing Policy Debate*, *6*(4), 867–904. https://doi.org/10.1080/10511482.1995.9521208

Lea, M. J. (1996). Innovation and the Cost of Mortgage Credit: A Historical Perspective. *Housing Policy Debate*, *7*(1), 147–174. https://doi.org/10.1080/10511482.1996.9521216

Liu, S., & Su, Y. (2021). The Impact of the COVID-19 Pandemic on The Demand for Density: Evidence from The U.S. Housing Market. *Economic Letters*, *207*, 110010. https://doi.org/10.1016/j.econlet.2021.110010

Lukas, J. A. (1985). *Common Ground*. Knopf.

Mallach, A. (2009). *Stabilizing Communities: A Federal Response to the Secondary Impacts of the Foreclosure Crisis*. Brookings Institution.

Marcuse, P. (1978). Housing Policy and the Myth of the Benevolent State. *Social Policy & Administration*, *8*, 21–26.

Marcuse, P., & Keating, D. (2006). The Permanent Housing Crisis: The Failures of Conservatism and the Limits of Liberalism. In R. Bratt, M. Stone, & C. Hartman (Eds.), *A Right to Housing: Foundation of a New Social Agenda* (pp. 139–162). Temple University Press.

Massey, D. S., & Denton, N. A. (1990). *American Apartheid: Segregation and the Making of the Underclass*. Harvard University Press.

McClure, K. (2000). The Low Income Housing Tax Credit as an Aid to Finance: How Well Has It Worked? *Housing Policy Debate*, *11*(1), 91–114.

McClure, K. (2008). Deconcentrating Poverty with Housing Programs. *Journal of the American Planning Association*, *74*(1), 90–99.

McClure, K., Schwartz, A. F., & Taghavi, L. B. (2015). Housing Choice Voucher Location Patterns a Decade Later. *Housing Policy Debate*, *25*(2), 215–233. https://doi.org/10.1080/10511482.2014.921223

McConahay, J. B. (1982). Self-Interest versus Racial Attitudes as Correlates of Anti-Busing Attitudes in Louisville: Is It the Buses or the Blacks? *The Journal of Politics*, *44*(3), 692–720.

Meyer, S. G. (2000). *As Long as They Don't Move Next Door: Segregation and Racial Conflict in American Neighborhoods*. Rowman & Littlefield.

Michney, T. M. (2021). How the City Survey's Redlining Maps Were Made: A Closer Look at HOLC's Mortgagee Rehabilitation Division. *Journal of Planning History*, *21*(4), 316–344. https://doi.org/10.1177/15385132211013361

Miller, B. (2018). A McMansion for the Suburban Mob Family: The Unfulfilling Single-Family Home of The Sopranos. *Journal of Popular Film and Television*, *46*(4), 207–218. https://doi.org/10.1080/01956051.2018.1499605

Mueller, E. J., & Schwartz, A. (2008). Reversing the Tide: Will State and Local Governments House the Poor as Federal Direct Subsidies Decline? *Journal of the American Planning Association*, *74*(1), 122–135.

Nelson, R., Winling, L., Marciano, R., & Connolly, N. (2022). *Mapping Inequality*. American Panorama. https://dsl.richmond.edu/panorama/redlining/#loc=5/39.1/-94.58

Nova, A. (2020, December). The CDC Banned Evictions. Tens Of Thousands Have Still Occurred. *CNBC*. https://www.cnbc.com/2020/12/05/why-home-evictions-are-still-happening-despite-cdc-ban.html

Oliver, M., & Shapiro, T. (2013). *Black Wealth/White Wealth: A New Perspective on Racial Inequality*. Routledge.
Olsen, E. O. (2001). *Housing Programs for Low-Income Households* (NBER Working Paper No. 8208). National Bureau of Economic Research.
Orlebeke, C. (2000). The Evolution of Low-Income Housing Policy, 1949 to 1999. *Housing Policy Debate*, *11*(2), 489–520.
Orfield, M. (2006). Land Use and Housing Policies to Reduce Concentrated Poverty and Racial Segregation. *Fordham Urban Law Journal*, *33*, 877.
Pogodzinski, J. M. (1991). The Effects of Fiscal and Exclusionary Zoning on Household Location: A Critical Review. *Journal of Housing Research*, *2*(2), 145–160.
Perin, C. (1977). *Everything in Its Place: Social Order and Land Use in America*. Princeton University Press.
Radford, G. (1996). *Modern Housing for America: Policy Struggles in the New Deal Era*. University of Chicago Press.
Rodríguez-Pose, A., & Storper, M. (2020). Housing, Urban Growth and Inequalities: The Limits to Deregulation and Upzoning in Reducing Economic and Spatial Inequality. *Urban Studies*, *57*(2), 223–248. https://doi.org/10.1177/0042098019859458
Rothstein, R. (2017). *The Color of Law*. Liveright Publishing Corporation.
Satter, B. (2009). *Family Properties: Race, Real Estate, and the Exploitation of Black Urban America*. Macmillan.
Schneider, A., & Ingram, H. (1993). Social Construction of Target Populations: Implications for Politics and Policy. *The American Political Science Review*, *87*(2), 334–347.
Schuman, H., & Bobo, L. (1988). Survey-Based Experiments on White Racial Attitudes Toward Residential Integration. *American Journal of Sociology*, *94*(2), 273–299. http://www.jstor.org/stable/2780776
Schwartz, A. F. (2014). *Housing Policy in the United States* (1st ed). Routledge.
Schwartz, A. F. (2021). *Housing Policy in the United States* (4th ed). Routledge.
Seitles, M. (1998). The Perpetuation of Residential Racial Segregation in America: Historical Discrimination, Modern Forms of Exclusion, and Inclusionary Remedies. *Florida State University Journal of Land Use & Environmental Law*, *14*(1), Article 3.
Stulz, R. M. (2010). Credit Default Swaps and the Credit Crisis. *Journal of Economic Perspectives*, *24*(1), 73–92. https://doi.org/10.1257/jep.24.1.73
Suglia, S. F., Duarte, C. S., & Sandel, M. T. (2011). Housing Quality, Housing Instability, and Maternal Mental Health. *Journal of Urban Health*, *88*(6), 1105–1116. https://doi.org/10.1007/s11524-011-9587-0
Taylor, K.-Y. (2012). Back Story to The Neoliberal Moment: Race Taxes and The Political Economy of Black Urban Housing in the 1960s. *Souls*, *14*(3–4), 185–206. https://doi.org/10.1080/10999949.2012.764836
Taylor, K.-Y. (2019). *Race for Profit: How Banks and the Real Estate Industry Undermined Black Homeownership*. The University of North Carolina Press.
Tighe, J. R. (2012). How Race and Class Stereotyping Shapes Attitudes Toward Affordable Housing. *Housing Studies*, *27*(7), 962–983. https://doi.org/10.2139/ssrn.2158851
Tighe, J. R., Hatch, M. E., & Mead, J. (2017). Source of Income Discrimination and Fair Housing Policy. *Journal of Planning Literature*, *32*(1), 3–15. https://doi.org/10.1177/0885412216670603
Tikkanen, R. S., & Schneider, E. C. (2020). Social Spending to Improve Population Health—Does the United States Spend as Wisely as Other Countries? *New England Journal of Medicine*, *382*(10), 885–887. https://doi.org/10.1056/NEJMp1916585
Trounstine, J. (2018). *Segregation by Design: Local Politics and Inequality in American Cities*. Cambridge University Press.
Tully, S. (2022, February 17). What slowdown? Home Prices are on Track to Soar 12% in 2022 says a Top Forecaster. *Fortune*. https://fortune.com/2022/02/17/real-estate-home-prices-climbing-prediction-2022/
Turner, M. A. (2003). *Strengths and Weaknesses of the Housing Voucher Program*. Congressional Testimony of Margery Austin Turner, Director, Metropolitan Housing and Communities Policy Center, The Urban Institute, prepared for the Committee on Financial Services, Subcommittee on Housing and Community Opportunity, United States House of Representatives, June 17, 2003
U.S. Census Bureau. (2000). *Historical Census of Housing Tables: Homeownership by Selected Demographic and Housing Characteristics*. https://www.census.gov/data/tables/time-series/dec/coh-ownerchar.html
U.S. Census Bureau. (2012). *United States Census Summary 2010: Population and Housing Counts* (CPH-2-1; p. 554). Department of Commerce.
U.S. Census Bureau. (2013). *United States 2010: Summary Population and Housing Characteristics* (CPH-1-1; p. 1036). Department of Commerce.

U.S. Census Bureau. (2021, August 12). *2020 Population and Housing State Data*. https://www.census.gov/library/visualizations/interactive/2020-population-and-housing-state-data.html

U.S. Federal Housing Administration. (1941). *Seventh Annual Report of the Federal Housing Administration* (Annual Report No. 7; p. 117).

U.S. Housing and Home Finance Agency. (1951). *Fourth Annual Report of the Housing and Home Finance Agency* (Annual Report No. 4; p. 489).

Vale, L. J. (2000). *From the Puritans to the Projects: Public Housing and Public Neighbors*. Harvard University Press.

Vale, L. J. (2002). *Reclaiming Public Housing: A Half Century of Struggle in Three Public Neighborhoods*. Harvard University Press.

von Hoffman, A. (1996). High Ambitions: The Past and Future of American Low-Income Housing Policy. *Housing Policy Debate*, *7*(3), 423–446.

Wallace, J. (1995). Financing Affordable Housing in the United States. *Housing Policy Debate*, *6*(4), 785–814.

Wang, R. (2018). Tracking "Choice" in the Housing Choice Voucher Program: The Relationship Between Neighborhood Preference and Locational Outcome. *Urban Affairs Review*, *54*(2), 302–331. https://doi.org/10.1177/1078087416646205

Wilson, G. (1996). Toward a Revised Framework for Examining Beliefs About the Causes of Poverty. *The Sociological Quarterly*, *37*(3), 413–428. https://doi.org/10.1111/j.1533-8525.1996.tb00746.x

Wilson, W. J. (1987). *The Truly Disadvantaged: The Inner City, the Underclass, and Public Policy*. University of Chicago Press.

Wolf, M. A. (2008). *The Zoning of America: Euclid v. Ambler*. University Press of Kansas.

Zucker, G. S., & Weiner, B. (1993). Conservatism and Perceptions of Poverty: An Attributional Analysis. *Journal of Applied Social Psychology*, *23*(12), 925–943. https://doi.org/10.1111/j.1559-1816.1993.tb01014.x

6

HOUSING AND WELFARE IN AUSTRALIA

Kath Hulse, Piret Veeroja and Zoë Goodall

The current state of housing in Australia

The state of Australian housing is shaped by the country's geography as the world's largest island (and sixth largest country), inhabited by only 26 million people, where 87 per cent of the population live on the coast (Clark et al., 2021) and 67 per cent live in capital cities (ABS, 2022a). Housing is also shaped by the country's history of colonization by the British in the late 18th century, leading to a housing system modelled on British laws and the subjugation of Aboriginal and Torres Strait Islander peoples, the land's traditional owners, who have lived in Australia for an estimated 65,000 years.[1] Post-WW2, the "great Australian dream" of individual households owning a detached house has been spread through political and cultural discourse (Burke et al., 2020). The prospect of owning a home as part of a better life attracted migrants from all over the world post-WW2 (Burke et al., 2022), with 30 per cent of people born overseas (2019), the second highest percentage of overseas-born people in the OECD (after Luxembourg) (OECD, 2022). Governments have consistently supported homeownership in a housing regime that was and remains somewhat distinct from the welfare regime.

Tenure overview

Housing in Australia is highly commodified – owned by occupiers or private landlords. In the most recent Census (August 2021), 31 per cent of private dwellings were owned outright; 35 per cent were owned with a mortgage; 27 per cent were rented from private landlords; and only 3.8 per cent rented from social landlords (ABS, 2022a, see also Table 6.1). Since 2001, the private rental sector (PRS) has been growing faster than the rate of all households (Hulse et al., 2019, p. 3). Increasing numbers of people are living in the PRS long-term (Pawson et al., 2017), including low-income and vulnerable households (PC, 2019). Properties move readily between owner-occupation and rental (Martin et al., 2018), in part due to the (world-first) introduction of strata title in 1961, which means that individuals can easily purchase units within multi-unit buildings. The majority of PRS stock is houses or units owned by individual landlords. While commercial landlords and build-to-rent developers have attempted to get a foothold in Australia, their impact remains small. Rental housing is regulated by state and territory governments under a consumer protection model known

 DOI: 10.1201/9781003212690-8

in each jurisdiction as the Residential Tenancies Act, but the rights granted to tenants vary (Martin et al., 2022). Social housing comprises public housing (managed by state/territory governments) and community housing (managed by not-for-profit organizations), with additional public and community housing specifically for Aboriginal and Torres Strait Islander peoples (AIHW, 2022). Public housing constitutes the largest share of social housing, at 68 per cent (Ibid.) but any growth in the sector in the 2000s has mainly been via community housing. Other tenure types, such as cooperative housing, are very small (Crabtree et al., 2021).

Assistance and affordability

Government housing assistance is split between the Australian (Commonwealth) government, six states, and two territories in Australia's federal system of governance. The Australian Government's main role is to provide rent assistance (RA), a supplementary payment to eligible private and community housing renters (but not public housing renters) receiving income support and family payments. Some states/territories have small-scale supplementary schemes to assist private renters, responding to differences in housing markets in such a geographically large country (AIHW, 2022). RA payments are mainly calculated based on household type and size. It is paid to households directly to increase their capacity to rent in the private market. RA payments have not kept pace with rising rents, and in 2020, 29.4 per cent of RA recipients still spent more than 30 per cent of their income on rent, using an often-cited benchmark (PC, 2021, Figure G2). The Australian Government also provides funding for supply-side measures (public and community housing) and homelessness services in collaboration with the states/territories via the National Housing and Homelessness Agreement (NHHA). Funding by governments since the mid-1990s has failed to address the extent of housing stress, and social housing is heavily rationed to those with the most urgent and complex needs (see "Rethinking policies on renting", below). Both federal and state/territory governments also periodically provide demand-side assistance by way of grants and stamp duty concessions to first home buyers, which have been criticized by Australia's Reserve Bank for further increasing house prices, which are already very high in a global context (Khadem, 2022). There are strong financial and tax incentives in Australia for homeowners and investors who buy properties, less so for renters (Eccleston et al., 2018).

Inequality and homelessness

The pace of house construction has increased in recent years (Shooshtarian et al., 2021, p. 2), (although there have been significant delays in 2021/2022 due to materials shortages), but this has not led to housing being more affordable to low-modest-income households. Inequalities between older homeowners and younger renters are a key source of tension in Australia. There is a growing discourse about young people being locked out of purchasing a home because older generations own multiple houses. However, there are other factors at play behind young people's difficulties with purchasing, such as labour market insecurity and increased casualization of jobs. Young people with access to family wealth are also less likely to face purchasing difficulties. Australia has no inheritance tax (abolished in the 1970s), so inequality in access to homeownership is intergenerational.

Homelessness is an ongoing problem in Australia, with an increase in homelessness among families with children, women and older people (Pawson et al., 2020). Reflecting Australia's history of colonization, according to the 2021 Census, Aboriginal and Torres Strait Islander peoples constitute 3.2 per cent of the Australian population (ABS, 2022a) but 20 per cent of people classified as homeless (ABS, 2018).[2] Aboriginal and Torres Strait Islander peoples also have lower rates of homeownership than non-Indigenous people and are over-represented in social housing (Moskos et al., 2022).

Environmental issues

Environmental issues are significant in Australia, as recent years have brought devastating natural disasters caused by climate change, including flooding and bushfires. The lack of Australian Government's support for climate change action has been a contentious issue and arguably led to the 2022 election of a more climate-focused federal government. Despite the environmental impact of construction, there has not been much movement to increase the sustainability of residential housing (Shooshtarian et al., 2021). The National Construction Code requires minimum energy efficiency standards only on new housing or major renovations, as measured by the Nationwide House Energy Rating Scheme (NatHERS, 2022). Federal and state governments have also created opt-in measures for households, such as financial incentives for adopting energy-efficient or renewable energy systems and appliances. Some state governments have also trialled measures to increase the energy efficiency of rented housing, for example by regulating that landlords install energy-efficient appliances.

A brief overview of the historical development of the housing regime

Development of housing as a field of politics and policy

The Australian Government has played a significant role in forming housing policy since WW2, although states have major responsibility for land use planning and housing matters (Pawson et al., 2020). Housing became a political and policy issue for the states in the Great Depression of the 1930s when community and church organizations campaigned against poor conditions and evictions associated with private renting, putting pressure on governments to become actively involved in providing public housing (Morris et al., 2021).

The "great Australian Dream" of homeownership post-WW2 was generated by rapid domestic population growth ("the baby boom"), waves of migration from European countries, and favourable institutional settings (Burke et al., 2020). The post-war institutional environment was characterized by the Keynesian economic system (Dufty-Jones & Rogers, 2016) and a booming economy, resulting in consistent growth of households' real incomes (Burke et al., 2020). The post-war housing regime included financial institutions advancing mortgages with low interest rates, selling public housing (mainly detached houses) to sitting tenants and offering capital grants to some first-time buyers (Berry, 1999). Cumulatively, these measures resulted in an increase in homeownership rates, from 53 per cent in 1947 to 70 per cent in 1961 (Burke et al., 2020; see Table 6.1).

"Golden age" of public housing

The Australian Government became involved in housing policy via ten successive intergovernmental agreements (Commonwealth-State Housing Agreements (CSHAs)) between 1945 (Berry, 1999) and 2003. The CSHAs provided funds to state housing authorities to build and manage public housing stock (Pawson et al., 2020), as well as assistance with homeownership. Kemeny (1983, p. 9) named the post-war decades the "golden age of public housing"

Table 6.1 Housing tenure by occupied private dwellings, Australia, 1947–2021[1]

	Australia								
Tenure type	*1947*	*1954*	*1961*	*1971*	*1981*	*1991*	*2001*	*2011*	*2021*
Owner	45.4%	48.1%	47.7%	68.8%[4]	34.2%	41.1%	41.7%	33.0%	31.6%
Purchaser	8.0%	15.2%	22.5%		36.0%	27.7%	27.8%	35.4%	35.1%
Owners total	*53.4%*	*63.3%*	*70.3%*		*70.1%*	*68.9%*	*69.5%*	*68.5%*	*66.7%*
Private and 'other landlord'[2]	44.0%[3]	30.0%	23.3%	22.3%	20.6%	22.1%	24.5%	24.6%	27.4%
Public landlord		4.3%	4.2%	5.6%	5.1%	5.3%	4.8%	4.8%	3.8%
Other	2.6%	2.4%	2.2%	3.3%	4.2%	3.6%	0.9%	2.1%[5]	2.1%[5]

Sources: Data from 1947 Census are from Kryger (2009), 1954–2001 Censuses from Hulse et al. (2012) and 2011 – 2021Censuses are original analyses (ABS, 2011; 2021).

Notes

1 "Tenure type not stated" was removed from the analysis due to inconsistent reporting.

2 From the 1954 Census, renters were classified whether they paid rent to a government landlord (public housing) or "other" landlord (that was not public housing; this is referred to as private renters and the "other" category, where it was not possible to distinguish landlord type).

3 Landlord type was not distinguished.

4 Owners and purchasers were not distinguishable.

5 Before the 2021 Census, people who selected "rent-free" were counted as renters, but they were placed in the "other" category in 2021. For the purposes of this analysis and to compare the most recent data, data from 2011 were categorized similarly. It was not possible in earlier years.

because the stock grew significantly, from 4,028 dwellings to about 100,000 dwellings and the CSHAs laid the foundation for the system (Hayward, 1996). Public housing was targeted to employed Australians on low and moderate incomes (Ibid.). Private renting was in decline up to the mid-1960s due to increased numbers of homeowners and people in public housing (Morris et al., 2021). Private renting was associated with (after-housing) poverty and poor housing conditions (e.g. crowding; limited facilities, such as social spaces, storage or laundry facilities; damp and mould; issues with plumbing) (Ibid.). Private rental accommodation was traditionally located in inner-city areas that were progressively gentrified, creating insecurity for renting households (Ibid.).

Suburbanization and gentrification of inner cities

Homeownership was supported by land planning systems operated by the states. Planning restrictions, for example, limit urban densification (Burke et al., 2020). Suburban areas were associated with good-quality housing that was located away from the polluted older industrial areas in the inner cities (Berry, 1999). Most people lived or aspired to live in single-family detached dwellings (Dufty-Jones & Rogers, 2016). Inner-city gentrification started in the 1950s and 1960s with waves of migrants arriving from Southern Europe and middle-class households buying and renovating inner-city dwellings (Morris et al., 2021). Additionally, during the 1960s and 1970s, state government housing commissions carried out inner city "slum clearance" by demolishing substandard housing (that was mainly occupied by private renters) and redeveloping the areas with public housing (Ibid.). These redevelopments were stopped by union and community campaigns by the mid-1970s (Ibid.).

Deregulation and market liberalism

The long boom in the Australian economy and housing was challenged by the oil shocks of the mid-1970s (Dufty-Jones & Rogers, 2016). In the 1980s, governance and finance markets were deregulated, and public policies transformed into a market-liberal system (Burke et al., 2020). The Australian finance market offered a wider range of financial products to owner-occupiers and investors, relaxed lending standards (Yates & Yanotti, 2016), and introduced lower deposit rates and longer lending periods (Burke et al., 2020). Australian housing policies were also heavily influenced by market liberalism (Ibid.). In the early 1990s, Australia went through a recession (Yates & Yanotti, 2016), but the economy then grew consistently until the COVID-19 pandemic of 2020, which led to border closures and lockdowns and a halt to migration from overseas. Since then, the economy has rebounded, with Gross Domestic Product (GDP) growth at 3.25 per cent in 2022 (RBA, 2022).[3] The rate of growth is expected to moderate due to a range of factors, including labour shortages (unemployment rate of 3.25 per cent at the end of 2022), material shortages, cost-of-living pressures and constraints on exports to Australia's major trading partner, China (RBA, 2022). Interest rates were reduced to record low rates in 2020–2021 to stimulate demand, resulting in an increase in home mortgages and a rapid increase in housing prices until interest rates started to rise again in mid-late 2022.

Continuing support for homeownership and private landlordism

The overall rate of homeownership has changed only slowly since the early 1970s, as shown in Table 6.1. There have been important compositional changes. While there remains a high rate of ownership among older people, homeownership rates among younger households (aged below 45) have steadily declined since the 1970s (Yates, 2015). In 2021, 83.4 per cent of adults aged 65 years or older owned their home (and 70.4 per cent of them were outright owners). The average age at which outright ownership is achieved, however, increased from 58 years in 1988 to 67 years in 2016/17 (Burke et al., 2020, p. 33), posing serious doubt about the link between homeownership and age pensions in the future (Coates, 2022). There is a growing proportion of private renters among the younger and middle-aged cohorts ("generation rent") who might never attain the dream of homeownership unless they have substantial help from their families (Burke et al., 2020).

From the mid-1980s, following the deregulation of financial services, including home mortgages, housing came to be seen as an asset and investment, not merely a place to live (Morris et al., 2021). Indirect housing assistance to homeowners and private landlords was increased through tax concessions such as the exemption of owner-occupiers from the capital gains tax introduced in 1986 (Yates & Yanotti, 2016). Rental property investments became increasingly popular but had greater advantages for people on higher incomes (Morris et al., 2021). Other forms of assistance for owners include exemption of the "family home" from state land tax (Ong et al., 2019). In terms of direct housing policies, several first homeowners' grants have been introduced, eliminated and reintroduced since 2000.

Since the late 1990s, Australian property prices have grown rapidly, especially in the most populous cities (Morris et al., 2021). According to one account, Australia's largest cities, Sydney and Melbourne, rank in the top five most unaffordable cities globally for housing (Demographia, 2022). This has made achieving homeownership more difficult for younger demographic groups, such as people on single incomes or low and moderate incomes (Morris et al., 2021). In terms of land use, the redevelopment of inner cities has increased land prices, pushing lower-income households to suburban areas where prices are lower (Dufty-Jones & Rogers, 2016).

Rethinking policies on renting

Strict entry requirements (means tests) were introduced in public housing (which had a near monopoly on social housing) in the early 1970s, heralding a move from housing workers on modest incomes towards those on social security payments (Hayward, 1996). Funding for public housing was boosted in the early 1980s (Ibid.), initially growing the stock but decreasing funding in real terms since the mid-1990s (with the exception of some one-off boosts). This means that public housing in the 2000s became accessible only to people with the most urgent and complex needs, usually on very low incomes (Morris et al., 2021). This had implications for the financial viability of public housing since most tenants pay rents geared to income (called rent rebates in Australia), and revenue declined as costs rose. During the 1980s, state governments started diversifying public housing via innovative community housing programmes, such as establishing rental associations and cooperatives run by community organizations (Hayward, 1996). Rents were set in a variety of ways, but in most cases, eligible tenants can access RA from the Australian Government (see below). Additionally, specific housing programmes for Aboriginal and Torres Strait Islander peoples were established (Pawson et al., 2020), but these programmes have remained relatively small scale.

Private renting increased slowly in 1981, and the Australian Government extended RA to more groups and changed the payment formula in response to increasing numbers of low-income private renters (Ong et al., 2020). There are no rent controls (other than in exceptional circumstances such as world wars), and there is no cap on the rate of increase in private rents, which are set at "market" level. In consequence, by the mid-1990s, expenditure on RA exceeded expenditure on public housing, a pattern that has become entrenched in the 2000s (Morris et al., 2021).

Aboriginal and Torres Strait Islander peoples

Compared to Australians of other ancestries, few Aboriginal and Torres Strait Islander people are homeowners, and many live in public housing (Dufty-Jones & Rogers, 2016). The first funding from the Australian Government for specific public housing and assisting Aboriginal and Torres Strait Islander peoples into homeownership was in 1988–1989, followed by government funding to Indigenous community organizations to deliver housing (Pawson et al., 2020). In the 2000s, there has been significant funding to increase and improve housing for Aboriginal and Torres Strait Islander peoples outside of capital cities. At the 2021 Census (ABS, 2021b), 62.9 per cent of people who identified as Aboriginal and Torres Strait Islander peoples lived in "rest of the state" areas (not in capital city areas).

The early 1990s saw important developments in addressing the dispossession of and discrimination against Aboriginal and Torres Strait Islander peoples. In June 1992, the High Court, in what became known as the "Mabo Decision", established that "native title rights" had not been extinguished by the British occupation of Australia and existed as part of Australian common law. A year later, the Native Title Act was passed, which enabled traditional owners to begin an often-protracted process to claim rights of occupancy and use of land without freehold title (AIATSIS, 2022).

The welfare regime and its development

Development of welfare regimes prior to WW2

Two key features of the Australian welfare regime had already been established in the early 1900s: wage regulation and social protection when not able to work. The Harvester Judgement of 1907 introduced the idea of a "living wage" for a male breadwinner supporting his family, and government pensions were introduced for those who could not work due to old age or 'invalidity' (disability) (Kewley, 1973). Housing was not an integral part of the institutional arrangements for social protection prior to WW2, due in part to relatively high homeownership rates in comparison with European countries early in the 20th century (Pawson et al., 2020).

The wage earners' welfare state: 1940 to the mid-1970s

In the 1940s, commitment to a "living" wage for male breadwinners continued, reflecting a high level of unionization of the post-war workforce and a shortage of labour; social protection was extended to include unemployment and sickness benefits. The 1940s also saw the introduction of the third element of the Australian welfare regime, progressive taxation, following the transfer of income tax powers from the states to the Australian government in 1942. The new taxation powers gave the federal government the means of funding national post-war reconstruction programmes in social services, health and housing. A successful national referendum in 1946 approved federal powers to make laws about social security matters and healthcare (Kewley, 1973). Although social security was extended, a national health service could not be established, leaving reliance on private health provision and private insurance. Federal funding of public housing was, by contrast, relatively uncontroversial in the context of acute post-WW2 housing shortages. However, by the time the CSHA was renegotiated in the mid-1950s, shortages were less pressing, and priority was given to homeownership assistance (Murphy, 1995). Some states embarked on extensive sales of public housing to tenants under very generous terms to promote homeownership.

The welfare regime in the decades after WW2 was characterized as the "wage earners' welfare state" (Castles, 1986). It was an era of near full employment for men, with a living wage that could support their families, non-contributory means-tested social security for defined groups, and progressive taxation. The regime relied on high rates of homeownership via a protected circuit of capital, as homeownership peaked at 72 per cent of households in 1966 (Berry, 1999). There was limited recognition of the plight of private renters with the introduction of a flat rate payment to single pensioners who rented in the PRS as cheaper than a general increase in the age pension (Kewley, 1973). Over the years, this payment (subsequently known as RA) became of increasing importance.

The wage earners' welfare state benefited white men in unionized industries who were living in traditional relationships. Its familial basis discriminated against working women living without a male partner and women in traditional relationships who lacked the financial capacity to leave the marital home for reasons including domestic and family violence. It also discriminated against Aboriginal and Torres Strait Islander peoples in Australia who did not benefit from high wages, were often discriminated against in industrial agreements, particularly for pastoral industries, and had low rates of housing ownership. Following a concerted campaign, in 1967 voters passed a referendum to give the Australian Government the power to make laws regarding Aboriginal and Torres Strait Islander peoples and require the Census to include Aboriginal and Torres Strait Islander peoples in the count. The successful referendum was important in addressing the housing and other disadvantages faced by Aboriginal and Torres Strait Islander peoples.

Challenging the male breadwinner model: Mid-1970s to early 1990s

By the early 1970s, it became clear that the post-WW2 welfare regime had left some groups living in poverty, notably older people, large families, and private renters (Commission of Inquiry into Poverty, 1975). The Labor Government (1972–1975) introduced a raft of national reforms to extend social protection and outlaw discrimination based on race and sex. In 1974, the "family" element of the minimum wage was abolished on the grounds that an industrial tribunal was not a social welfare agency and support for families was a matter for governments. The reforms marked a fundamental shift in theory, if not yet in practice, from the patriarchal systems that supported the male breadwinner model (Castles & Mitchell, 1993) towards a more progressive and inclusive model that supported female participation in paid work (with equal pay) and an increased diversity of household types. Funding for healthcare remained a contested issue in Australia due to opposition from private health providers and insurance funds until 1983, when a national health system was established (Medicare) as an integral part of Australia's welfare regime.

The oil price shocks of the mid-1970s, together with other demographic and economic changes, prompted further changes to the Australian welfare regime. Financial services were deregulated, including housing finance, and the economy opened up to global flows of capital and labour in the mid-1980s. These changes put greater emphasis on markets and competition, although prices and income accords were negotiated with employers and unions through the 1980s. The union movement was also involved in the introduction of compulsory "superannuation" in 1993, a system of retirement accounts primarily funded by employer contributions based on wages. It was expected that over time, older Australians would be less dependent on the age pension.

Labour market deregulation and welfare: Mid-1990s to present

The biggest change to wage earners' welfare state since the mid-1990s has been the limited regulation of wages. Trade unionism fell from 45.6 per cent of all employees in 1986 to 14.3 per cent in 2020 (ABS, 2020), reflecting the decline in manufacturing industries employing male labour and the rise of the service sector, such as education and health, which employ predominantly female labour. The result has been an increase in contract, casual and part-time work and wage increases significantly below the rate of house price increases (Ryan-Collins & Murray, 2021). The Coalition (conservative) federal government (1996–2007) tried to introduce individual work contracts as the norm, although this was reversed after 2007 and, for the most part, "enterprise bargaining agreements" remain in place. The institutional infrastructure to provide minimum (not living) wages remains via the Fair Work Commission. However, notwithstanding increased female participation for many years and 50 years since the introduction of the Equal Pay Act, there is still a gender pay difference that cannot be accounted for by different types of work (KPMG, 2022).

The 2000s have seen a tightening of some social security measures, influenced by neo-liberal ideas about the primacy of market competition and smaller government. Social protection for working-aged people has borne the brunt of this, with various jobseeker (unemployment) payments kept at levels well below the poverty line to "encourage" people into paid work. These changes brought Australia nearer to the "liberal" welfare regime identified by Esping-Andersen (1990) in his seminal work on welfare regimes.[4]

The third feature of the workers' welfare state, progressive taxation via income tax, remains but has been somewhat diluted by the Goods and Services Tax introduced in Australia in 2001 and extensive tax expenditures (concessions). The biggest tax expenditures are for housing, as outlined earlier, and superannuation (Stebbing & Spies-Butcher, 2010). Such tax concessions benefit older, wealthier Australians, contributing to increasing inequality and friction between generations (Wood & Griffiths, 2019). The Gini coefficient for income inequality after taxes and transfers was 0.33 in 2017–2018, lower than many European countries but close to the OECD average of 0.34 (PC 2018). There is greater inequality in the distribution of wealth compared to income, but

Australia ranks eighth most equal among 28 OECD countries, as measured by the Gini coefficient of net household worth (0.62 in 2017–2018) (Productivity Commission, 2018, p. 3) in part due to prevailing high homeownership rates.

The interrelation between housing and welfare

The historical role of housing in the welfare regime

Housing was not historically regarded as part of the broader welfare regime but had its own funding and institutional arrangements. Access to housing assistance through these arrangements was administered by the states/territories and rationed. The welfare regime, by contrast, was a national system with social security payments funded from general taxation and administered by the Australian Government. The underlying principles of the latter were entitlement subject to means testing and payments that varied according to household type and size but not location.

Despite these differences in governance, funding and underlying principles, homeownership (with no mortgage) has long been regarded a key aspect of security in old age (PC, 2015). It was the fourth pillar of Australia's retirement system, alongside the means-tested age pension, superannuation and individual savings, enabling older homeowners to have sufficient income for other essentials after housing costs (Ong et al., 2019). It has long been asserted that high rates of social expenditures seen in Europe were not required in Australia due to the "social wage" for working people and low housing costs for retirees who owned their homes outright (no debt) (Castles, 1998).

RA, introduced earlier, is an uneasy hybrid between welfare and housing. Developing from an income supplement for some pensioners to a widespread payment to assist with affordability for private renters (and some community renters), it reflects the features of the national social security scheme rather than housing assistance but has been increasingly relied upon as the key plank of national housing policy.

Social rental housing in Australia was originally for working families, but since the 1970s, it has moved towards a welfare model (Hayward, 1996). Not only were entry means tested, but, from the 1990s on, there was also more and more tight targeting of a very small stock of housing compared to many developed countries. Since some of this housing could be distinguished from other housing, this targeting led to the stigmatization of social housing and its tenants, and it has been difficult to get political and community support for additional funding for social housing construction.

Housing and the welfare regime: Contemporary developments

The increase in private renting has placed a degree of importance on RA that was never intended, and the hybrid nature of the scheme poses difficulties. Low-paid workers without dependent children are not eligible for RA, and payments do not address affordability issues in high-rent cities such as Sydney and Melbourne (Ong et al., 2020). Attempts to reform the RA scheme to make it more sensitive to different housing markets have failed. There is conceptual disagreement about whether the problem is one of affordability or lack of affordable housing, i.e., income or housing. More pragmatically, since RA, unlike other forms of housing assistance, is a demand-driven programme, governments are nervous about the cost blowouts associated with any reforms.

The low level of social housing is still a big concern, notwithstanding a broader-based national agreement (NAHA) replacing the CSHA in 2009.[5] Funding levels under this agreement have not kept pace with rising needs or even population growth. In addition, state governments have been very slow to reform other areas that affect renters, including using land management powers to elicit affordable rental supply and residential tenancy legislation that could enable greater affordability and security for renters (Martin et al., 2022).

Although historically distinct even from housing, as it emerged from the welfare sector, policy on homelessness does provide something of a bridge between housing and welfare in contemporary Australia. The NAHA, established in 2009, included homelessness for the first time (together with social housing and Indigenous housing). Research into homelessness indicates that there is a significant underlying problem of a shortage of affordable and available housing (Hulse et al., 2019), but there are also clearly connected problems with the welfare, health and justice systems (Batterham, 2018).

Governments, markets and housing disadvantage

The role of the market in housing, and increasingly in welfare, is very strong in Australia. At a national level, the main lever for government housing assistance is via social security and income tax powers, which have supported the current emphasis on RA as a means of improving affordability for private renters. At times, the Australian Government has taken a more direct role in funding not-for-profit organizations to provide accessible and more affordable housing for older people and those with disabilities. More recently, it has allocated significant specific-purpose funding, particularly for Aboriginal and Torres Strait Islander peoples living in regional and remote areas. The Australian Government also continues to provide some funding to the states/territories for social housing and homelessness.

The Australian Government has powers specified in the 1901 Constitution, as amended by (only) eight successful referenda. State governments retain other powers, including land use planning, development, and property taxes, all of which affect the supply and price of housing coming onto the market. The states also have legislation on residential tenancies, which is rather light-touch by the standards of many developed countries (Martin et al., 2018). Market stakeholders, notably developers, builders and their unionized workforces, are influential in policy-setting in these areas. For example, there have been only a few examples of using the planning system to elicit more affordable rental housing (unlike, for example, in the UK). States/territories are responsible for assessing the needs of different housing groups and allocating housing accordingly. In recent years, they have given priority to people who are homeless with complex needs and women who have experienced family and domestic violence.

Local government in Australia is generally small and weak. It was not recognized in the 1901 Constitution, and subsequent referenda to do so in 1974 and 1988 were unsuccessful. Many Australians feel that they are over-governed, which perhaps explains these results. Local government operates under delegated state powers, with decisions subject to being overruled by state government ministers and administrative tribunals. Despite this impediment and very real funding constraints, some local governments have become involved in the provision of affordable housing for groups such as the elderly, usually in conjunction with not-for-profit organizations operating in this space.

In sum, the rather complex intergovernmental arrangements that attempt to deal with the division of responsibilities across different levels of government in Australia's federal system, quite distinct geographic housing markets due to patterns of settlement, and the politics of federal governance make it difficult to make radical changes to the status quo on housing policies.

Housing: Market commodity or social right?

Australian governments have responded to new social risks through support for markets and targeting housing assistance, moving away from the workers' welfare state towards the liberal welfare regime articulated by Esping-Andersen (1990). It has proved difficult to effect changes to the established housing-welfare regime nexus even when there is an increasing disjuncture between existing institutional arrangements and emerging risk profiles (Esping-Andersen, 1999, p. 172). In Australia, these new risks have centred on young and low/middle-income households unable to access home ownership, difficulties accessing and sustaining private rental, blocked access to social housing, and increasing homelessness.

Continuing support for markets and market-driven solutions is at odds with a rights-based approach that could address some of these risks. Attempts to draw on housing as a human right (Morris, 2010), particularly in relation to homelessness, have not been successful to date. Although Australia is signatory to some international treaties that include the right to housing (e.g., International Covenant on Economic, Social and Cultural Rights Article 11), there is no Charter of Human Rights in Australia, and international obligations are not enshrined in domestic legislation. Indeed, the UN Special Rapporteur on Housing (2007) condemned Australia for failing to implement the human right to adequate housing, with groups including Aboriginal and Torres Strait Islander peoples and women missing out on safe and affordable housing.

The ethos of property rights is very strong in Australia and often trumps other considerations. Housing affordability has been an issue in Australia for many years and is now affecting children in middle-class families who are unable to buy homes in high-priced markets. While the lack of affordability in rental housing evokes some concern, so too does the plight of "rental investors", who are ascribed with an everyman archetype (Hulse et al., 2020). Social housing is more of a niche concern with a somewhat vexed history. It has been the concern of welfare, homelessness and housing organizations that lobby for more action from governments. This lobbying has at times run into stiff opposition from property interests and lobbyists, as in attempts to reform residential tenancy legislation.

Regional outlook

Australia is both a country and a continent. Its geographic neighbours are the Pacific Islands, which constitute Melanesia, Micronesia and Polynesia. The Pacific Island region is vast, diverse and not easily comparable to Australia. One similarity between the Pacific Islands and Australia is concern over climate change's impacts on housing and the environment, including increased natural disasters and rising sea levels.

More thorough comparisons can be made between Australia and New Zealand (also known as Aotearoa, the current Māori-language name), a country located in Polynesia. Although much smaller than Australia, with a population of just over 5 million (Stats NZ, 2022), New Zealand was also colonized by the British, and, as in Australia, people of European heritage are currently the majority of the population (Stats NZ, 2019). New Zealand's housing system is also based on British laws, and the Indigenous people, the Māori, have experienced the damaging impacts of colonization.

There are key similarities between Australia and New Zealand in the realm of housing and welfare. Firstly, New Zealand, along with Australia, was characterized as the workers' welfare state (Castles, 1986), with a different model from the high social expenditure states of many other developed countries (Castles, 1996). Secondly, New Zealand also valorized homeownership as part of its national "dream". The most recent NZ Census (2018) showed the homeownership rate had fallen to

64.5 per cent, the lowest rate since 1951, from its peak of 73.8 per cent in the early 1990s (Stats NZ, 2020, p. 28). Otherwise, the housing tenure situation is somewhat similar to Australia, with 32 per cent of households renting, more than four in five of those in the PRS (Stats NZ, 2020, pp. 36–37).

There are also significant differences between Australia and New Zealand's housing and welfare systems, notably in governance. As indicated earlier, Australia has a complex multi-level governance system, which makes a comprehensive housing and land management policy difficult to achieve. New Zealand has a unitary government system (no state or territory governments) and therefore has centralized housing policymaking (Pawson et al., 2020) that enables greater policy change. In the 1990s, New Zealand embarked on a radical experiment according to a neo-liberal prescription of corporatization and commercialization of government (Cheyne et al., 2008). In this process, market rents were introduced for public housing, and an income-related housing payment (accommodation supplement) was paid to eligible people to assist with rents. This policy led to some adverse consequences in terms of overcrowding and the incidence of infectious diseases (Baker et al., 2013). Māori and Pacific Islander communities were disproportionately affected, and regulated rents and secure tenancies were subsequently restored to social housing in 2001 (Murphy, 2003).

Similar to Aboriginal and Torres Strait Islander peoples (Dufty-Jones & Rogers, 2016), the homeownership rate among Māori is lower compared to people of other ancestries, and Māori are more reliant on social housing (Schrader, 2013). Indigenous people in Australia and New Zealand are also more likely to live in substandard housing and experience crowding (Dufty-Jones & Rogers, 2016; Schrader, 2013). However, there are some differences in policy action relating to Indigenous peoples' housing. In late 1980, the New Zealand government gave some ownership and control of housing resources to Māori organizations, and different schemes were introduced to increase homeownership rates among Māori, such as lending money to build on multiple-title land and partnerships between Māori trusts and the government (Ibid.). In 2010, the New Zealand government also implemented the Whānau Ora (well families) scheme that aimed to reduce welfare dependency among Māori (Ibid.).

A further difference between Australia and New Zealand concerns policy priorities. While Australia is in the early stages of understanding the conditions of its housing stock (Baker et al., 2019), the New Zealand government and research sectors have invested in studying housing stock, with a focus on the relationship between housing conditions and various physical and mental health outcomes. Dampness and mould were found to be common in New Zealand housing, especially in rental housing (Stats NZ, 2020). As a response, in 2019, Healthy Home Standards became a law in New Zealand. The law sets minimum standards for heating, insulation, ventilation, dampness and drainage and draught stopping in private rental housing (Ministry of Housing and Urban Development, nd.).

Concluding discussion

This chapter has illustrated the complexity of housing and welfare regimes in Australia and their points of connection and distance. Once established, institutional arrangements have proved difficult to change in Australia's federal system of government. This is of concern in view of three critical and ongoing challenges that we contend are the most important for the future of Australian housing: (i) lack of housing affordability and its widespread consequences; (ii) dealing with the legacy of colonization in which many Aboriginal and Torres Strait Islander peoples do not have adequate and affordable housing in line with their cultural preferences and connection to country; and (iii) addressing the effects of climate change and severe weather events through more resilient housing systems.

Affordability and affordable housing

The main housing policy challenge in Australia is the lack of affordability, a situation also faced in New Zealand. Lack of affordability is a problem for buyers and renters in light of recent interest rate rises affecting home mortgages. However, affordability pressures are most acute for private renters under housing stress who have no other viable alternative. Living in capital cities (e.g., Sydney and Melbourne) has become increasingly unaffordable for low- and even moderate-income households (Hulse et al., 2019). This, in turn, adds demand pressure to social housing, which falls far short of meeting current needs and will not do so under any likely future scenario. Recent government policies have sought to provide less security in social housing to encourage people to move on when their lives have stabilized. This has been relatively unsuccessful and only deals with the microlevel. Other solutions put forward to address affordability problems include arrangements to purchase in which a government takes an equity share, thereby minimizing loan size and debt repayments, with an option to increase equity subsequently; increasing the level of social and affordable rental housing with some regulation of rents; a very significant increase to the level of RA (40 per cent or more) particularly in high-price markets; and ending homelessness (achieved somewhat with COVID-19 emergency measures). Broader solutions include changes in land use and infrastructure planning (e.g., intensification of activity centres and transport corridors), which could help address the supply side of the unaffordability problem.

Housing for aboriginal and Torres Strait Islander peoples

A major challenge affecting Australia, New Zealand and the wider Pacific Island region is enabling Indigenous people to access and remain in decent housing that takes into account differences in laws, cultures, family structures and social practices. Key to this is building on customary forms of ownership of land that predate colonization in the region and the introduction of European statutory laws about land tenure (Juswanto & Kelkar, 2019). Financial institutions modelled on European ones generally lend to individuals/households secured against titles created by statutory law and not against other forms of customary ownership, including common ownership. This contributes to lower rates of homeownership in Australia for Aboriginal and Torres Strait Islander peoples compared to the general population and significantly higher rates of renting (similar to New Zealand). The challenge is to secure funding for housing on land with customary ownership, such as multiple titles, in partnership with Aboriginal and Torres Strait Islander peoples, with housing designed and managed to suit the needs of local communities rather than a one-size-fits-all approach. This has created friction in Australia's sparsely populated country away from the coast, where the costs of providing infrastructure associated with housing can be very high. Nevertheless, unless this challenge is met, housing for Aboriginal and Torres Strait Islander peoples outside of major cities will continue to be of low standard and lack the necessary infrastructure for health and wellbeing.

Addressing climate change

The impacts of climate change have been a serious concern in Australia and are believed to have led to the election of a more climate-focused federal government in 2022. Catastrophic bushfires and floods have affected Australia significantly in the last three years and have led to homes becoming destroyed or severely damaged. Rising sea levels will also impact housing, as coastal areas will become riskier to inhabit. In 2021, the Reserve Bank of Australia (RBA) flagged that properties in locations more at risk of climate change could see a decline in prices due to rising insurance costs (Bellrose et al., 2021). As temperatures rise, the ability to live in climate-resistant, comfortable housing will be limited to wealthier Australians. The new Australian Government has enshrined in law a commitment to reduce emissions by 43 per cent by 2030 and have net zero emissions by 2050. However, evidence such as the State of the Environment Report (Department of Agriculture, Water and the Environment, 2022), which confirmed the serious ongoing deterioration of Australia's environment, demonstrates that further government action is necessary. Recent citizen responses include people from the Torres Strait taking legal action against the (previous) federal government for lack of climate change action because climate change is already damaging the low-lying islands (see Our Islands, Our Home, 2022; The Australian Climate Case, 2022). The impacts of climate change on housing conditions, prices and construction may be the most pressing issue for the future of Australia's housing system.

Notes

1 Aboriginal peoples are from the mainland of Australia, while Torres Strait Islander peoples are from the Torres Strait, a strait between Australia and Papua New Guinea that contains many islands.

2 The Australian Census is conducted every five years, and effort is made to include every person and household in Australia. In 2021, over 96 per cent of dwellings completed the Census, so it is extremely comprehensive (ABS, 2022b).

3 Further lockdowns in major cities including Sydney and Melbourne in mid-2021 saw one more quarter of GDP decline followed by a further rebound, technically avoiding a recession.

4 Esping-Andersen in his 1990 work asserted that Australia along with the United States and Canada were 'the archetypical examples' of the liberal welfare regime (Esping-Andersen, 1990, p. 27), later extended to include New Zealand (along with Ireland and the UK) (Esping-Andersen, 1999, p. 75). Australia scholars rejected this categorisation asserting that Australia (along with New Zealand) represented a 'fourth world' of welfare capitalism (Castles & Mitchell, 1993) based on regulation of minimum wages, social guarantees and progressive taxation.

5 The National Affordable Housing Agreement 2009–2018 became the National Housing and Homelessness Agreement from 1 July 2018.

References

ABS. (2011) *2011 Census - Counting Dwellingss, Place of Enumeration* [Census TableBuilder] [Accessed: 15 September 2022].

ABS. (2017, June 26). *2001 Census Data*. https://www.abs.gov.au/websitedbs/censushome.nsf/home/historicaldata2001?opendocument&navpos=280

ABS. (2018, March 14). *Census of Population and Housing: Estimating Homelessness*. https://www.abs.gov.au/statistics/people/housing/census-population-and-housing-estimating-homelessness/2016

ABS. (2020). *Trade Union Membership*. Released 11/12/2020. https://www.abs.gov.au/statistics/labour/earnings-and-working-conditions/trade-union-membership/latest-release

ABS. (2021a). *2021 Census - Counting Dwellings, Place of Enumeration* [Census TableBuilder] [Accessed: 27 September 2022].

ABS. (2021b). *2021 Census - Counting Persons, Place of Enumeration* [Census TableBuilder] [Accessed: 27 September 2022].

ABS (Australian Bureau of Statistics). (2022a). *Snapshot of Australia*. Released 28/06/2022. https://www.abs.gov.au/statistics/people/people-and-communities/snapshot-australia/2021#data-download

ABS. (2022b). *Participation in the 2021 Census*. https://www.abs.gov.au/census/about-census/2021-census-overview/participation-2021-census.

ABS (2022c). *21. Historical Censuses (Pre 1996)*. https://www.abs.gov.au/AUSSTATS/abs@.nsf/ViewContent?readform&view=ProductsbyCatalogue&Action=Expand&Num=2.2Australian Institute of Aboriginal and Torres Strait Islander Studies (AIATSIS). (2022). *The Mabo Case*. AIATSIS. https://aiatsis.gov.au/explore/mabo-case [Accessed: 15 August 2022].

Australian Institute of Health and Welfare (AIHW). (2022). *Housing Assistance*. https://www.aihw.gov.au/reports-data/health-welfare-services/housing-assistance/overview

Baker, E., Beer, A., Zillante, G., London, K., Bentley, R., Hulse, K., Pawson, H., Randolph, B., Stone, W., & Rajagopolan, P. (2019). *The Australian Housing Conditions Dataset* (ADA Dataverse, V1) [Data set]. https://doi.org/10.26193/RDMRD3

Baker, M. G., McDonald, A., Zhang, J. & Howden-Chapman, P. (2013). *Infectious Diseases Attributable to Household Crowding in New Zealand: A Systematic Review and Burden of Disease Estimate*. Dunedin: He Kainga Oranga/Housing and Health Research Programme. https://www.healthyhousing.org.nz/sites/default/files/2022-02/HH-Crowding-ID-Burden-25-May-2013.pdf

Batterham, D. (2018). Homelessness as Capability Deprivation: A Conceptual Model. Housing. *Theory and Society*, *36*(3), 274–297. https://doi.org/10.1080/14036096.2018.1481142

Bellrose, K., Norman, D., & Royters, M. (2021). *Climate Change Risks to Australian Banks*. Reserve Bank of Australia. https://www.rba.gov.au/publications/bulletin/2021/sep/pdf/climate-change-risks-to-australian-banks.pdf

Berry, M. (1999). Unravelling the "Australian Housing Solution": The Post-War Years. *Housing, Theory and Society*, *16*(3), 106–123. https://doi.org/10.1080/14036099950149974

Burke, T., Hulse, K., & Ralston, L. (2022). 'I Will Call Australia My Home': Migration and Housing Tenure. In I. Levin, C. Nygaard, P. Newton, & S. Gifford (Eds.), *Migration and Urban Transitions in Australia: Then, Now and in the Future* (pp. 127–148). Palgrave Macmillan.

Burke, T., Nygaard, C., & Ralston, L. (2020). *Australian Home Ownership: Past Reflections, Future Directions* (AHURI Final Report, 328). https://doi.org/10.18408/ahuri-5119801

Castles, F. G. (1986). *The Working Class and Welfare: Reflections on the Political Development of the Welfare State in Australia and New Zealand, 1890-1980*. Allen & Unwin in association with Port Nicholson Press.

Castles, F. G. (1996). Needs-Based Strategies of Social Protection in Australia and New Zealand. In G. Esping-Andersen (Ed.), *Welfare States in Transition: National Adaptations in Global Economies* (pp. 88–115). Sage.

Castles, F. G. (1998). The Really Big Trade-Off: Homeownership and the Welfare State in the New World and the Old. *Acta Politica*, *33*(1), 5–19.

Castles, F. G., & Mitchell, D. (1993). Worlds of Welfare and Families of Nations. In F. G. Castles (Ed.), *Families of Nations: Patterns of Public Policy in Western Democracies* (pp. 93–128). Dartmouth.

Cheyne, C., O'Brien, M., & Belgrave, M. (2008). *Social Policy in Aotearoa New Zealand* (4th ed.). Oxford University Press.

Clark, G., Fischer, M., & Hunter, C. (2021). *Coasts, in Australia: State of the Environment 2021*. Department of Agriculture, Water and the Environment. https://soe.dcceew.gov.au/coasts/introduction

Commission of Inquiry into Poverty, & Henderson, R. F. (1975). *Poverty in Australia: first Main Report, April 1975*. Australian Government Publishing Service. https://nla.gov.au/nla.obj-1357060179

Crabtree, L., Perry, N., Grimstad, S., & McNeill, J. (2021). Impediments and Opportunities for Growing the Cooperative Housing Sector: An Australian Case Study. *International Journal of Housing Policy*, *21*(1), 138–152. https://doi.org/10.1080/19491247.2019.1658916

Demographia. (2022). *Demographia International Housing Affordability 2022 Edition*. Urban Reform Institute and the Frontier Centre for Public Policy. http://www.demographia.com/dhi.pdf

Department of Agriculture, Water and the Environment. (2022). *Australia: State of the Environment 2021*. https://soe.dcceew.gov.au/

Dufty-Jones, R., & Rogers, D. (2016). Housing in Australia: A New Century. In R. Dufty-Jones & D. Rogers (Eds.), *Housing in 21st Century Australia: People, Practices and Policies* (pp. 1–18). Routledge. https://doi.org/10.4324/9781315587110

Eccleston, R., Verdouw, J., Flanagan, K., Warren, N., Duncan, A., Ong, R., Whelan, S., & Atalay, K. (2018). *Pathways to Housing Tax Reform* (AHURI Final Report, 301). https://doi.org/10.18408/ahuri-4111001

Esping-Andersen, G. (1990). *The Three Worlds of Welfare Capitalism*. Polity Press.

Esping-Andersen, G. (1999) *Social Foundations of Postindustrial Economies*. Oxford University Press.

Hayward, D. (1996). The Reluctant Landlords? A History of Public Housing in Australia. *Urban Policy and Research*, *14*(1), 5–35. https://doi.org/10.1080/08111149608551610

Hulse, K., Burke, T., Ralston, L., & Stone, W. (2012). *The Australian Private Rental Sector: Changes and Challenges*. AHURI Positioning Paper, 149. https://www.ahuri.edu.au/sites/default/files/migration/documents/AHURI_Positioning_Paper_No149_The-Australian-private-rental-sector-changes-and-challenges.pdf

Hulse, K., Reynolds, M., Nygaard, C., Parkinson, S., & Yates, J. (2019). *The Supply of Affordable Private Rental Housing in Australian Cities: Short-Term and Longer-Term Changes* (AHURI Final Report, 323). https://doi.org/10.18408/ahuri-5120101

Hulse, K., Reynolds, M., & Martin, C. (2020). The Everyman Archetype: Discursive Reframing of Private Landlords in the Financialization of Rental Housing. *Housing Studies*, *35*(6), 981–1003. https://doi.org/10.1080/02673037.2019.1644297

Juswanto, W., & Kelkar, V. (2019, February). *The Dynamics of Urbanization, Housing, and Land Provision in the Pacific Island Countries*. Asia Development Bank Institute, Policy Brief, No. 2019-1. https://www.adb.org/sites/default/files/publication/488221/adbi-pb2019-1.pdf

Kemeny, J. (1983). *The Great Australian Nightmare*. Georgian House.

Kewley, T. H. (1973). *Social Security in Australia 1900-1972* (2nd ed.). Sydney University Press.

Khadem, N. (2022, May 25). HomeBuilder and First Home Buyer Grants Made Housing Less Affordable Says Reserve Bank. *ABC News*. https://www.abc.net.au/news/2022-05-25/homebuilder-first-home-buyer-grants-housing-affordability-rba/101097376

KPMG. (2022). *She's Price(d)less: The Economics of the Gender Pay Gap*. KPMG, Sydney. https://assets.kpmg/content/dam/kpmg/au/pdf/2022/kpmg-shes-priced-less-2022.pdf

Kryger, T. (2009). *Homeownership in Australia Data and Trends*, Research Paper no 21 2008-09 (11/02/2009) for Parliament of Australia. https://www.aph.gov.au/About_Parliament/Parliamentary_Departments/Parliamentary_Library/pubs/rp/rp0809/09rp21#_Toc222125572

Martin, C., Hulse, K., & Pawson, H. (2018). *The Changing Institutions of Private Rental Housing: An International Review* (AHURI Final Report, 292). https://www.ahuri.edu.au/research/final-reports/292; https://doi.org/10.18408/ahuri-7112201

Martin, C., Hulse, K., Ghasri, M., Ralston, L., Crommelin, L., Goodall, Z., Parkinson, S., & O'Brien Webb, E. (2022). *Regulation of Residential Tenancies and Impacts on Investment* (AHURI Final Report No. 391), Australian Housing and Urban Research Institute Limited, Melbourne. https://www.ahuri.edu.au/research/finalreports/391; https://doi.org/10.18408/ahuri7124801.

Ministry of Housing and Urban Development (NZ). (nd). *Healthy Homes Standards*. https://www.hud.govt.nz/our-work/healthy-homes-standards/

Morris, A. (2010). The Lack of a Right to Housing and Its Implications in Australia. *Journal of Australian Political Economy*, (65), 28–57. https://search.informit.org/doi/10.3316/informit.388386634519941

Morris, A., Hulse, K., & Pawson, H. (2021). *The Private Rental Sector in Australia Living with Uncertainty*. Springer.

Moskos, M., Isherwood, L., Dockery, M., Baker, E., & Pham, A. (2022). *'What Works' to Sustain Indigenous Tenancies in Australia* (AHURI Final Report, 374). https://doi.org/10.18408/ahuri3122901

Murphy, L. (2003). Reasserting the 'Social' in Social Rented Housing, Housing Policy and Housing Reforms in New Zealand. *International Journal of Urban & Regional Research*, *27*(1), 90–101. https://doi.org/10.1111/1468-2427.00433

Murphy, J. (1995). *The Commonwealth-State Housing Agreement of 1956 and the Politics of Homeownership in the Cold War*, Urban Research Program, Working Paper, 50, Research School of Social Sciences, Australian National University, Canberra

NatHERS (Nationwide House Energy Rating Scheme). (2022). *What Is NatHERS?* Department of Climate Change, Energy, the Environment and Water. https://www.nathers.gov.au/

OECD (2022) *OECD Data: Foreign Born Population 2019*. https://data.oecd.org/migration/foreign-born-population.htm

Ong, R., Pawson, H., Singh, R., & Martin, C. (2020). *Demandside Assistance in Australia's Rental Housing Market: Exploring Reform Option* (AHURI Final Report, 342). https://doi.org/10.18408/ahuri8120801

Ong, R., Wood, G., Cigdem-Bayram, M., & Salazar, S. (2019). *Mortgage Stress and Precarious Home Ownership: Implications for Older Australians* (AHURI Final Report, 319). https://doi.org/10.18408/ahuri-8118901

Our Islands Our Home. (2022). *Home Page*. https://ourislandsourhome.com.au/ [Acessed September 28, 2022].

Pawson, H., Hulse, K., & Morris, A. (2017). Interpreting the Rise of Long-Term Private Renting in a Liberal Welfare Regime Context. *Housing Studies*, *32*(8), 1062–1084. https://doi.org/10.1080/02673037.2017.1301400

Pawson, H., Milligan, V., & Yates, J. (2020). *Housing Policy in Australia: A Case for System Reform*. Palgrave Macmillan. https://doi.org/10.1007/978-981-15-0780-9_10

PC. (2015). *Housing Decisions of Older Australians*. Commission Research Paper. Productivity Commission.
PC. (2018). *Rising Inequality? A Stocktake of the Evidence,* Productivity Commission.
PC. (2019). *Vulnerable Private Renters: Evidence and Options, Commission Research Paper.* Canberra: Australian Government. https://www.pc.gov.au/research/completed/renters
PC (Productivity Commission). (2021). *Report on Government Services 2021 – Part G: Housing and Homelessness*. Australian Government. https://www.pc.gov.au/research/ongoing/report-on-government-services/2021/housing-and-homelessness
RBA (Reserve Bank of Australia). (2022) *Statement on Monetary Policy August 2022, 5*. Economic Outlook. https://www.rba.gov.au/publications/smp/2022/aug/economic-outlook.html
Ryan-Collins, J., & Murray, C. (2021). When Homes Earn More Than Jobs: The Rentierization of the Australian Housing Market. *Housing Studies*. https://doi.org/10.1080/02673037.2021.2004091
Schrader, B. (2023). 'Māori housing – te noho whare', Te Ara - the Encyclopedia of New Zealand, http://www.TeAra.govt.nz/en/maori-housing-te-noho-whare [Accessed 18 September 2023].
Shooshtarian, S., Hosseini, M. R., Martek, I., Shrestha, A., Arashpour, M., Costin, G., & Seaton, S. (2021). Australia's Push to Make Residential Housing Sustainable - Do End-Users Care? *Habitat International, 114*. https://doi.org/10.1016/j.habitatint.2021.102384
Stats NZ. (2019, September 23). *New Zealand's Population Reflects Growing Diversity*. https://www.stats.govt.nz/news/new-zealands-population-reflects-growing-diversity [Acessed 28 September 2022].
Stats NZ. (2020, December 8). *Housing in Aotearoa: 2020*. https://www.stats.govt.nz/reports/housing-in-aotearoa-2020 [Acessed 28 September 2022].
Stats NZ. (2022, June). *Homepage*. https://www.stats.govt.nz/ [Acessed 28 September 2022].
Stebbing, A., & Spies-Butcher, B. (2010). Universal Welfare by 'Other Means'? Social Tax Expenditures and the Australian Dual Welfare State, *Journal of Social Policy, 39*(4), 585–606.
The Australian Climate Case. (2022). *Home Page*. https://australianclimatecase.org.au/
UN Special Rapporteur on Housing. (2007). *Report of the Special Rapporteur on Adequate Housing as a Component of the Right to an Adequate Standard of Living, and on the Right to Non-Discrimination in This Context*. Miloon Kothari: Addendum. https://digitallibrary.un.org/record/649445
Wood, D., & Griffiths, K. (2019). *Generation Gap, Ensuring a Fair Go for Younger Australians*. Grattan Institute. https://grattan.edu.au/report/generation-gap/
Yates, J. (2015). *Trends in Home Ownership: Causes, Consequences and Solutions.* Submission to the Standing Committee on Economics Inquiry into Home Ownership. https://www.google.com/url?sa=t&rct=j&q=&esrc=s&source=web&cd=&ved=2ahUKEwjd1prx_Lr6AhXxaGwGHT27DtYQFnoECAMQAQ&url=https%3A%2F%2Fwww.aph.gov.au%2FDocumentStore.ashx%3Fid%3D09aeea3d-4921-4e78-8df1-ee83fb39801b%26subId%3D352865&usg=AOvVaw00PK_KiqlyTReFg-bExTIG
Yates, J., & Yanotti, M. B. (2016). *Australia's 25 Years With a Deregulated Housing Finance System, Milestones in European Housing Finance* (pp. 37–53). John Wiley & Sons.

7

HOUSING AND WELFARE IN SWEDEN, NORWAY AND THE WIDER NORDIC REGION

Jardar Sørvoll, Carina Listerborn and Matilda Annemark Sandberg

In this chapter, we will direct our attention to the four largest Nordic countries – that is Denmark, Finland, Norway, and Sweden. To give a brief introduction, the population (year 2020 Q4) in Denmark is 5.8 million (Statistics Denmark, 2022), 5.5 in Finland (Statistics Finland, 2021), 5.4 in Norway (Statistics Norway, 2021) and 10.4 Sweden (Statistics Sweden, 2022). The capital cities are, similarly to the populations in total, comparatively small. The largest capital is Stockholm, Sweden, with 2.4 million inhabitants in its metropolitan area (Boverket, 2021a). Next comes Copenhagen, with 1.3 million in the metropolitan area (Nordic Co-operation, 2022b).

The living and housing standards in the Nordic countries are some of the highest in Europe, and the rate of "severe housing deprivation"[1] is lower than in most other countries (Borg & Guio, 2021). When it comes to GDP per capita (2020), oil-rich Norway leads the field with 67,300 US dollars. The Danish GDP per capita is 61,100, followed by Sweden with 52,300, and then Finland with 48,800. For reference, the average in the European Union (EU) is 34,100 and the global average is 10,900 USD (World Bank, 2022a). Denmark, Finland, and Sweden are part of the EU. Norway is part of the European Economic Area. The Nordic countries are known for significant natural resources, and the living standard is in general high; for example, extensive health care services for all are provided by the state.

Historically, particularly the Swedish postwar housing regime has been singled out for praise by outside observers (see Blackwell, 2021; Christophers, 2013, for examples from the literature). However, the Nordic world of housing has never been the Panglossian best of all possible worlds. The present era is no exception to this rule. Spiralling property prices and household debt levels, segregation, speculation, growing inequality and precarious rental housing to varying degrees characterize all four mainland Nordic housing regimes.

In this chapter, we discuss the historical development and the current state of the housing and welfare regimes in the Nordic countries, using Sweden and Norway as our main cases. The two neighbouring countries on the Scandinavian peninsula are often classified as generous social democratic welfare regimes with a high level of universalism and de-commodification (Esping-Andersen, 1990). In the first postwar decades, however, Sweden and Norway chose different paths in the sphere of housing policy. The Swedish government subsidized the construction of rented-, owner-occupied- and cooperative housing, but many local governments gave the highest priority to the universal municipal public rented sector (*allmännyttan,* trans. "for the benefit of everyone"

DOI: 10.1201/9781003212690-9

in Grander, 2018). In contrast, the Norwegian governments post-1945 shunned rental housing, followed the Swedish ideal of tenure neutrality with no specific form of "social housing" directed to low-income groups, and supported the mass construction of owner-occupied and collectively owned cooperative housing (Bengtsson, 2013). It is an overstatement to claim that this divergence made Sweden a nation of tenants and Norway a nation of homeowners, as homeownership was always a major feature of the social and political reality in Sweden. In the first three postwar decades, moreover, the institutional differences between Swedish public rented housing and Swedish and Norwegian cooperative housing were not enormous; both were state-subsidized and highly regulated tenures (Sørvoll, 2014). Nonetheless, many scholars have rightly emphasized the differences between the two housing regimes in the past and present: the political organizations and government regulations in the rental market are still much stronger in Sweden than in Norway (Bengtsson, 2013; Jensen, 2013; Sejersted, 2011). In 2022, the Swedish housing regime will have one of the world's largest tenants' movements with over half a million members, institutionalized collective bargaining between tenants and landlords, and the use-value system of rent determination. Since 1945, moreover, political majorities in Sweden have always rejected "social housing" targeted at low-income groups in favour of universal public rented housing (Bengtsson & Grander, 2023). The Norwegian housing regime has none of these unique Swedish features.

In this chapter, we argue that it is still broadly correct to regard Sweden, Norway, Finland and Denmark as four distinct housing regimes with major differences in terms of public policies and housing market characteristics. On the other hand, the contemporary Nordic housing regimes have many similar features and challenges, not least connected to the debt-fuelled escalation of residential property prices (Blackwell, 2018; Kohl & Sørvoll, 2021). Our comparative discussion of Norway and Sweden seeks to illustrate how common challenges related to globalization, financialization, affordability and inequality manifest themselves and are dealt with in two unique housing regimes in the Nordic region. In our concluding discussion, we briefly discuss the prospects of political initiatives to counter major obstacles to the goals of achieving fairness, freedom and well-being for all in the welfare state's "wobbly pillar" (Torgersen, 1987).

The current state of housing

Sweden

The contemporary Swedish housing regime is characterized by, on the one hand, rent negotiations between tenants and landlords and security of tenure, and, on the other hand, a pronounced marketization, which caused Brett Christophers to characterize it as being a "monstrous hybrid", as these parallel tendencies "serve as a decisive mechanism for the creation, reproduction and intensification of socio-economic inequalities" (2013, p. 888). The benefits of owning compared to the high costs of renting have been essential in reinforcing gaps in wealth (Christophers, 2021). The historically strong position of the rental sector in terms of market share and appeal to consumers has been gradually weakened in tandem with the growing economic privileges of ownership, but the rental sector is still protected by extensive regulation.

In Sweden, there are four main forms of tenure. Ownership rights (owner-occupied housing) are mainly enjoyed in single-family houses. Until 2009, it was not possible to build owner-occupied apartments, and today they represent only 0.5 per cent of the housing market. Tenant-owner right (*bostadsrätt*) means that the tenant is a member of a tenant-owner association (*bostadsrättsförening*), which owns the property, and each member has the exclusive right to occupy an apartment. Cooperative rental rights (*kooperativ hyresrätt*) are similar in the sense that an association owns – or rents – a property, and the association members rent their apartment from the association with some form of deposit, but the apartment cannot be sold. This is a small segment of the housing market. Rental rights cover around one-third of the housing market, and about half of these properties are owned by municipal public housing companies (*allmännyttan*) (Boverket, 2022a). See Table 7.1 for an overview of the market shares of the tenure forms and Table 7.2 for a comparison with Norway.

Ownership rights and tenant-owner rights have been profitable from a household perspective as prices have increased. Between 1996 and 2017, real prices increased by 232 per cent (annual increase of 5.9 per cent), stimulated by interest rate payments being tax deductible and tax on capital gains being rolled over if the profit is reinvested in more expensive housing, the latter thus encouraging households to "switch up". The wealth tax was removed in 2007, which then contributed to a 10 per cent increase in housing prices. The loan-to-value ratio for home buyers has increased from 59 per cent in 2002 to 71 per cent in 2010 (Englund 2016; see also Christophers, 2021). This reflects the financialized character of Swedish homeownership since the deregulation of mortgage lending in the 1980s. In 2018, 74 per cent of homeowners held a mortgage (Aarland & Sørvoll, 2021).

Tenants in the rental sector are protected through security of tenure, which refers to the tenant's right to have the lease renewed when the landlord wishes to terminate it if rents have been paid and no other rules have been broken. Unique to Sweden is that the rents are set through collective negotiations between landlords and the Swedish Union of Tenants, based on the use-value of the apartment's standard, quality and service. The annual rent increase must be "equitable" (*skälig*), and for the last ten years, that has, according to Statistics Sweden (2021), meant between 0.8 per cent and 2.8 per cent. But, if the landlords make improvements, such as kitchen, bathroom, and water pipe replacements, then rents are allowed to increase by up to 50–60 per cent in accordance with their improvements. However, such major improvements need to be accepted by the tenants. The tenants can object and bring the question of renovation to the Rent Tribunal, but they are seldom successful (Polanska & Axen, 2021). This contentious state of affairs has led scholars and activists to talk about "renovictions" (renovation leading to eviction), a term coined by activists in Canada that highlights the threat of displacement with serious social and health consequences for the tenants (Baeten et al., 2017; Polanska & Richard, 2019; Pull, 2020). According to a study by the National Board of Housing, Building and Planning (*Boverket*), up to 25 per cent of tenants never move back into renovated buildings (Boverket, 2014). The lure of extensive rent increases through "renoviction" has also attracted international investment companies, such as Blackstone, to purchase Swedish rental housing from private and public landlords. Global investment and real estate companies are quite new in Sweden but have had a major impact on contemporary housing in terms of increased distance to service and implementing renovation schemes (Gustafsson, 2021, 2022; Kadioglu et al., 2022).

Homelessness in Sweden stands out in a Nordic context (FEANTSA & Foundation Abbé Pierre, 2020). Municipal social services (as opposed to housing authorities) are responsible for homelessness and can provide different forms of housing, which will be discussed further in the section "The interrelation between housing and welfare". For people unable to enter the rental market or to buy housing, the sublet market in all its variations is the most common solution, leading to a precarious housing market (Listerborn, 2021). Younger people increasingly tend to stay with parents as they face difficulties entering the housing market (Grander, 2023), and overcrowding is strongly associated with low incomes (Grander & Sandberg, 2022). The shortage of affordable housing, exacerbated by the very low levels of municipal new build since subsidies were cut in the 1990s and the sale of public housing, in combination with the increased inequalities on the housing market, has led to this situation (Grander, 2018; Listerborn, 2021). The annual housing market survey conducted by *Boverket* (2021b) found that 207 (of 290) municipalities reported housing shortages, and this concerns in particular young people, students, newly arrived refugees, people with disabilities and older people who want or need to move homes.

Since the early 2010s, there has been a rapid increase in housing construction (Boverket, 2019). Construction is financed by private companies and bank loans. During 2017–2021, a state investment fund (*investeringsstöd*) was initiated to support the construction of general rental housing and student housing, but it was removed by the conservatives. There remains a shortage of affordable housing. The major restructuring of the housing market in the early 1990s (Christophers, 2013; Grundström & Molina, 2016; Hedman, 1994; Listerborn & Molina, 2022) has led to considerable difficulties for low-income households and an emerging resistance and social movements around housing issues (Gustafsson et al., 2019; Listerborn et al., 2020).

Norway

The Norwegian housing regime is a liberal and selective sector of what is still a generous welfare regime with many universal benefits and services (Bengtsson et al., 2017; Stamsø, 2009). Housing construction is predominantly planned and financed by private actors; there is no rent control or price regulation in the housing market; economic support and social rented housing are targeted at low-income groups; and homeownership is championed by the government through tax subsidies and state-backed mortgages for disadvantaged households. The long-standing goal of the government is to expand homeownership to all social strata, even low-income households (Aarland & Reid, 2019; Sørvoll & Nordvik, 2020). Norway has one of the highest homeownership rates in Western Europe: 76.4 per cent of households are either owner-occupiers (62.7 per cent) or tenant-homeowners (13.7 per cent) (SSB, 2021). The latter tenure is a form of shareholder housing almost identical to Swedish tenant ownership with roots dating back to the 1920s (Sørvoll, 2014).

Akin to the Swedish case, homeownership in Norway is currently highly financialized. A large majority of homeowners have mortgages with commercial banks, and only one-third of homeowners own their homes outright (Aarland & Sørvoll, 2021). The rich stream of capital flowing into the housing market has fuelled the parallel growth of household debt and residential property prices since the 1980s (Sørvoll, 2021).

Whereas homeownership is the cultural norm (Vassenden, 2014) and favoured across the ideological spectrum, rental housing has limited political support and the tenants' movement is very weak compared to the Swedish case. The government has pledged to promote security for tenants (KMD., 2020), but so far this rhetoric has not been accompanied by substantial policy changes. The Norwegian rental sector is small and characterized by weak security of tenure and lower housing standards than the owner-occupied sector (FR, 2021). Even though the rental sector has grown slightly in recent years, it still houses only 23.6 per cent of all households (SSB, 2021). Furthermore, the social rented sector in Norway is one of the more means-tested and market-oriented in Europe (Sørvoll, 2019). In stark contrast to the Swedish case, the Norwegian municipal public rented sector accounts for a mere four per cent of the housing stock and is almost exclusively targeted at the most disadvantaged households, such as low-income families with children, refugees, individuals struggling with addiction and mental illness, and other households in danger of becoming homeless (Aarland & Sørvoll, 2021).

The means-tested and market-oriented social rented sector reflects the general liberal and selective character of the Norwegian housing regime. Policy documents published in recent years have emphasized the virtues of the free market, personal responsibility and the need to prioritize the homeless and other disadvantaged groups when allocating social housing and state-backed mortgages. The latter has probably contributed to a substantial reduction in the number of homeless individuals in the years between 2012 and 2020 (Dyb, 2017; Sørvoll, 2021). Along with Finland, Norway is the only European country that has succeeded in lowering the homelessness rate in recent years (Benjaminsen et al., 2020).

However, the contemporary Norwegian housing regime is not met with approval in all quarters. According to vocal critics from the left and centre-left, the predominantly market-based production, exchange, and distribution of housing has significantly contributed to greater economic inequality and led to a reduction in the number of affordable homes for low- and medium-income households in urban areas (cf. Gitmark, 2020; Tranøy et al., 2020). The latest empirical research suggests that the critics have a point (Aaberge et al., 2021; Eggum & Larsen, 2021), and it is therefore not surprising that the central state and local governments have recently tried to respond to calls for more affordable homes in the owner-occupied and rental sectors. According to Kjærås and Haarstad (2022), there has been a comprehensive re-politicization of housing in Oslo in recent years; alternative models of affordable rental and owner-occupied housing are on the agenda of both political parties and civil society actors. To date, however, this re-politicization has not changed the core features or even the finer details of the Norwegian housing regime.

A brief overview of the historical development of the housing regimes

Along with the workers' and women's movements in the early 20th century, the struggles for decent housing were fundamental to the development and improvement of housing conditions in both countries (Rolf, 2020). In Sweden, a major government-led inquiry aiming for improved housing standards (*Bostadssociala utredningen* 1933–1947) laid the groundwork for the birth of the municipal housing companies (*allmännyttan*) and for the universal housing regime that survived at least until the early 1990s. The minimum housing space standard was elevated to a maximum of two people per room, excluding the kitchen. The ratio between housing costs and income was set at a maximum of 20 per cent of the household income. Family housing allowances directed at poor households and building allowances directed to construction companies were introduced, and the public took responsibility for financing housing construction. The state would be responsible for risks and subsidies, while the municipalities would organize, plan, take building initiatives and mediate state loans.

Both the Swedish and Norwegian postwar governments were allied with vibrant social democratic housing movements and sought to eliminate the postwar housing shortage through the ambitious mass construction of housing backed by brick-and-mortar subsidies. To limit speculation and ensure a fair distribution of resources, housing was also subject to rent control, price regulation and municipal allocation. Moreover, cooperative building companies based on non-profit principles, such as *HSB* and *Riksbyggen* in Sweden and *OBOS* in Norway, were important providers of housing on both sides of the border. To simplify somewhat, before the 1980s, housing was regarded more as a social right provided and distributed with the aid of the government than a market commodity primarily produced and allocated by profit-seeking companies (Sørvoll, 2014). In the postwar era, the Swedish and Norwegian housing regimes may be described as universal in the sense that the main government subsidies and state-supported housing forms were directed at the entire population (Bengtsson et al., 2017). However, the two housing regimes differed in many other significant respects.

In Norway, a "social homeownership model" (Sandlie & Gulbrandsen, 2017) was developed based on the cheap mortgages of the State's Housing Bank (*Husbanken*) and politically controlled interest rates. In contrast to the Swedish case, the dominant party of government until the 1980s, the Social Democrats, were strongly pro-homeownership in the single-family and cooperative tenant-ownership varieties. This unusual ideological position was based on anti-market and anti-landlord sentiments. Tenant ownership was accordingly price-regulated to inhibit speculation and supplied by non-commercial cooperative building companies (Bengtsson et al., 2017). Rental housing was also heavily regulated but not in any way promoted by the government. In the decades after 1945, the total stock of rental housing in Norway was significantly reduced through government-supported conversions to tenant-ownership and freehold flats. Social rental housing, moreover, has always been quite marginal and residual in Norway (Sørvoll, 2014).

In the housing studies literature, the Swedish post-war housing regime is understandably best known for its large and universal public rental sector and the determination of rent levels by way of corporatist negotiation between the representatives of landlords and a comparatively strong tenants' movement (cf. Kemeny, 2006). Even though price-regulated tenant ownership produced by non-profit cooperative providers expanded on both sides of the border and both governments intervened strongly in the planning, building and distribution of housing (Sørvoll & Bengtsson, 2018), there were still clear differences between the Swedish and Norwegian housing regimes before the era of retrenchment and deregulation commenced gradually from the late 1970s onwards. The large universal public rented sector, the ideal of tenure neutrality, negotiated rents and strong security of tenure are features of the Swedish housing regime that distinguished it from the Norwegian case.

As was the case in many other European countries, the private rental sector shrank substantially in the golden years of Scandinavian post-war housing policy. In Norway, tenant ownership was the main beneficiary of the decline of landlord-owned housing. Across the border, tenant ownership also expanded, but it had to play second fiddle to the municipal housing companies (Sørvoll, 2014). Between 1945 and 1980, the public rented sector's share of the housing stock grew from 6 to 24 per cent (see Tables 7.1 and 7.2). In Sweden, 53 per cent of the population lived in public or private rental housing in the 1970s, compared with today's 38 per cent (Löfgren, 2021).

Sweden and Norway both experienced a gradual "system shift" in housing policy in the 1980s, 1990s and after the millennium. Large-scale government subsidies for housing construction and comprehensive state regulation of housing markets were scaled back or abolished and replaced with modest economic support to disadvantage groups and more liberal market regulations (Holmqvist & Turner, 2014; Stamsø, 2009). In Sweden, the liberal conservative government of Carl Bildt (1991–1994) even dismantled the Ministry of Housing (Lindbom, 2001). However, it was arguably the Norwegian housing regime that changed most dramatically. Starting from the late 1970s, a universal housing policy aimed at all households featuring a vast array of subsidies and regulations was abolished in favour of free markets, the principle of personal responsibility and selective state support for low-income households. Most significantly, the price regulation on cooperative tenant-ownership was *de facto* abolished during the 1980s, and the State's Housing Bank, hitherto so important for the mass construction of affordable housing, was transformed into the government's main policy instrument in the fight against homelessness and other forms of disadvantage in the first years of the new millennium (Sørvoll, 2021).

In Sweden, the central postwar institutions of housing policy – such as tenant ownership, the municipal housing companies and the use-value system of rent determination – have all changed in notable ways during the last forty years. The price regulation on cooperative tenant-ownership was lifted in 1968, something that paved the way for the tenant-ownership housing boom in the era of financial deregulation in the 1980s. Much like its Norwegian counterpart, the Swedish originally cooperative tenure of tenant ownership was transformed from a regulated, non-commercial and semi-public tenure to a commercial, privatized and financialized form of housing. By the 1990s, the cooperative providers of tenant-ownership, such as *HSB* in Sweden and *OBOS* in Norway, had shed the non-profit ideology of cost-based pricing and acted like any other commercial housing company, and the asking prices and mortgage levels in the cooperative tenant-ownership sector have soared almost unabated in the last twenty years (Sørvoll & Bengtsson, 2018). In parallel with this commercialization and financialization, there were several waves of mass conversions from private and public rented housing to tenant-owner cooperatives, most notably in the most attractive residential areas in Stockholm and other major cities. In Stockholm, the proportion of rental flats fell from 73 per cent to 36 per cent between 1990 and 2010 as a direct result of political decisions. Moreover, in the inner parts of the Swedish capital, tenant ownership increased its share of the housing stock from 26 to 62 per cent between 1990 and 2010 (Andersson & Turner, 2014).

Liberal-conservative parties cheered on the conversion of public rented to tenant ownership as part of a wider strategy of promoting homeownership. The Swedish government, headed by the conservative Fredrik Reinfeldt (2006–2014), legalized owner-occupied flats and abolished the property tax to stimulate the growth and welfare of homeowners (Holmqvist & Turner, 2014). The latter decidedly signalled the end of any pretence of tenure neutrality, a concept that was always more of a rhetorical construct than a social and political reality since homeowners in single-family housing had always been favoured by the tax system (cf. Lundqvist, 1987).

During the last 30 years, moreover, the large public rented sector has lost many of its distinguishing features and is no longer the indisputable institutional pillar of the Swedish housing regime. In the early 1990s, the generous government subsidies to public rented housing were abolished, and thereafter, the municipal housing companies were left to fend for themselves in the housing market (Gustafsson, 2021). Significant changes in the public rented sector also occurred in the wake of adjustments to the EU's laws regarding competition and state subsidies to market actors (Elsinga & Lind, 2013). In 2011, parliament stipulated that municipal housing companies should be governed according to business-like principles, meaning that profit-seeking, at least in the long term, should justify all the companies' decisions (Grander, 2017). Furthermore, the new legislation meant the end of the norm-setting role of public housing; from then on, rent negotiations were done independently for private and public housing companies (Bengtsson & Jensen, 2021). In addition, the public rented sector has undergone a process of residualization in recent decades. Even though it is still formally universal, it has become "social by default" (Magnusson & Turner, 2008) in the sense that middle-class households have flocked to the owner-occupied sector and the average public tenant has become poorer (Andersson & Turner, 2014; Borg, 2019).

With the above changes noted, the Swedish housing regime in the years after the 1980s was also characterized by almost surprising elements of stability (Bengtsson & Jensen, 2021; Blackwell & Bengtsson, 2021). In 2021, the formal universality of the public rented sector, negotiated rents and the soft rent control inherent in the use-value system were still features of the Swedish housing regime. Certainly, local governments offer fixed-term so-called "social contracts" to tenants unable to afford private or public rented housing (Grander, 2017), but there is still no Norwegian-style means-tested social housing sector directed at the most disadvantaged groups in Sweden. This institutional stability has led some scholars to argue that Sweden remains a universal housing regime despite all the notable changes in the era of retrenchment and deregulation (Bengtsson et al., 2017). In contrast, the Norwegian case is a less complicated and unambiguously selective housing regime based on means-testing, free markets, and homeownership (Aarland & Sørvoll, 2021). However, the comparative perspective adopted in this chapter also helps us zoom in on important similarities between our two cases. The most notable of these are the socio-economic and spatial inequalities caused by the parallel debt and price growth in the owner-occupied sector

Table 7.1 Tenures in Sweden 1945–2020 (%), measured in dwellings

	Sweden				
	1945	*1960*	*1980*	*2000*	*2020*
Homeownership	38	34	41	39	38
Tenant ownership	4	9	14	17	24
Private rented	52	43	21	21	17
Public rented	6	14	24	23	16
Other	–	–	–	–	5

Source: Bengtsson and Grander, 2023, p. 96.

Table 7.2 Tenures in Norway 1945–2020 (%), measured in households

	Norway				
	1945	*1960*	*1980*	*2000*	*2020*
Homeownership	53	53	56	62	63
Cooperative ownership	NA.	11	17	14	14
Private Rented	41	36	27	23	19
Public rented	NA.	NA.	NA.	NA.	NA.

Sources: SSB, 2021; Sørvoll, 2011, 2014.

on both sides of the border (Tranøy et al., 2020). In both Sweden and Norway, the financialization of tenant ownership from the 1980s onwards was a key driver behind this development (Kohl & Sørvoll, 2021).

Furthermore, the conversion of rented housing to tenant ownership gradually moved the Swedish homeownership rate closer to the level of its neighbour in the West. In the years from 2000 to 2020, the Swedish homeownership rate (including tenant-ownership) increased from 56 per cent to 62 per cent, a surge that narrowed the owner-occupation gap with Norway from approximately 20 to 15 per cent (see Tables 7.1 and 7.2).

The welfare regime and its development

In the decades after 1945, Sweden and Norway developed some of the most comprehensive welfare states in Europe (Sejersted, 2011). When Esping-Andersen published his classic book The Three Worlds of Welfare Capitalism in 1990, both countries were classified as social democratic welfare regimes with high levels of de-commodification and generous tax-financed universal benefits. In the comparative literature, the "Nordic model" is also associated with high levels of employment, gender equality, trade union power and low levels of poverty and inequality. Briefly put, in the postwar era, the Nordic states took a high degree of responsibility for the welfare of all citizens from the cradle to the grave in the spheres of health, care, education, pensions and housing. The role of the family, the market and private insurance in the provision of welfare was quite limited, seen from an international comparative perspective. In 2021, this was still the case to some extent, but that being noted, Sweden, Norway and the other Nordic welfare states have not been immune to major changes since the 1980s, in the years of new public management, neoliberalism and retrenchment (Greve, 2007; Kangas & Kvist, 2018; Pedersen & Kuhnle, 2017).

Starting in the 1990s, moreover, inequality started to increase substantially in the hitherto uniquely egalitarian Nordics, particularly in Sweden. From 1987 to 2013, the level of income inequality, as measured by the Gini index, increased by 22 per cent in Norway and 45 per cent in Sweden (Barth et al., 2021). In oil-rich Norway, the internal pressure for cutbacks in welfare benefits and services has been relatively weak. The economic downturn of the early 1990s was less dramatic than the Swedish experience, and a low rate of unemployment was maintained until the COVID crisis of 2020. Whereas Sweden moved much closer to the OECD average in terms of the generosity of social security benefits paid to the ill, unemployed, and old after 1990, Norway has kept clear of the pack and is near the top of the league in this area (Pedersen & Kuhnle, 2017). Nonetheless, it was not possible to escape entirely from the ideological winds of welfare conditionality and workfare in Norway. In 2008, the government implemented a milder version of the Swedish pension reform of 1994 with the motive of curtailing public expenditure in response to the ageing of society (Ellingsæter et al., 2020). In a comparative perspective, however, the social democratic model of welfare has survived relatively unscathed in Norway; the term "austerity" does not have a well-worn Norwegian counterpart.

Compared to the case of Norway, the Swedish welfare state has undergone profound changes in the last thirty years (Hort, 2014). Cutbacks have been more severe and the privatization of public services more comprehensive, but Sweden has maintained a high level of social investment in child care, parental leave, and other forms of public services, boosting the level of participation and competence in the labour force (Ferragina, 2022; Hort et al., 2019). The third (voluntary) sector is relatively small in both Norway and Sweden but has increasingly played a role in the fight against homelessness. In the Swedish National City Mission Report (Sveriges Stadsmissioner, 2021), the third sector criticize the municipalities for lacking a long-term plan to deal with homelessness and for their failure to develop any effective models for collaborating with the non-profit sector.

The interrelation between housing and welfare

In this section, we discuss the development of the housing regime in relation to the wider welfare regime. The Swedish constitution states that "the public institutions shall secure the right to employment, housing and education and shall promote social care and social security, as well as favourable conditions for good health" (SFS, 1974 p. 152, The Instrument of Government, 1 Chapter 2 Art.). This does not mean that housing is a legal right for individuals, but it should rather be understood as a "social right" in the sense that the provision of decent housing is an area of concern for the welfare state (Bengtsson, 2001). Rather like the Swedish case, housing has never been an individual right in Norway, but municipalities are legally responsible for providing temporary shelter to households unable to secure a satisfactory dwelling themselves (Solstad, 2021).

Even though the contemporary welfare state has been altered by privatization and retrenchment, particularly in Sweden (Hort, 2014), housing is currently something of a marketized exception in the two welfare regimes in question. Compared to the long post-war years between 1945 and 1990, housing is currently much more of a financialized and marketized commodity produced and distributed according to the logic of the market. According to Tranøy et al. (2020, p. 406), the Scandinavian countries "still decommodify human lives with respect to pensions, health, life risks in general and education, but no longer when it comes to housing". Housing is, on the one hand, a responsibility of the state and, on the other hand, primarily provided and allocated through market

Table 7.3 Social policy instruments designed to "correct markets" and satisfy the basic housing needs of disadvantaged households in Sweden and Norway

Sweden	*Norway*
Housing allowances	Housing allowances
Public rented housing	Public social rented housing
Housing for homeless organized by the social services, e.g. social contracts	Start-up mortgages to encourage low-income homeownership
Government grants for housing construction (periodically)	Government loans for housing construction

Source: Authors.

mechanisms. The state's role is partly to "correct" or mitigate the welfare outcomes produced by the market (Bengtsson, 2001; Sørvoll, 2021). In both countries, there exists many policy instruments designed to "correct markets" and satisfy the basic housing needs of vulnerable groups, as shown in Table 7.3.

The Swedish Union of Tenants (Hyresgästföreningen) (Hyresgästföreningen, 2021) estimates that in 2020, Norway spent 1.6 per cent of GDP on support to the housing sector, compared to 1 per cent in Sweden. These expenditure estimates include housing allowances, interest deductions, support for construction, etc. The expenditure gap between the two countries is partly explained by generous Norwegian tax and grant subsidies that stimulate homeownership.

However, the clearest chasm between the two housing regimes is arguably the approach to low-income households and other disadvantaged groups in the housing market. In Norway, national and municipal housing policy has prioritized the most vulnerable groups for the last 20 years, and the small public rented sector is almost purely devoted to catering to the housing needs of this group. This may partly explain why Norway, unlike Sweden, has experienced a significant fall in the number and share of homeless people since 2012 (cf. Benjaminsen et al., 2020; Dyb, 2017). In Norway, moreover, the government's promotion of homeownership also extends to households with incomes way below the national average through so-called startup loans from the state Housing Bank (Aarland & Reid, 2019). This is another contrast to the Swedish case that underscores the selective and owner-oriented character of the Norwegian housing regime.

That said, it should not be forgotten that selective housing allowances is a long-standing feature of both housing regimes. The government's spending on housing allowances (*bostøtte*) in the selective Norwegian housing regime is actually lower per capita than the combined cost of housing supplements (*bostadstillägg*) and housing benefits (*bostadsbidrag*) in Sweden (OECD, 2021). In Norway, almost every household with (very) low incomes and high housing expenses, as defined by the government, may claim housing allowances. Across the border, the approach to selective housing support is more targeted: only families with children and people under the age of 29 are eligible for housing allowances, and only pensioners and people receiving sickness or activity compensation may receive housing supplements (Aarland & Sørvoll, 2021; Nordic Co-Operation, 2022a).

In Sweden, public rented housing is still universal in the sense that it is formally open to all strata of the population and not directed towards disadvantaged groups. On the other hand, new stricter rental policies implemented in the aftermath of the 2011 reform and "social contracts" for low-income households have weakened the principle of universalism (Grander, 2017). The so-called social contracts, as well as other short-term solutions, are part of the needs-tested housing solutions that the municipal social services can organize for homeless households (see e.g., Wirehag, 2021). The contracts can generally be described as forming a "secondary housing market" where the municipality rents dwellings to the homeless (Sahlin, 1996). Often, the municipality holds a first-hand lease on the regular rental market (public or private) and then sublets it to the household. While the epithets as well as characteristics of the contracts vary, they are normally somewhat temporary without security of tenure and may entail other conditions such as supervision by the social services or (very) short notice periods (Boverket, 2022b; Knutagård, 2018; Wirehag, 2021). The social contracts through the social services now equal over 1 per cent of the Swedish rental market (Boverket, 2022b). However, those who lack housing but are not considered to have any additional social problems are not always seen as part of the target group of the social services (Sahlin, 2020; Wirehag, 2021).

Another notable similarity between the two countries is the contribution of the uneven distribution of housing wealth to the general increase in economic inequality noted in the previous section (Christophers & O'Sullivan, 2019; Sørvoll & Nordvik, 2020). For instance, Eggum and Larsen (2021) document how housing wealth in Norway has exacerbated inequality between segments of the population and parts of the country. Homeowners in the most attractive urban locations reaped a highly disproportionate share of the wealth emanating from housing from 2007 to 2019. In this period, the top five percentiles of homeowners made 61 per cent more capital gains than the next five percentiles, and the 10 per cent of owner-occupiers that made the largest capital gains in Oslo increased their housing wealth with a staggering 3.35 million, or around 355,000 Euros (Eggum & Larsen, 2021). This illustrates that even though housing wealth is more evenly distributed than other sources of capital, the unregulated housing market is still one of the major drivers of inequality in a Nordic reality of increasing economic differences. A recent study by Aaberge et al. (2021) shows that data and methods traditionally used to measure inequality dramatically underestimate economic disparities in Norway, in large part because the value of homeowners' imputed rent and capital gains are not included.

The growth of inequality in Norway and Sweden, including in the housing market, raises serious questions in a context traditionally described as "social democratic" welfare regimes. The unique development of housing compared to other pillars of welfare in Sweden and Norway is arguably not surprising given that it has always been the "market commodity of the welfare state" (Bengtsson, 1995). Public ownership of housing was never the norm, and the state could therefore withdraw relatively quickly when the pressure for deregulation and retrenchment increased from segments of the population and right-wing politicians. Moreover, the extensive bricks and mortar subsidies that existed in the first postwar decades (1945–1990) were gradually phased out in both countries when political and bureaucratic elites judged them to be an unsustainable burden on public budgets (Borg, 2004; Sørvoll, 2021). Theoretical arguments may also help explain the deregulation and retrenchment in the housing sector in the last few decades. Whereas the benefits of deregulation to tenant-owners were highly concentrated and visible, in the form of increased market values and capital gains, the costs of retrenchment and liberalization were not as obvious to consumers and mainly affected the housing expenditures of homebuyers and tenants in the long term (cf. Lindbom, 2001; Pierson, 1996). This may have contributed to making housing a fertile ground for politicians looking to curry favour with tenant owners or limit public spending.

At the same time, Sweden and Norway have a strong progressive social democratic tradition, and concepts such as "just distribution", "fairness", "equality" and "the welfare state" are still very much a part of the political culture. Thus, the welfare outcomes produced by market-oriented housing policies and precarious rental housing have sparked protests from social movements and political parties.

So far, the consequences of the changes in the Nordic welfare regimes since the 1980s for the housing outcomes of individuals have not been studied systematically. However, it is safe to assume that growing disparities in housing wealth have contributed to wider inequalities between segments of the population. At the same time, cuts in welfare benefits, particularly in Sweden, must have harmed the prospects of low-income groups in the housing market.

Regional outlook

The housing regimes of the four largest Nordic countries share many of the same challenges. Apart from a period after the 2008 financial crisis, residential housing markets in urban areas have boomed in parallel with growing levels of household debt since the mid-1990s. House prices increased by approximately 147 (Sweden), 109 (Norway), 49 (Denmark) and 27 per cent (Finland) in real terms from 2000 to 2019 (Anundsen, 2021). In 2020, Denmark and Norway topped the rankings in the EU when it came to the total level of household debt, and Sweden and Finland placed fourth and sixth, respectively (OECD, 2022). The high level of mortgaged homeownership is a common feature of all Nordic housing regimes; in the EU only, the Netherlands can compete when it comes to the financialization of homeownership (Table 7.4).

The expansion of mortgage debt in the Nordics is partly driven by financial market deregulation and the removal or relaxation of price controls on tenant ownership in Sweden (1968), Norway (1980s) and Denmark (post-2000). This deregulation transformed highly state-regulated tenures into fully fledged examples of financialized homeownership (Kohl & Sørvoll, 2021).

OECD, IMF, and policy experts have warned that the interrelated growth of house prices and debt constitutes a major threat to the economies of the Nordic nations. In a recent study, however, Anundsen (2021) shows that Nordic housing prices have rarely exceeded the levels expected by assessing the fundamentals (disposable income per capita, housing stock per capita, real interest rate) by a great margin and finds no evidence of housing bubbles in Oslo, Helsinki, Copenhagen or Stockholm. Nonetheless, the phenomenon of booming markets for owner-occupied housing in various forms is a significant common feature of the Nordic housing regimes. Like most of Europe, with the exception of the German-speaking countries, homeowners are in a clear majority in all the Nordic nations (Kohl & Sørvoll, 2021).

Another shared feature of all the Nordics, except Denmark, is that they are sparsely populated compared to the average in the EU (World Bank, 2022b). Urban-rural differences are important to consider in the Nordic housing regimes. In the Swedish case, the municipalities with a housing

Table 7.4 Mortgaged ownership in the Nordic countries and reference countries (% individuals living in households)

	2015	*2020 (or nearest)*
Denmark	48.6	45.2
Finland	42.4	40.6
Sweden	59.1	52.1
Norway	61.7	61.5
Iceland	62.8	57.4
EU-27	26.8	26.6
Netherlands	60.1	60.7

Source: EU-SILC.

market characterized by a balance between demand and supply are almost exclusively located outside the three largest metropolitan regions, and the over 70 per cent of local governments that report housing shortages account for 90 per cent of the total population. Those who report housing surpluses are all smaller municipalities with fewer than 25,000 inhabitants and a population decline (Boverket, 2021c). Moreover, throughout the Nordic countries, it tends to be more difficult to build new housing in rural locations because low market values are anathema to private construction companies and financial institutions. In a recent report on housing in rural areas in the Nordic region, the challenge of "securing a loan for construction, refurbishing, or just buying existing houses" is recognized "as a barrier to development in rural locations", as it "enforces the trends towards urbanization and a declining rural population" (Eliasen et al., 2020, p. 3).

In urban housing markets, however, it is rather the high cost of housing that creates social problems. The market-based allocation of owner-occupied housing (including tenant ownership) strengthens pre-existing patterns of economic inequality, segregation and residualization of the rental sector (cf. Christophers, 2021; Sørvoll et al., 2020). Moreover, the promise of growing property values and high rent yields makes rental housing in urban areas attractive for private capital. As previously touched on in one of the sections on Sweden, transnational investment companies, such as Black Rock, have increasingly invested in private rental housing and other property categories in the Nordic region. Nordic property companies, such as Norwegian *OBOS* and *Heimstaden*, also buy and build residential properties throughout the region (Tellefsen, 2021). This blurs the boundaries between the four Nordic housing regimes: construction firms and property investors increasingly operate and invest throughout the region.

Despite common challenges and the influence of transnational investment companies, it is still fair to speak of four distinct and (mostly) separate Nordic housing regimes. The Danish regime is similar to the Swedish case in the sense that it features housing associations providing rental housing open to the entire population (almene boliger). In contrast to Sweden, however, Danish municipalities have the right to allocate up to 25 per cent of all housing provided by non-profit associations to vulnerable households (Nielsen & Haagerup, 2017). The Finnish regime is closer to the Norwegian case, as homeownership is considered the norm and government-subsidized rental housing is directed at disadvantaged low-income groups. Compared to the Norwegian case, Finnish social rented housing accounts for a greater share of the housing stock (12 per cent), but the size of this sector has diminished in recent years due to the general retrenchment of housing policy (Ruonavaara, 2017).

Concluding discussion

Since the 1990s, the Nordic housing regimes have gradually gone from being rather local and welfare-oriented, to becoming a part of a globally financialized housing market. *Vonovia*, a listed company based in Germany, is today one of the largest private landlords in Sweden and the owner of several Swedish real estate companies. This increased social and geographical distance between owner and tenants has not only led to confusion among tenants about who is the owner of their housing but has also created fertile ground for conflicts (Kadioglu et al., 2022). At the same time, the Swedish company *Akelius* (later purchased by *Heimstaden*) has been the target of anti-gentrification campaigns in Berlin and Toronto. In 2021, the owner declared that they would leave Sweden, Denmark, and Germany for more profitable and less regulated markets outside the EU (Von Bahr, 2021). To define housing markets as enclosed seems increasingly redundant. Still, local financial and political structures create specific contexts and preconditions; the institutions and political debates in national housing regimes still matter.

For several decades, the housing question was dormant in Sweden, but due to a shortage of affordable housing, research and activism have increased around housing issues in recent years. However, within national politics, the housing question is still not prioritized. An exception to this occurred in 2021, when the government lost support and was forced to resign by the left-wing party *Vänsterpartiet*, because it proposed to introduce market-based rents. The opposition to market-based rents was supported by a nationwide social movement called *Nej till marknadshyra!* (No to market rent!) organized by the *18 april-kommittén* and other organizations. Housing movements are organized in local networks related to local conflicts, often concerning threats of "renoviction", but also on a national level through nationwide networks like *Bostadsvrålet* (The Housing Scream), *Alla-ska-kunna-bo-kvar*, (Everybody-should-be-able-to-stay-put), *jagvillhabostad.nu* (Iwanthousing.now) or *Ort-till-ort* (Neighbourhood-to-neighbourhood). Several of these organizations and networks were set up as a response to the lack of engagement from the Swedish Union of Tenants, which was described as sleepy, old, and passive. The new housing movements also championed a wider set of issues, like antiracism, anti-gentrification, and environmental concerns. These networks are increasingly involved with researcher and scholar activism (Listerborn et al., 2020).

Organized social protests against the conditions in the housing market have also occurred in Norway in recent years. Significantly, activists have taken the initiative to establish *Min drittleilighet* (My crappy apartment) and *Reduser husleia* (Reduce the rent) – two related social media campaigns reacting against the bad conditions and high prices of rental housing and the lack of affordable housing in urban areas (Kjæras & Haarstad, 2022).

The organized housing protests mentioned above suggest that there is currently no Fukuyama-like "end of history" moment in the two housing regimes covered in this chapter. Even though local and national governments, particularly in Sweden, have been slow to react to the demands of activists, protest movements may shape Nordic housing regimes in the future. It also seems likely that the increasingly market-based provision of housing and challenges related to housing affordability and housing inequality in the Nordic countries will continue to spark heated debate in countries with a strong passion for "equality" and "welfare for all", at least in large sections of the population.

Note

1 The EU definition of "severe housing deprivation" refers to households living in a dwelling that is both overcrowded (according to the EU standards) and has at least one of the following problems with housing deprivation: a leaking roof, damp walls, floors or foundation, or rot in window frames or floor; no bath/no shower, nor indoor toilet; darkness (Borg & Guio, 2021, p. 206).

References

Aaberge, R., Mogstad, M., Vestad, O. L., & Vestre, A. (2021). *Økonomisk Ulikhet i Norge i Det 21. Århundre* (33/2021; SSB-report). SBB. https://www.ssb.no/inntekt-og-forbruk/inntekt-og-formue/artikler/okonomisk-ulikhet-i-norge-i-det-21.arhundre/_/attachment/inline/46945fe1-533f-45b3-9ef2-cde52936f6fc:eaf3f053ce1e5878a6fb288814b428703665122e/RAPP2021-33.pdf

Aarland, K., & Reid, C. (2019). Homeownership and Residential Stability: Does Tenure Really Make a Difference? *International Journal of Housing Policy*, *19*(2), 165–191. https://doi.org/10.1080/19491247.2017.1397927

Aarland, K., & Sørvoll, J. (2021). *Norsk Boligpolitikk i Internasjonalt Perspektiv* (7/2021; NOVA-report). NOVA. https://oda.oslomet.no/oda-xmlui/bitstream/handle/11250/2835038/NOVA-Rapport-7-2021.pdf?sequence=1&isAllowed=y

Andersson, R., & Turner, L. M. (2014). Segregation, Gentrification, and Residualisation: From Public Housing to Market-Driven Housing Allocation in Inner City Stockholm. *International Journal of Housing Policy, 14*(1), 3–29. https://doi.org/10.1080/14616718.2013.872949

Anundsen, A. (2021). House Price Bubbles in the Nordic Countries. *Nordic Economic Policy Review*. Nordic Council of Ministers. https://hdl.handle.net/11250/2991820

Baeten, G., Westin, S., Pull, E., & Molina, I. (2017). Pressure and Violence: Housing Renovation and Displacement in Sweden. *Environment and Planning A, 49*(3), 631–651. https://doi.org/10.1177/0308518X16676271

Barth, E., Moene, K., & West Pedersen, A. (2021). Rising Inequality in the Egalitarian Nordics. In G. Fischer & R. Strauss (Eds.), *Europe's Income, Wealth, Consumption, and Inequality* (pp. 219–246). Oxford University Press.

Bengtsson, B. (1995). *Bostaden – Välfärdsstatens Marknadsvara* [Doctoral dissertation, Uppsala University].

Bengtsson, B. (2001). Housing as a Social Right: Implications for Welfare State Theory. *Scandinavian Political Studies, 24*(4), 255–275.

Bengtsson, B. (Ed.) (2013). *Varför så Olika? Nordisk Bostadspolitik i Jämförande Historiskt Ljus*. Égalité.

Bengtsson, B., & Grander, M. (2023). *Bostadsfrågan Som Politik Och Intressekamp*. Égalité.

Bengtsson, B., & Jensen, L. (2021). Unitary Housing Regimes in Transition - Comparing Denmark and Sweden from the Perspective of Path Dependence and Change. In J. Erola, P. Naumanen, H. Kettunen, & V.-M. Paasivaara (Eds.), *Norms, Social Structures and Policy Changes: Essays in Honour of Hannu Ruonavaara* (pp. 181–201). Acta Universitatis Upsaliensis.

Bengtsson, B., Ruonavaara, H., & Sørvoll, J. (2017). Home Ownership, Housing Policy and Path Dependence in Finland, Norway and Sweden. In R. Ronald & C. Dewilde (Eds.), *Housing Wealth and Welfare* (pp. 60–84). Edward Elgar Publishing.

Benjaminsen, L., Dhalmann, H., Dyb, E., Knutagård, M., & Linden, J. (2020). Measurement of homelessness in the Nordic Countries. *European Journal of Homelessness, 14*(3), 159–180.

Blackwell, T. (2018). *Manufacturing Debt: The Co-evolution of Housing and Finance Systems in Sweden* [Doctoral dissertation, University of Sussex].

Blackwell, T. (2021). Power, Production and Disorder: The Decline of Sweden's Housing Industrial Complex and the Origins of the Present Housing Discontents. *European Urban and Regional Studies, 28*(4), 338–352. https://doi.org/10.1177/09697764211009570

Blackwell, T., & Bengtsson, B. (2021). The Resilience of Social Rental Housing in the United Kingdom, Sweden and Denmark. How institutions matter. *Housing Studies, 38*(2), 269–289. https://doi.org/10.1080/02673037.2021.1879996

Borg, P. (2004). *Systemskifte: En Studie av Tröghet vid Fyra Brytpunkter Inom Svensk Välfärdspolitik* [Doctoral dissertation, Stockholm University].

Borg, I. (2019). Universalism Lost? The Magnitude and Spatial Pattern of Residualisation in the Public Housing Sector in Sweden 1993–2012. *Journal of Housing and the Built Environment, 34,* 405–424. https://doi.org/10.1007/s10901-018-09638-8

Borg, I., & Guio, A.-C. (2021). Improving our knowledge of Housing Conditions at EU Level. In A.-C. Guio, E. Marlier, & B. Nolan (Eds.), *Improving the Understanding of Poverty and Social Exclusion in Europe* (pp. 203–216). Publications Office of the European Union.

Boverket. (2014). *Flyttmönster till Följd av Omfattande Renoveringar* (2014:34). https://www.boverket.se/globalassets/publikationer/dokument/2014/flyttmonster-omfattande-renoveringar.pdf

Boverket. (2019). *Boverkets Indikatorer: Analys av Utvecklingen på bygg- och Bostadsmarknaden med Byggprognos* (No. 1, juni 2019). https://www.boverket.se/globalassets/publikationer/dokument/2019/boverkets-indikatorer-juni-2019.pdf

Boverket. (2021a). *Bostadsmarknaden i Regioner och Kommuner*. https://www.boverket.se/sv/samhallsplanering/bostadsmarknad/bostadsmarknaden/bostadsmarknadsenkaten/region-kommun/

Boverket. (2021b). *Bostadsmarknadsenkäten.* https://www.boverket.se/sv/samhallsplanering/bostadsmarknad/bostadsmarknaden/bostadsmarknadsenkaten/

Boverket. (2021c). Läget på Bostadsmarknaden i Riket. https://www.boverket.se/sv/samhallsplanering/bostadsmarknad/bostadsmarknaden/bostadsmarknadsenkaten/region-kommun/riket/

Boverket. (2022a). *Analyse Construction and Housing Market.* https://www.boverket.se/en/start/building-in-sweden/swedish-market/housing-market/

Boverket. (2022b). Kommunernas Särskilda Boendelösningar. https://www.boverket.se/sv/samhallsplanering/bostadsmarknad/olika-grupper/hemlosa/kommunernas-boendelosningar/

Christophers, B. (2013). A Monstrous Hybrid: The Political Economy of Housing in Early Twenty-First Century Sweden. *New Political Economy*, *18*(6), 885–911. https://doi.org/10.1080/13563467.2012.753521

Christophers, B. (2021). A Tale of Two Inequalities: Housing-Wealth Inequality and Tenure Inequality. *Environment and Planning A: Economy and Space*, *53*(3), 573–594. https://doi.org/10.1177/0308518X19876946

Christophers, B., & O'Sullivan, D. (2019). Intersections of Inequality in Homeownership in Sweden. *Housing Studies*, *34*(6), 897–924. https://doi.org/10.1080/02673037.2018.1495695

Dyb, E. (2017). Counting Homelessness and Politics: The Case of Norway. *European Journal of Homelessness*, *11*(2), 15–37.

Eggum, T., & Larsen, T. R. (2021). *Is the Housing Market an Inequality Generator?* (2/2021; Housing Lab working paper series). Oslo Metropolitan University. https://housinglab.oslomet.no/wp-content/uploads/2021/06/HLWP2021_2_new.pdf

Eliasen, S. Q., Vestergaard, L., Sigurjonsdottir, H., Turunen, E., & Penje, O. (2020). *Rural Housing Challenges in the Nordic Region* (Report 2020:7). Nordregio. https://vbn.aau.dk/ws/portalfiles/portal/402017903/rural_housing.pdf

Ellingsæter, A. L., Hatland, A., Haave, P., & Stjernø, S. (2020). *Den Nye Velferdsstatens Historie. Ekspansjon og Omdanning Etter 1966*. Gyldendal.

Elsinga, M., & Lind, H. (2013). The Effect of EU-Legislation on Rental Systems in Sweden and the Netherlands. *Housing Studies*, *28*(7), 960–970. https://doi.org/10.1080/02673037.2013.803044

Englund, P. (2016). Milestones in Swedish Housing Finance. In J. Lunde & C. Whitehead (Eds.), *Milestones in European Housing Finance* (pp. 375–391). Wiley Blackwell.

Esping-Andersen, G. (1990). *Three Worlds of Welfare Capitalism*. Princeton University Press.

FR. (2021). *Å Leie Bolig*. Forbrukerrådet. https://fil.forbrukerradet.no/wp-content/uploads/2021/03/a-leie-bolig-i-norge-1.pdf

FEANTSA & Foundation Abbé Pierre. (2020). Fifth Overview of Housing Exclusion in Europe 2020. FEANTSA.

Ferragina, E. (2022). Welfare State Change As a Double Movement: Four Decades of Retrenchment and Expansion in Compensatory and Employment-Oriented Policies Across 21 High-Income Countries. *Social Policy & Administration*, *56*(5), 705–725. https://doi.org/10.1111/spol.12789

Gitmark, H. 2020. *Det Norske Hjem. Fra Velferdsgode til Spekulasjonsobjekt*. Res Publica.

Grander, M. (2017). New Public Housing: A Selective Model Disguised As Universal? Implications of the Market Adaptation of Swedish Public Housing. *International Journal of Housing Policy*, *17*(3), 335–352. https://doi.org/10.1080/19491247.2016.1265266

Grander, M. (2018). *For the Benefit of Everyone? Explaining the Significance of Swedish Public Housing for Urban Housing Inequality* [Doctoral dissertation, Malmö University].

Grander, M. (2023). The Inbetweeners of the Housing Markets: Young Adults Facing Housing Inequality in Malmö, Sweden. *Housing Studies*, *38*(3), 502–522. https://doi.org/10.1080/02673037.2021.1893278

Grander, M., & Sandberg, M. (2022). En God Bostad? Om Bostadsyta som Kvalitet i en Tid av Ojämlikhet. In B. Bengtsson, M. Holdo, & E. Holmqvist (Eds.), *Allas Rätt Till Bostad: Marknadens Begränsningar och Samhällets Ansvar* (pp. 147–160). Daidalos.

Greve, B. (2007). What Characterise the Nordic Welfare State Model. *Journal of Social Sciences*, *3*(2), 43–51. http://www.scipub.org/fulltext/jss/jss3243-51.pdf

Grundström, K., & Molina, I. (2016). From Folkhem to Life-style Housing in Sweden: Segregation and Urban Form, 1930s – 2010s. *International Journal for Housing Policy*, *16*(3), 316–336. https://doi.org/10.1080/14616718.2015.1122695

Gustafsson, J. (2021). Renovations as an Investment Strategy: Circumscribing the Right to Housing in Sweden. *Housing Studies*. https://doi.org/10.1080/02673037.2021.1982872

Gustafsson, J. (2022). "They Had Already Sold": Uncovering Relations Among the Local State, the Market and the Public in the Case of Municipal Housing Privatization, in Rosengård, Sweden. *Environment and Planning*, *54*(2), 247–264. https://doi.org/10.1177/0308518X211053017

Gustafsson, J., Hellström, E., Richard, Å., & Springfeldt, S. (2019). The Right to Stay Put: Resistance and Organizing in the in the Wake of Changing Housing Policies in Sweden. *Radical Housing Journal*, *1*(2), 191–200. https://doi.org/10.54825/MKVM1056

Hedman, E. (Ed.) (1994). *Bostadsmarknaden och 90-talets Förändringar* (1994:1). Boverket.

Holmqvist, E., & Turner, L. M. (2014). Swedish Welfare State and Housing Markets: Under Economic and Political Pressure. *Journal of Housing and the Built Environment*, *29*, 237–254. https://doi.org/10.1007/s10901-013-9391-0

Hort, S. E. O. (2014). *Social Policy, Welfare State, and Civil Society in Sweden. Volume II: The Lost World of Social Democracy 1988-2015*. Arkiv förlag.

Hort, S. E. O., Kings, L., & Kravchenko, Z. (2019). The Swedish Welfare System. In S. Blum, J. Kuhlmann, & K. Schubert (Eds.), *Routledge Handbook of European Welfare Systems*. Routledge (pp. 466–482). https://doi.org/10.4324/9780429290510-26

Hyresgästföreningen. (2021). *Vad kan Svensk Bostadspolitik Lära av: Norge*. https://mb.cision.com/Public/19081/3466736/95ad65e8414d0751.pdf

Jensen, L. (2013). Housing Welfare Policies in Scandinavia: A Comparative Perspective on a Transition Era. *LHI Journal*, *4*(2), 133–144. http://doi.org/10.5804/LHIJ.2013.4.2.133

Kadioglu, D., Kellecioglu, I., & Listerborn, C. (2022). Transnationella bostadsbolag, Ägarbyten Och Lokal Organisering Bland Hyresgäster. In B. Bengtsson, M. Holdo, & E. Holmqvist (Eds.), *Allas Rätt Till Bostad: Marknadens Begränsningar Och Samhällets Ansvar* (pp. 275–291). Daidalos.

Kangas, O., & Kvist, J. (2018). Nordic Welfare States. In B. Greve (Ed.), *Routledge Handbook of the Welfare State* (pp. 124–136). Routledge.

Kemeny, J. (2006). Corporatism and Housing Regimes. *Housing, Theory and Society*, *23*(01), 1–18. https://doi.org/10.1080/14036090500375423

Kjærås, K., & Haarstad, H. (2022). A Geography of Repoliticisation: Popularising Alternative Housing Models in Oslo. *Political Geography, 94*. https://doi.org/10.1016/j.polgeo.2021.102577

KMD. (2020). *Alle Trenger et Trygt Hjem. Nasjonal Strategi for den Sosiale Boligpolitikken*. Kommunal- og moderniseringsdepartementet. https://www.regjeringen.no/contentassets/c2d6de6c12d5484495d4ddeb-7d103ad5/alle-trenger-et-trygt-hjem-nasjonal-strategi-for-den-sosiale-boligpolitikken-2021-2024-2.des.20.pdf

Knutagård, M. (2018). Homelessness and Housing Exclusion in Sweden. *European Journal of Homelessness, 12*(2), 103–119.

Kohl, S., & Sørvoll, J. (2021). Varieties of Social Democracy and Cooperativism: Explaining the Historical Divergence between Housing Regimes in the Nordic and German-Speaking countries. *Social Science History*, *45*(3), 561–587. https://doi.org/10.1017/ssh.2021.16

Lindbom, A. (2001). Dismantling Swedish Housing Policy. *Governance*, *14*(4), 503–526. https://doi.org/10.1111/0952-1895.00171

Listerborn, C. (2021). The New Housing Precariat: Experiences of Precarious Housing in Malmö, Sweden. *Housing Studies, 38*(7), 1304–1322. https://doi.org/10.1080/02673037.2021.1935775

Listerborn, C., Molina, I., & Richard, Å. (2020). Claiming the Right to Dignity: New Organizations for Housing Justice in Neoliberal Sweden. *Radical Housing Journal*, *20*(1), 197–137. https://doi.org/10.54825/DBXL1532

Listerborn, C., & Molina, I. (2022). Rätten Till Bostad Eller Kris i Bostadsfrågan. In M. Dahlstedt & N. Altermark (Eds.), *Bortom Systemskiftet. Mot en ny Gemenskap* (pp. 492–515). Verbal förlag.

Löfgren, H. (2021). *Det Legitima Ägandet: Politiska Konstruktioner av Allmännyttans Privatisering i Stockholms stad 1990–2015* [Doctoral dissertation, Södertörn högskola].

Lundqvist, L. J. (1987). Sweden's Housing Policy and the Quest for Tenure Neutrality: An Evaluation of Goal Achievement. *Scandinavian Housing and Planning Research*, *4*(2), 119–133.

Magnusson, L., & Turner, B. (2008). Municipal Housing Companies in Sweden – Social by Default. *Housing, Theory and Society*, *25*(4), 275–296. https://doi.org/10.1080/14036090701657397

Nielsen, R. S., & Haagerup, C. D. (2017). The Danish social housing sector: Recent changes and future challenges. *Critical Housing Analysis*, *4*(1), 142–149. https://doi.org/10.13060/23362839.2017.4.1.333

Nordic Co-operation. (2022a). *Housing Allowance in Sweden*. https://www.norden.org/en/info-norden/housing-allowance-sweden

Nordic Co-operation. (2022b). *Population*. https://www.norden.org/en/information/population

OECD. (2021). *PH3.1 Public Spending on Housing Allowances*. OECD Directorate of Employment, Labour and Social Affairs - Social Policy Division. https://www.oecd.org/els/family/PH3-1-Public-spending-on-housing-allowances.pdf

OECD. (2022). *Household Debt (indicator)*. https://doi.org/10.1787/f03b6469-en

Pedersen, A. W., & Kuhnle, S. (2017). The Nordic Welfare State Model. In O. P. Knutsen (Ed.), *The Nordic Models in Political Science: Challenged, but Still Viable?* (pp. 249–272). Fagbokforlaget.

Pierson, P. (1996). The New Politics of the Welfare State. *World Politics*, *48*(2), 143–179.

Polanska, D. V., & Axen, H. (2021). *Den Orättvisa Rättvisan: En Rättspraxis som Renovräker*. Verbal.

Polanska, D. V., & Richard, Å. (2019). Narratives of Fractured Trust in the Swedish Model: Tenants' Emotions of Renovation. *Culture Unbound*, *11*(1), 141–164. https://doi.org/10.3384/cu.2000.1525.2019111141

Pull, E. (2020). *The Original Sin: On Displacement Through Renoviction in Sweden* [Doctoral dissertation, Malmö University; Roskilde University].

Rolf, H. (2020). *En Fackförening för Hemmen: Kollektiv Mobilisering, Hyresgästorganisering och Maktkamp på Hyresmarknaden i Stockholm och Göteborg 1875–1942* [Doctoral dissertation, Ersta Sköndal Bräcke Högskola].

Ruonavaara, H. (2017). Retrenchment and Social Housing: The Case of Finland. *Critical Housing Analysis*, *4*(2), 8–18. https://doi.org/10.13060/23362839.2017.4.2.382

Sahlin, I. (1996). *På Gränsen Till Bostad: Avvisning, Utvisning, Specialkontrakt* [Doctoral dissertation, Lund University].

Sahlin, I. (2020). Moving Targets: On Reducing Public Responsibilities through Re-Categorising Homeless People and Refugees. *European Journal of Homelessness*, *14*(1), 27–54.

Sandlie, H. C., & Gulbrandsen, L. (2017). The Social Home Ownership Model – the Case of Norway. *Critical Housing Analysis*, *4*(1), 52–60. https://doi.org/10.13060/23362839.2017.4.1.324

Sejersted, F. (2011). *The Age of Social Democracy. Norway and Sweden in the Twentieth Century*. Princeton University Press.

SFS 1974:152, *The Instrument of Government* [*Regeringsformen*]. https://www.riksdagen.se/globalassets/07.-dokument-lagar/regeringsformen-eng-2021.pdf

Solstad, A. (2021). Boligsosialt Arbeid. In *Boligsosialt Arbeid* (pp. 24–46). Universitetsforlaget.

SSB. (2021). *Boforhold, Registerbasert. Andelen Hushold som eier Egen Bolig*. https://www.ssb.no/bygg-bolig-og-eiendom/bolig-og-boforhold/statistikk/boforhold-registerbasert

Stamsø, M. A. (2009). Housing and the Welfare State in Norway. *Scandinavian Political Studies*, *32*(2), 195–220. https://doi.org/10.1111/j.1467-9477.2008.00223.x

Statistics Denmark. (2022). *FOLK1A: Population at the First Day of the Quarter by Region, Sex, Age and Marital Status*. https://www.statbank.dk/statbank5a/selectvarval/define.asp?PLanguage=1&subword=tabsel&MainTable=FOLK1A&PXSId=199113&tablestyle=&ST=SD&buttons=0

Statistics Finland. (2021). *Finland in Figures 2021*. https://www.stat.fi/tup/julkaisut/tiedostot/julkaisuluettelo/yyti_fif_202100_2021_23492_net_p2.pdf

Statistics Norway. (2021a). *Population. 01222: Population and Changes During the Quarter, by Region, Contents and Quarter*. https://www.ssb.no/en/statbank/table/01222/

Statistics Sweden. (2021b). *Lägre Hyreshöjning 2021*. https://www.scb.se/hitta-statistik/statistik-efter-amne/boende-byggande-och-bebyggelse/bostads-och-hyresuppgifter/hyror-i-bostadslagenheter/pong/statistiknyhet/hyror-i-bostadslagenheter-2021/

Statistics Sweden. (2022). *Population in the Country, Counties and Municipalities on 31 December 2020 and Population Change in 2020*. https://www.scb.se/en/finding-statistics/statistics-by-subject-area/population/population-composition/population-statistics/pong/tables-and-graphs/population-statistics-year/population-in-the-country-counties-and-municipalities-on-31-december-2020-and-population-change-in-2020/

Sveriges Stadsmissioner. (2021). *Hemlös 2021 – Tema: Är Kommunernas Arbete mot Hemlöshet Effektivt?* https://www.sverigesstadsmissioner.se/files/5216/3828/8124/Stadsmissionen_Hemlos2021_webb.pdf

Sørvoll, J. (2011). *Norsk Boligpolitikk i Forandring 1970-2010. Dokumentasjon og Debatt* (Report No. 16/11). NOVA. https://oda.oslomet.no/oda-xmlui/bitstream/handle/20.500.12199/3361/4799_1.pdf?sequence=1

Sørvoll, J. (2014). *The Politics of Cooperative Housing in Norway and Sweden 1960-1990 (1945-2013) - The Swedish Deregulation of 1968 and the Norwegian Liberalization of the 1980s* [Doctoral dissertation, The University of Oslo].

Sørvoll, J. (2019). The Dilemmas of Means-tested and Market-Oriented Social Rental Housing: Municipal Housing in Norway 1945-2019. *Critical Housing Analysis*, *6*(1), 51–60. https://doi.org/10.13060/23362839.2019.6.1.460

Sørvoll, J. (2021). *Husbanken og Boligpolitikkens Historie 1996–2021*. En jubileumsbok. Cappelen Damm Akademisk.

Sørvoll, J., & Bengtsson, B. (2018). The Pyrrhic Victory of Civil Society Housing? Co-operative Housing in Sweden and Norway. *International Journal of Housing Policy*, *18*(1), 124–142. https://doi.org/10.1080/14616718.2016.1162078

Sørvoll, J., & Nordvik, V. (2020). Social Citizenship, Inequality and Homeownership. Postwar Perspectives from the North of Europe. *Social Policy and Society*, *19*(2), 293–306. https://doi.org/10.1017/S1474746419000447

Sørvoll, J., Nordvik, V., & Turner, L. M. (2020). Bostedssegregasjon og Politiske Strategier. Erfaringer fra Norge og Oslo. *Samfundsøkonomen, 38*(3), 65–78. https://doi.org/10.7146/samfundsokonomen.v0i3.122597

Tellefsen, T. K. (2021, January 14). *Norges Største Utleiekonger*. Kapital. https://kapital.no/reportasjer/naeringsliv/2021/01/14/7605477/norges-storste-utleiekonger

Torgersen, U. (1987). Housing: The Wobbly Pillar Under the Welfare State. In B. Turner, J. Kemeny, & L. Lundqvist (Eds.), *Between State and Market: Housing in the Post-Industrial Era*. Almqvist & Wiksell.

Tranøy, B. S., Stamsø, M. A., & Hjertaker. I. (2020). Equality as a Driver of Inequality? Universalistic Welfare, Generalised Creditworthiness and Financialised Housing Markets. *West European Politics*, *43*(2), 390–411. https://doi.org/10.1080/01402382.2019.1612161

Vassenden, A. (2014). Homeownership and Symbolic Boundaries: Exclusion of Disadvantaged Non-Homeowners in the Homeowner Nation of Norway. *Housing Studies 29*(6), 760–780. https://doi.org/10.1080/02673037.2014.898249

Von Bahr, O. (2021 September 7). *Roger Akelius: "Vi ska Lämna Sverige"*. Fastighetsnytt. https://www.fastighetsnytt.se/fastighetsmarknad/bostader/roger-akelius-vi-ska-lamna-sverige/

Wirehag, M. (2021). Gatekeeping, Managing Homelessness and Administrating Housing for the Poor: The Three Functions of Local Housing Services for People Experiencing Homelessness in Sweden. *European Journal of Homelessness*, *15*(1), 83–105.

World Bank. (2022a). *World Bank Open Data. GDP Per Capita (current US$)*. https://data.worldbank.org/indicator/NY.GDP.PCAP.CD?locations=DK-FI-IS-NO-SE-EU-1W

World Bank. (2022b). *World Bank Open Data. Population Density (people per sq. km of land area)*. https://data.worldbank.org/indicator/EN.POP.DNST?locations=DK-FI-IS-NO-SE-EU

8

HOUSING AND WELFARE IN CONSERVATIVE CENTRAL EUROPE

The cases of Germany, Austria and Switzerland

Steffen Wetzstein

The current state of housing in the region

The central European region – consisting of Germany (population of 84 million), Austria and Switzerland (9 million each) – shares borders, history, language and culture in the heart of Europe (Figure 8.1). The area is also referred to as DACH region (Germany (D), Austria (A) and Switzerland (CH)). It is characterized by stable economies and well-functioning political systems and rose to economic power during industrialization in the 19th and 20th centuries. Germany and Austria underwent rapid recovery after the Second World War whereas Switzerland played an important part in facilitating European recovery through its competitive banking system. In the 21st century all three countries have experienced relatively strong and stable economic growth and overcame the challenge of the Great Financial Crisis – unlike many neighbouring countries – relatively unscathed. A captivating question is to what degree the DACH countries can continue to set trends for European progress as they have done for most of the past.

All three countries represent so-called *Bismarckian Welfare Regimes*. These continental European nations are bound together by a distinct model of public welfare organization named after German Chancellor Otto von Bismarck that can be traced back to the heights of 19th-century industrialization. It is defined by life course risk management consisting of public and private institutions and actors that seek to protect and advance the well-being, health, education and material base of citizens and workers. This 'social insurance state' had been explicitly designed to support and accommodate labour during the rise of industrial capitalism. It is therefore much less concerned with reducing inequality than with providing workers with security, and to conserve status.

Housing provision – or the process through which housing becomes available to users involving promotion and financing of housing production, as well as the building, distribution and management of housing units (Ruonavaara, 2020) – features an intriguing mix of commercial, for-profit, non-profit/limited-profit and state-led provision in the DACH region. Provision is strongly governed by principles and structures of corporatist institutions and conservative attitudes towards target groups, inclusiveness as well as actor- and gender relations. At the same time, broadly interpreted welfare principles such as equality, harmony and poverty reduction have shaped housing system developments. Conservative housing finance structures and structural reforms have scaled down the public support for mortgage debt, such as tax subsidies for homeowners, with the goals

 DOI: 10.1201/9781003212690-10

Table 8.1 Housing tenure distribution (households) in Germany, Austria and Switzerland (%, 2020)

Tenure type/country	*Germany*	*Austria*	*Switzerland*
Outright owner	26	31	5
Owner with mortgage	19	18	35
Tenants private	50	31	55
Tenants subsidized	4	13	3
Other	1	7	2

Source: OECD Affordable Housing Database (OECD - Social Policy Division - Directorate of Employment, Labour and Social Affairs; http://oe.cd/ahd; data available at https://www.oecd.org/els/family/HM1-3-Housing-tenures.pdf).

of balancing budgets, reviving competitiveness and reducing market distortions that channel investments away from production.

The conservative and top-down negotiated provision system is reflected in distinct tenure arrangements. Table 8.1 presents the intra-regional tenure split. One notes that the region's housing systems are relatively similar and exhibit a remarkedly balanced tenure distribution. A combination of large, relatively strongly regulated private rental sectors (PRSs) and smaller alternative sectors as constitutive elements in – initially – unitary rental systems are complemented by comparatively small homeownership sectors. Austria features a sizeable social housing sub-system, Germany a much smaller social sector and Switzerland a small cooperative sector. Both the large and unrivalled PRSs in Switzerland and Germany as well as Austria's remarkable social housing system are exemplary and often used as international references for good practice and progressive policy. With pronounced urban-rural differences the three national housing systems exhibit a noteworthy spatialized character with rental, subsidized and alternative tenures being much more prevalent in cities and urban agglomerations. Referring to Kemeny's theory of dualist and unitary/integrated housing systems (Kemeny, 1995), all three countries might be grouped into unitary rental markets originally but have progressed towards dualization. This process features increasing attempts to regulate affordability and tenure security and a propensity for cross-subsidization housing business models rather than state-subsidized provision. Kemeny's integrated rental market thesis is at least modestly applicable to Austria where the subsidized sectors exert demonstrable price pressure on the for-profit market and therefore achieve measurable cooling effects.

The DACH region – in general – has a solid reputation for looking after households from lower socio-economic backgrounds. Three housing-mediated factors can be highlighted. First, as quintessential renter nations, the tenant-landlord power balance is of extreme importance as it affects PRS price setting, refurbishment conditions, termination of contracts and evictions. Second, all forms of subsidized social and communal housing share a concern with the lower socio-economic strata of society as the target group, although the middle classes are significant too. Third, poor tenants' affordability problems are confronted by the state at large by helping individuals on the demand side via housing benefits and rental assistance.

Social housing in the region takes on different forms. In Germany, limited social housing is available to people who possess a Social Housing Permit (Wohnberechtigungsschein). Eligibility depends on having residency in the country for at least one year and earning below income thresholds that differ from state to state. In Austria, widespread social/subsidized housing is provided via different housing sectors conditional on residence permits (between two and five years) and earning below generous income thresholds. Social housing in Switzerland varies across regional states (cantons), both in terms of availability and application procedures. Cantonal authorities or local housing associations typically set rents at a percentage of market value. In the absence of a national low-cost housing policy, every growing city has to design its own social housing strategy and program, including means-tested forms of provision (Glaser, 2017).

One structural challenge concerns the adaptation of the built environment. With mixed construction rate performances (Austria and Switzerland very high; Germany below average) building and modernization activity face massive challenges now in terms of adapting the existing built environment to the special requirements of demographic transition and climate change. In terms of age- and disability-aligned housing planning and construction, for example, more attention needs to be paid to elevator access, wheelchair-proof spaces and noise reduction. Planet/climate- and environment-aligned built environment adaptations should focus on de-carbonising and retrofitting the existing housing stock. Moreover, energy and heating efficiency and reliability should be combined with quality residential neighbourhood design and attractive spaces for work and employment.

A structural pan-regional challenge is *diminishing housing affordability*. The emerging gap between income/wealth and housing/residential energy expenses (Wetzstein, 2017) has even affected the otherwise well-regarded Austrian housing system because the demand for social and affordable housing has outgrown supply. In Germany, after an extended period of rather balanced urban housing markets, declining subsidies and a continuously shrinking stock of social housing have contributed to a growing number of tight urban housing markets in large, medium-sized and university cities. The German urban housing shortage matters doubly as the country is one of the largest receiving countries in terms of refugees and migrants, with a net migration surplus of 6 million between 2000 and 2021, and new pressures now stemming from the effects of Russia's War against Ukraine.

Arguably the biggest issue facing housing and construction today is the *"perfect storm" for inflation*. More costly loans due to interest rate rises and additional costs for materials and workers mean socially sensitive business models for housing are rapidly becoming uneconomical without greater state support. Private developers can hardly pass on extra costs to people struggling with the cost-of-living crisis. Under public finance arrangements, for example in Austria, many housing providers are no longer able to build at a price that falls within the permitted threshold (Koessl, 2022). For Germany, this new challenge compounds pre-existing mismatches between the federal government's construction targets (400,000 apartments/year), and outcomes (294,000 in 2021). Consequently, industry advocates now call to suspend the German stamp duty land (transaction) tax, reduce the goods and services/VAT tax for social housing and embark on a massive €10 billion support package for newly built affordable housing from 2023 (FAZ, 2022).

A brief overview of the historical development of the housing regimes

The structures of the DACH-region housing systems can be traced back to rapid 19th-century industrialization and urbanization. By housing the rapidly growing workforces they supported social and economic reproduction and thus indirectly stabilized economic production and promoted industrial expansion. In addition to cultural preferences, socio-demographic factors and economic drivers it is historical influences, especially ones concerning urban space and form, which have skewed Germanic housing tenures in favour of rental markets and against homeownership. Crucial in this context were strict city limits, an absolutist city-planning style and an apartment-living middle-class of state employees (Blackwell & Kohl, 2019). Sizable private rental-based urban housing sectors emerged in early unitary rental markets in the 19th century where non-profit organizations competed with for-profit ones (Kemeny, 2005). In fact, private rental has continued to represent an attractive and accessible alternative to homeownership (Kemp, 2015). In other words, DACH countries embarked early on what could be called a social policy alternative to the recently proliferating property-debt-driven approaches that have favoured household indebtedness as a general growth strategy and thus relied on a privatized Keynesian demand stimulus regime (Crouch, 2009).

A key hallmark of the *German* housing market from a comparative perspective is the size of the PRS. Individual private landlords rather than property companies and financial institutions dominate this housing tenure type that features relative tenure neutrality and – these days – no strong fiscal support for either owner-occupied housing or the rental sector (Kofner, 2017). The building-up of a large rental housing stock via a combination of regulation, housing subsidies and, in the past, generous tax treatment has created the basis for a stable tenure. This situation contrasts with a fragile system of temporary subsidies for social housing at moderate profits, which eventually transfers the dwellings to the private market. In recent years a severe market division between existing and new PRS tenancies in urban growth regions emerged. The reason lies in marked rental price increases despite cooling measures such as the "rent break"- legislation and a tendency for shifting to inflation-coupled "index rental contracts".

Following World War 2 rebuilding of housing in both parts of the now divided country became a high priority. This momentous task was achieved with strong private sector involvement in the West and considerable state control in the East. Germany's unification meant integrating the East German housing system into the well-functioning West German system. Post-unification housing policy channelled large subsidies into the housing stock at a time of strong outmigration from East to West, leading to excess supply in some Eastern regions. In the early 21st century, housing policy was devolved from the federal administration to federal states causing underinvestment in social housing and privatization of subsidized and communal housing stock. These developments have fuelled gentrification (Bernt & Holm, 2009) and further marginalized social housing within the restructured housing welfare regime (Schönig, 2020). Unsurprisingly, urban housing affordability pressures have grown, driven by tighter urban land markets, growing population, increasing numbers of single-person households and the high costs of commuting (Egner et al., 2018). The key political response had been the introduction of the "rent brake" legislation that sought to slow down rent increases in urban areas but had – initially – little demonstrable impact (Cholodilin et al., 2016) due to flaws in data methodology, uneven landlords-tenant relations in tight urban markets and too many exceptions that watered down the intervention (Wetzstein, 2021).

Housing in *Austria* is embedded in both the strong corporatist tradition ("social partnership") and the well-established conservative welfare regime (Reinprecht, 2014). The political consensus sees housing as a basic human right that should be provided by society at large, and not just via markets. There are two different types of owner-occupied and three types of rented tenures (Koessl, 2022). First, owner-occupation means ownership of flats in multi-family homes (leasehold) and single-family homes (freehold). Rental tenures can be divided by municipal rented housing in public ownership but often managed by arms-length organizations, for-profit rentals provided either by private landlords or commercial housing organizations, and Limited-Profit Housing Associations (LPHA) that charge rents at cost levels. Of countries with initially unitary rental markets, the Austrian share of social rental dwellings is among the highest in Europe (Oosterlynck et al., 2015). The state of housing is (still) good in Austria if assessed against household expenditure, available living space, quality of the housing stock and segregation/ghettoization (Amann & Mundt, 2005).

Austria's post-World War 2 housing path has been well-aligned with the nation's welfare system. Yet, the European Commission's competition law that makes it more difficult to subsidize social housing has put political pressure on the country in recent decades (Mundt & Amann, 2010).

Following the Global Financial Crisis (GFC) of 2008 house prices started to soar in Austria and outstripped all other Euro member states (Oosterlynck et al., 2015). Austria is increasingly facing challenges similar to those European nations that also have a long history of social housing (Mundt, 2018). The debate currently concerns the ongoing liberalization of housing markets, immigration and the risk of segregation, as well as the increasing polarization in society that housing markets exacerbate (Reinprecht, 2014). Post-GFC house price rises, further boosted by the COVID-19 pandemic, have added to growing housing inequalities, both in terms of housing wealth and in terms of housing affordability and quality (Koessl, 2022).

A special place in the DACH-region housing narrative is reserved for Austria's capital, Vienna. After World War 1 no longer the imperial capital of one of the longest dynasties in human history, Vienna was reduced to being the only large city in a small country (Reinprecht, 2014). Inheriting strong housing deprivation but having strong control over its own taxation the Social Democratic city administration promoted the famous "Red Vienna" political agenda and invested massively in public housing (Matznetter, 2002). Notwithstanding high annual costs for ongoing financial support the Viennese housing system performs well given that a century later two-thirds of all people rent subsidized units mostly from institutional landlords of large communal and cooperative housing organizations; making Vienna Europe's unofficial capital of social housing. Enviable affordable housing access has come at the cost of a bifurcated housing system that reproduces strong dichotomies between lucky insiders (secure tenure and affordable costs) and those outsiders forced to rent expensive private accommodation (Wetzstein, 2021). These diverging housing outcomes largely follow from the applicability rules of the City of Vienna that currently state that people have to be registered in Austria's Capital for at least two years before qualifying for subsidized housing.

Switzerland is one of the most politically and fiscally decentralized countries in the world, with a flexible zoning system and direct democracy at federal, regional (cantons) and local (municipal) levels (Hilber & Schöni, 2016). The country features the lowest homeownership rate across the whole OECD; followed by Germany. High house prices and the taxation of imputed rent constitute two key factors in limiting homeownership (Bourassa & Hoesli, 2010). Housing policy is divided between tenancy regulation and the promotion of housing construction (Cuennet et al., 2002). The important landlord-tenant relationship is regulated at the federal level whereas housing construction is a shared responsibility between all three tiers of government. Second-home investors and settling foreign buyers (for tax and quality-of-life benefits) have put sustained pressure on the Swiss housing market (Hilber & Schöni, 2016). Moreover, natural features such as lakes and mountains strongly impede urban expansion and thus reduce the elasticity of the housing supply.

Non-profit cooperative providers are responsible for much of the country's affordable housing. Switzerland's non-profit housing has existed in large urban areas for more than a century and promotion strategies have mediated very different movements in public policy from municipal socialism, and the cooperative movement to privatization and monetarism (Lawson, 2009). Today there exist more than 1,500 non-profit housing providers, and cooperatives own around 170,000 apartments. Concerning housing cooperatives' contribution to the supply of homes and the conditions explaining their contribution, Koch (2021) distinguishes four historical phases in their development: emergence and survival (1860–1918), establishment and expansion (1918–1950), consolidation (1950–1990) and re-politicization (1990–2020). Nine out of twenty-six federal states (cantons) engage in active affordable housing policies by supporting (1) non-profit housing developers, (2) individual non-profit housing projects, (3) tenants, (4) public housing and (5) zoning (Balmer & Gerber, 2018). In their ambiguous position between self-organization and being entangled with state practices, cooperatives can be characterized as representing meaningful policy instruments aiming to make contemporary urbanization governable (Barenstein et al., 2022).

Concerning the politicization of housing, dominant themes in Germany cover intra-state contentiousness concerning federal and state housing subsidies, unaffordable urban housing markets, controversial "rent brake" policies, and – after almost half a century – the re-introduction of the limited-profit housing sector. Radical demands for expropriation directed at large commercial landlords have made headlines in Berlin in recent years. More specifically, strong bottom-up concern about unaffordable living conditions triggered the 2021 Berlin referendum "Expropriate Deutsche Wohnen & Co". Voters endorsed the expropriation of the property of private real-estate companies with 3,000 or more units; affecting almost a quarter of a million rental apartments out of 1.5 million total apartments through public purchases by the Berlin state government. An Expert Commission that advises the Berlin Government on implementation currently works on questions of constitutional alignment and compensation.

In Austria, political tensions about the differing regulatory treatment of old and new rental contracts, migration-related housing responses under populist political agendas and not least the ongoing conservative and right-wing political challenges addressed at the Social Democratic Viennese housing system are significant. The first issue has become a country-wide topic for debate. In Austria, a tenant with a new contract (under two years old) pays almost double the amount compared to a tenant who rents with a contract older than 30 years. This stark difference applies especially to urban housing across all rental sectors but is biggest for the private rental market and smallest for limited-profit apartments and municipal housing. Because privileged old contracts can be passed on within families there now exists strong discrimination against young families and housing market newcomers.

In Switzerland, key topics have been the ongoing adaptation of rent control policies as well as housing affordability (largely via non-profit cooperatives) for lower-income households in bigger cities. However, the main housing-related policy issue in the recent past has been the limitation of rural sprawl – or the growth of towns and villages in rural areas – and the preservation of a liveable countryside. More specifically, following two referendums on second-home restrictions and contained rural development, the 2010s saw an increasing political divide between urban populations - that are increasingly migrating towards rural landscapes - and rural regions that want to obtain a similar level of living as the urban dwellers (Leimgruber, 2014). Since then, this topic has become a key challenge not just for national cohesion but also for the effectiveness of direct democracy.

Historicising pan-regional housing policy development, and adapting Bengtsson and Jensen (2013), five historical stages can be identified post-1945. In the immediate aftermath of the war, policies were introduced with the explicit goal of alleviating housing shortages. Some of these policies represented a continuation of previous models, but others (notably in East Germany) marked a clearer historical break. An intense investment and construction period followed that helped to relieve acute housing needs in inner cities. Later construction of larger, green-field estates with the help of industrial construction methods sought to produce accessible modernist neighbourhoods. The 1970s and 1980s saw a stronger management and quality focus and housing policy devolution while neoliberal policies increasingly advocated for a reduction in public funding for housing. At this point, governmental support policies also shifted from object to subject-subsidies – although Austria to an extent continued with the previous model. The current fifth stage is characterized by growing and vexed contradictions that have arisen under the global market-oriented model and cheap-money-based policy framework advocated by international finance institutions and central banks. Consequently, renewed emphasis – especially by progressive movements, housing advocates and often municipal state factions under social pressure – has been put on de-commodification, rent control, strategic land use policies and affordable urban housing solutions.

The welfare regimes and their development

Welfare regimes comprise state entities and well-established networks of social institutions that protect and promote the economic and social well-being of its citizens. They are rooted in principles of equal opportunity, equitable distribution of wealth and public responsibility for citizens unable to avail themselves of the minimal provisions for a good life. Welfare states, generally, achieve their objectives by applying a range of techniques that can be grouped into four basic tiers of a social security system (Wang & Aspalter, 2006): (1) *Social Insurance* (including pension insurance, health care insurance, accident insurance and unemployment), (2) *Social Assistance* (including children subsidies and housing subsidies), (3) *General Universal Welfare Benefits* (e.g. family allowance, minimum guaranteed income, child-care allowance and tax credit for children) and (4) *Social Savings/Provident Fund Systems* (state-regulated mandatory personal savings accounts).

The *Bismarckian Conservative Continental Welfare Model* pursues 'prosperity for all' so all citizens can participate in the material wealth the national economy produces (Friedrich-Ebert-Stiftung, 2022). Such "social insurance states" share common features: access to benefits related to work position, most of the benefits are in cash and are contributory, much of the social costs are financed via earnings-related contributions and the management of schemes is corporatist. Welfare is organized on the basis of occupational categories and the male breadwinner. Importantly, welfare in the German Reich and the Habsburg Empire was largely about controlling the working classes. In this model, catholic reform ethos geared at reconciling class differences had been married with the era's neo-absolutist ideal of an absolute monarchy and a hierarchically organized state-bureaucratic system (Esping-Andersen & Rainwater, 1987). Outside the DACH region France, the Netherlands, the Czech Republic and others have followed the Bismarckian route too.

Three different phases of welfare state development are commonly identified. The first period of emergence that ranging from the 19th century to the Second World War was succeeded by its 'golden age' – the period of steady growth from 1945 to the mid-1970s. This epoch was in turn superseded by an era of difficulties (Palier, 2006). The latter period is characterized by reform efforts across countries. These, Esping-Andersen (2010) notes, feature a striking degree of convergence with respect to the Bismarckian nation adaptation profiles. At the heart of fundamental reform strategies in Bismarckian welfare states were employment-facilitating responses to combat labour-shedding outcomes from the 1970s to 1980s and confronting the increasing burden of non-wage labour costs, e.g. social insurance contributions, for employers (Hemerijck & Eichhorst, 2009).

In the early 21st century, Bismarckian systems challenged policymakers in four domains. First, there is an *international competition* imperative to redirect the welfare effort, redesign institutions and implement new principles of social justice and redistribution. Second, *human capital-biased patterns of social exclusion* result from ageing populations, skills-biased technological change, changing gender roles in households because of women's growing labour force participation and a post-industrial shift to service economies. Third, long-standing *social policy commitments* resurface in the areas of unemployment and pensions which have been undermined in a period of austerity. Fourth, *European political and economic integration* since the 1980s has recast the boundaries of national systems of employment regulation and social protection, both by constraining autonomy for domestic policy and opening opportunities for EU-led social and employment coordination (Hemerijck & Eichhorst, 2009).

Germany's welfare state tradition is deeply connected with Reich Chancellor Otto von Bismarck. Under his leadership health insurance for workers was first introduced in 1883, followed by relatively comprehensive social legislation. Central is the logic of insurance: benefits are proportional to contributions, and only the financial risk of becoming dependent on benefits is borne by the whole community of contributors. This way, those who become dependent are subsidized by those who do not. From an institutional point of view, the German welfare state rests on five pillars: (1) public and semi-public insurance against vital risks such as old age, (2) direct and indirect tax-financed income subsidies for persons with insufficient own means, (3) mandatory social expenses of employers (including six weeks sick pay), (4) free education including at universities for all residents, and (5) subsidies and tax exemptions for large segments of the population (e.g. families with children, savers, buyers of homes) (Friedrich-Ebert-Stiftung, 2022).

The recent evolution of the German welfare state has been guided by the desire to respond to inner problems (unemployment, lower productivity, social cost explosion) and outer challenges (international competitiveness, European integration). According to Blank (2020), the first period (1998–2008) saw transformative reforms, among them retrenchment and privatization measures but also qualitative changes in fields of pension policies, unemployment protection/labour market policies and family policies. Reforms in areas of healthcare, long-term care and social assistance did not match the reforms in the other fields. The second period (2009–2018) was distinguished by relative stability in social policymaking. As in other developed democracies, Germany's social spending represents the largest individual item of public spending today. In 2018, for example, with almost 1 trillion EUR roughly one-quarter of the gross domestic product had been committed to social affairs (Friedrich-Ebert-Stiftung, 2022).

The above-mentioned deep-cutting welfare state restructurings of the early 2000s have become known as Hartz reforms. As part of Germany's Centre-Left Government's "Agenda 2010" the goal was to tackle high levels of long-term unemployment by a combination of creating a modern administration (including the renaming of the Federal Labour Agency and introduction of local Jobcentres), flexibilitizing of work arrangements by introducing part-time work models with lower or gradually rising taxes and insurance payments as well as providing start-up grants for aspiring entrepreneurs. At the heart of the reforms were the controversial Hartz-IV provisions that lowered social entitlements for unemployed people and linked them to legally binding obligations to improve their job situation. While the reforms added to national economic competitiveness they have been criticized for stigmatization and impoverishment. In 2019, the ruling of the Federal Constitutional Court prohibited controversial harsh sanctions against benefit recipients, and in 2023 Hartz-IV was replaced with a so-called Bürgergeld which increased the standard unemployment benefit and reduced conditionalities.

Austria builds thoroughly on conservative catholic values by acknowledging a strong role for the family. Welfare, including housing, is a decidedly middle-class state project. As an archetypal conservative-corporatist welfare state (Esping-Andersen, 1990) its key elements are (1) a mixture of insurance-based and universal benefits, (2) (male) labour market participation as key to social citizenship and (3) the family as the key unit of care (Mayrhuber et al., 2015). The latter principle means that family members can claim benefits without paying contributions (Österle & Heitzmann, 2009). The institutional make-up of the welfare system is very fragmented and the whole stakeholder system evolves via incremental - and not radical - changes. The Austrian social security system is made up of three main tiers: (1) social insurance such as health insurance, pension insurance, work-accident insurance and unemployment insurance, (2) social assistance, only accessible if the asset- and means-tested income is below a certain limit and (3) general universal welfare (Aspalter, 2006).

The Austrian system has shown strong path dependency even under the current pressures of neo-liberalization (Oosterlynck et al., 2015). In this context, Österle and Heitzmann (2020) noted an 'Austrification' of the welfare state as policymakers only picked certain ideas of ideologies such as activation, cost containment, familialism, harmonization or deservingness, and applied them in a distinct, conservative way. The EU membership in 1995 changed the Austrian corporatist system more than domestic politics. The role of lobby organizations and the power of larger firms increased (Oosterlynck et al., 2015) and for-profit organizations have increasingly been promoted even in social policy (Heitzmann & Wukovitsch, 2015). Yet, Austria also emerged as a new leader in universalism in social security in the continental welfare state systems in areas of pensions and general universal welfare (Wang & Aspalter, 2006).

Switzerland's welfare system relies on strong social security systems coupled with decent social assistance. Swiss welfare does not apply large systems of general universal benefits, or social savings accounts. Healthcare insurance is rather unusual (Daley et al., 2012) as public, subsidized private and fully private elements have been intertwined (Aspalter, 2006). In addition, decentralization has given rise to significant differences between regional states in terms of healthcare spending and provision (Huber & Orosz, 2003). A highly redistributive basic pension system guarantees effective poverty reduction and prevention among the working poor and their families (Aspalter, 2006). Institutionally, a highly decentralized and bottom-up oriented country of twenty-six federal states features large spatial variation in welfare state policies while the sub-regimes are not functionally connected (Armingeon et al., 2004). Like in its neighbouring countries, Swiss Bismarckian micro-institutions have clashed with the structural imperative surrounding austerity and post-industrialism in recent times.

Switzerland has developed a multi-tiered system of social protection, where financing, decision-making and implementation are shared between different government levels. Noteworthy is a shift in core policy areas such as old age pensions, unemployment insurance or health insurance from the cantonal to the federal government (Armingeon et al., 2004). Due to the Swiss consensus democracy, the welfare system has evolved slowly and in a modest and fragmented manner. It has combined different policy logics such as insurance, capitalization, means-testing and private health care from the very start (Häusermann, 2009). As every reform can be challenged in a popular referendum, negotiation is the main mechanism of reform. Interestingly, over decades the Swiss welfare state has continuously increased social spending to become a leading welfare state in Europe (Aspalter, 2006).

Lately, traditional institutions and instruments protecting European citizens from social risks have been weakening as a consequence of tremendous economic, demographic, social and political changes (Matznetter, 2002). In response, similar-style retrenchment and modernization across Bismarckian countries have included cost containment policies in pensions and unemployment insurance (Häusermann, 2009). A recurring claim is that welfare costs too much and that the growing financial burden it imposes jeopardizes national competitiveness and discourages entrepreneurial activity (Friedrich-Ebert-Stiftung, 2022). As a result, Bismarckian states have all developed a stronger focus on individual activation and female labour market participation in recent years in order to refuel the engines of economic prosperity. Tight to specific political agendas, welfare regime performance across the insurance, savings, assistance and benefits axes can become under scrutiny in terms of cost, quality of protection, questions of inclusiveness, economic benefits and general acceptance in society.

The interrelation between housing and welfare

Functionally, policy-institutionally and materially housing and welfare are intertwined in particular ways that include direct (sector- and target-group-specific effects) and indirect (consequential, follow-on or outcome-based effects). This mutuality of relations, flows and impacts can be conceptualized as *Housing – Welfare Interdependency (HWI)* (Figure 8.1). This concept distinguishes

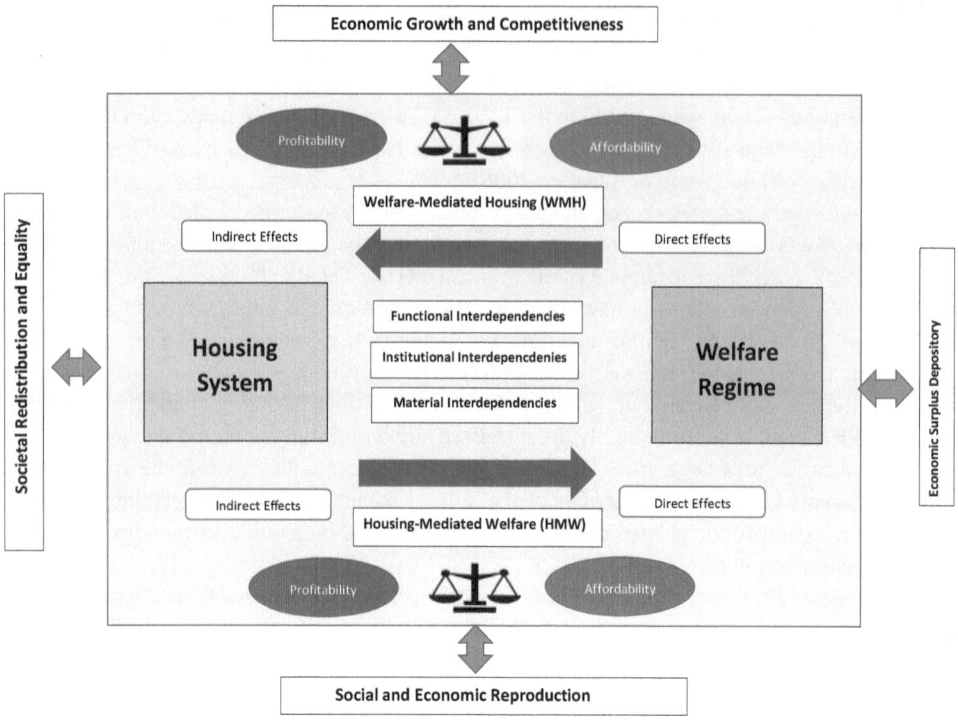

Figure 8.1 Housing – welfare interdependency model.
Source: Author.

between (1) welfare-mediated housing (WMH), where the welfare system, directly and indirectly, supports the housing system and actors, and (2) housing-mediated welfare (HMW), where the housing system produces welfare effects directly to target groups that otherwise would have to receive welfare, and by saving overall welfare system costs. The ongoing balance between profitability and affordability can be understood as both a chief aspiration and a central concern of HWI dynamics.

This HWI model also positions changing national housing-welfare dynamics within the changing and contested agendas of political reform efforts. In other words, light can be shed on how housing-welfare relations are being (re)shaped by four meta-programmatic (national) aspirations. These contextual drivers for change are economic production/growth/competitiveness, social and economic reproduction, societal redistribution and equality as well as societal cohesion and citizenship. This broader explanatory frame allows political attempts for housing-welfare system recalibrations to be linked to structurally embedded longitudinal development characteristics (e.g., nation-building, productive viability preservation, consumption demand stabilization). By inserting political economy and political theory in the analysis this conceptual devise expands welfare-housing relations beyond narrow transactional and simplifying sectoral frames. This model will assist in interpreting the interesting empirics of the DACH region below.

The contested position of *housing in the German welfare regime* makes for a fascinating historical account of path-continuing and path-breaking interchanges at particular historical moments.

Subsidized social and state housing efforts and a functioning PRS focused on post-war rebuilding while subsidy-led private rental investment dominated East Germany post-unification. Social concerns changed from the creation of affordable housing stock to promoting owner-occupation amongst the middle classes (West), to promoting a balanced PRS after unification, to focussing on the quality of the existing stock. Marginalized social housing has been criticized recently for the temporary nature of subsidization, the shortening of binding commitment periods to social housing letting, the reduction of federal subsidies, the importance of home ownership subsidy priorities and the fact that certain actors are preying on cheap social rental housing privatization opportunities (Kofner, 2017). Welfare, post-1945, has been concerned with the reconstruction of the Bismarckian social insurance system (West) and the creation of a quasi-universal social assistance system (East). Post-1989, the absorption of East German households into the German welfare infrastructure became paramount, before the above-outlined neoliberal Hartz-reforms of the early 2000s created Europe's largest low-wage employment sector. Current efforts are trying to tackle the problematic outcomes of these stigma-producing Hartz-IV-reforms.

Applying the proposed HWI model, WMH produces direct effects via rental assistance, 'cost of accommodation'-payments (Hartz-IV initiated housing and heating cost coverage for long-term unemployed and social beneficiaries) and construction children support. Indirect effects include the promotion of affordability and associated tenure continuity for renters but also rental income stability for private landlords. WMH-effects also underwrite new and controversial avenues for making money like the Hartz-IV business model (Bernt et al., 2017). In terms of HMW, direct effects include state-subsidized housing and often municipality-centred cross-subsidized and cost-rental housing and tax paid on rental income that boosts fiscal and thus welfare budgets. Indirect effects are affordability produced via balanced supply-demand in PRSs and thus cost savings for the wider welfare regime.

Housing–welfare dynamics are strongly governed by a productivist societal orientation where productive economic growth and social and economic reproduction take centre stage in political and administrative agendas rather than inequality reduction per se (Hassel et al., 2020). WMH has always been important for Germany - underwriting both profitability and affordability for a long time. One finds relative functional autonomy of the housing and welfare institutional spheres but intertwined institutional and policy practices. Key political projects in each domain – for example, the end of the Limited-Profit housing provision in 1980s West Germany and the comprehensive Hartz-IV-reforms of the 2000s – happen largely independently of one another yet always link up with broader structural agendas. For highly urbanized Germany urban policy is the domain where housing considerations are strongly intertwined with planning and design, and via municipal and communal governance also with social and welfare domains. In the global age, where capitalist urban housing has produced more unbalanced profitability-affordability outcome patterns, welfare has to compensate even more than before in response to increasingly failing spatial economic and housing systems. Thus, WMH utterly matters in Germany at the moment.

The *welfare-housing relation in Austria* can also be traced back to the late 19th century and the vision of great Austrian reformer Karl von Vogelsang and the actions of paternalistic and authoritarian politician Eduard von Taaffe (Wang & Aspalter, 2006). Austrian housing policy is a much younger creation. During reconstruction after World War II grand new schemes for welfare and housing were rejected and many pre-fascist arrangements were re-activated and re-institutionalized. For housing, Vienna council housing was re-instated, limited-profit cooperatives were re-installed and limited-profit housing companies continued their operation. Self-developed and self-built homes received institutional and public funding. Importantly, rent regulation was continued (Matznetter, 2002). Because of decades of ongoing and cumulative investment, the well-performing Austrian housing system today is comparatively cheap. As a result of the long-term government support for housing, every sixth inhabitant of Austria now lives in an apartment built and/or managed by a LPHA today (Co-operative Housing, 2022).

Matznetter (2002) detects a clear similarity between housing structures and policies, and wider welfare system features. Particular attributes of the country's quintessential conservative and familialistic welfare regime such as fragmentation, corporatism, familialism and immobilism do have their repercussions within Austrian housing and housing policy. He claims that housing policy has been fragmented from the beginning, but along lines other than social insurance but between types of tenure, types of developer and building cohorts on the one hand, and between the nine provinces on the other. In fact, provincial housing policy offers an excellent example of corporatist networks mediating interests between developers and political parties.

History firmly matters for Austria's housing and welfare systems, albeit via path-continuing trajectories and significantly positive cumulative effects. Actually, very cost-efficient direct HMW effects became possible precisely because of basically uninterrupted institutional and financial trajectories in subsidized and cost-rental housing development over decades. PRS outcomes differ between indirect effects for regulated older housing stock and increasing delivery gaps for newer housing stock. WHM is less important but growing in significance now because of urban affordability pressures in the PRS segment. From a policy-institutional perspective, housing and welfare are organizationally and intricately intertwined. In terms of links to broader agendas, it appears that state and stakeholders are particularly driven by aspirations of social cohesion and universal welfare objectives – perhaps reflecting most strongly the inherited catholic social value system. The dominant conservative family orientation is well-aligned with the guiding rural mindset in a country with few bigger cities. In conclusion, and in part because of longitudinally coherent policy and support frameworks, Austria can (still) rely on strong, effective and cost-efficient housing to produce important welfare effects that are unmatched by the other two DACH countries. In short, HMW features prominently.

The *Swiss welfare-housing experience* is different again. Does history matter as path-continuing housing policy? has erected a long-standing and sophisticated rent regulation framework that successfully underwrites both affordability and profitability. A smaller yet direct and actor-mediated link between housing and welfare occurs via the burgeoning cooperative sector. Especially in urban areas such as Zurich, Bern and Basle, it is this non-profit segment that ensures urban affordability and accessibility for lower-income households. The reasons behind their success are complex but relying on private initiative and targeting the middle class has contributed to their popularity. Intriguingly, Swiss non-profit cooperatives stand for the paradox of neoliberal state withdrawal in social policies promoting housing that is the most "common property"-based and de-commodified (Balmer & Gerber, 2018). Nevertheless, while an important tenure mechanism non-profit renting in Switzerland is only weakly influencing the overall rental market (Kemeny et al., 2005).

This regulated private rental market is indeed the heart of Swiss housing. Tightly interwoven financial, fiscal and housing arrangements have produced noteworthy welfare effects via affordable and secure private rental in most parts of the country. Put differently, the two main effects of rent control in Switzerland appear to be the reduction of housing costs for long-standing tenants and the prevention of abuse through arbitrary eviction and exploitation of temporary shortages (Werczberger, 1997). Growth effects and prosperity gains, in contrast, are ensured through the attraction of people and money to competing localities in the context of a local competitive agenda that actually somewhat resembles a local growth machine logic like in the United States. The promotion of housing construction via relaxed regulations and incentives guarantees that a relatively high proportion of new supply is coming to the market every year. Supply is thus responding quite well to demand increases under local growth agenda conditions. Intriguingly, well-developed democratic localism acts as institutional "checks and balances", a safety net for keeping welfare, housing and growth well-aligned.

On reflection, like in Austria, we observe more of a rural mindset with an active countryside preservation agenda that aims to safeguard against sprawling settlements in a pristine landscape. Like in Germany, a strongly regulated private rental system constitutes the basis for a balanced housing system that ensures both profitability for housing providers and – here in combination with the more urban tenure alternative of cooperative housing - affordability for the population. In Switzerland, we note a relative functional autonomy of the housing and welfare institutional spheres but somewhat intertwined institutional practices via fiscal-financial and political means. The intriguing symbiosis of growth/profitability with welfare/affordability is held together by the flagship model of intense democratic localism. Switzerland still seems to strike a working balance between HMW and WMH.

In conclusion, DACH-region welfare seems to act not foremost as a societal tool against immediate poverty or inequality but at large as an underwriting mechanism for productivist economic expansion, effective social reproduction and social cohesion based on conservative family and/or rural values. Housing mirrors welfare in this respect. Its main instrument is the strongly regulated private rental system complemented by auxiliary housing sectors that are more explicitly positioned to produce direct welfare effects. Germany uses its now inadequate social housing sector, Switzerland utilizes its cooperatives mainly in cities, and Austria works with well-regarded limited-profit and municipal housing sectors. Vienna constitutes a leading global housing-welfare brand in itself. In recent times common challenges have affected all countries; especially devolution tendencies, pro-market mechanisms and affordability concerns but also rising tensions in and between housing and welfare, growing status-quo contestation and a re-politicization of housing. One consequence of increasingly disruptive political economies is compounding housing-welfare inequalities for lower-income urban residents as well as a worrying new gap opening up for middle-income households. This significant welfare/housing constituency group may (still) do quite well in the labour market but struggle to find adequate and affordable urban housing in non-subsidized sectors.

Regional outlook

Concerning housing and welfare, what tends to be common across this continental European case study region? A first shared characteristic is the relatively highly developed economic base and thus the prosperity people enjoy. This high standard of living is mirrored in housing consumption, for example, in the available residential space for households, the quality of housing and the share of independently living young people. Germany, Austria and Switzerland have built prosperity on the back of housing that is facilitative towards the productive economy. On the one hand, it has ensured adequate social and economic reproduction based on power-balanced housing markets, general affordability and tenure security. On the other hand, affordable housing has lowered firms' labour costs and thus boosted competitiveness. Consequently, housing can be viewed as a foundational pillar in the national productivist economic systems that promote – especially in Germany – high-quality value-added manufacturing and export activity (Reisenbichler, 2021). From this perspective, the DACH region is arguably much more aligned with the East-Asian economic and welfare systems than with its continental neighbours. Austria, however, demonstrates how over many decades housing in the conservative European context can also function as a formidable, generative and sustainable welfare producer.

The region also shares a rich corporatist tradition. Public welfare is deeply grounded in dialogue, negotiation and interest reconciliation. A strong conservative culture – both catholic- and protestant-based – had originally been influential in accommodating rising industrial capitalism. The dominance of (later regulated) private rental systems can be traced back to this early cultural-religious contract as can housing system principles that – overall – are more loser-focussed (within limits) and have less stigma attached to housing-deprived people than most other housing systems. To deliver on these principles, however, means ongoing reliance on a well-functioning, productive, profitable and inclusive economy with well-paying labour markets and overall modest prices.

An important common feature is the strong PRS-tenure share. In fact, Switzerland and Germany lead the world in this statistic. The sector still produces relatively affordable housing and a high-quality residential built environment. Furthermore, and in contrast to many homeownership societies, it still delivers relatively well on human rights by offering a considerable degree of independence, dignity and plannability for tenants. Together with (residualizing) social housing (Germany), a well-functioning limited-profit sector (cooperatives, companies) in Austria but also Switzerland, and municipal housing in most cities of the region housing systems can still reconcile to some degree private, for-profit business models with wider societal goals of affordability, security of tenure, accessibility and – more recently – pressing planetary imperatives such as decarbonization. Also, the region has repeatedly shown a high absorption capacity for refugees and migrants as a response to ethical, demographic and growth imperatives.

Yet, the three countries also differ significantly. *Austria* features a mosaic of tenures that together ensure an overall high level of affordable housing that is retained for long periods. Vienna is a notable outlier even on a global stage and is considered exemplary. Strong conservatism has ensured path-dependency over a very long time and has created an actor landscape that is well accustomed and resourced to deliver WHM. *Switzerland* is a fascinating conglomerate of core Bismarckian welfare aspects (regulation, corporate negotiation) married with local growth agenda dimensions, relatively relaxed land-use regulation and the facilitation of international capital and banking exchange. Swiss cooperative housing, the strongly rent-controlled PRS, intense local democracy and anti-rural sprawl movement make for a creative housing policy case. *Germany*, the region's economic and population powerhouse, has become famous for its integration of social housing into private business models. Reconciling security of tenure, affordability and business profitability has worked for a long time. Recent structural challenges have destabilized the PRS-based balance, especially in cities, however.

Generally, housing policy in DACH countries is often an implicit part of broader social, fiscal or spatial policy development. Entangled with welfare considerations, housing policy may focus more on urban dynamics (Germany), rural sprawl protection (Switzerland) and explicit social concerns for residents (Austria). Poverty is considered a rather implicit target to be tackled via pre-emptive interventions and arrangements although homelessness has become an important housing intervention domain in itself ("Housing First"-initiatives). The intertwining, embedding and partial subordination of housing policy in relation to social policy has - possibly with the exception of Austria - undermined the creation of a strong professional, intellectual and institutional presence domestically and internationally. It may, for example, explain the under-representation of German-speaking researchers at the European Network for Housing Research. This trend has been amplified by a strong "housing-as-business"-focus on research and teaching that can be related to the dominance of 'for-profit' stakeholders, especially in PRSs.

How does the DACH region compare to other parts of Europe? On the one hand, one finds relatively low levels of financialized housing that can be attributed to a continuing regional productivist growth model (Reisenbichler, 2021) favouring manufacturing exports over inflationary housing markets. However, Wijburg and Aalbers (2017) point to the possibility of an alternative German financialization model based on the financialized privatization of many public housing associations and the recent market entry of listed real estate companies. On the other hand, highly developed welfare functions contrast with a housing policy that has remained a minor and often tacit activity of the state and a hybrid kind of policy directed at different goals simultaneously (Matznetter, 2002). Obviously, this differentiates this region from European housing policy heartlands such as the Netherlands or the Nordic countries. In sum, the particular symbiosis of housing and welfare, and its spatio-temporal and institutional expressions, make this region both a globally distinctive outlier region yet also an exemplary European welfare region. In fact, DACH countries contribute considerably to the European "Third Way" imaginary in-between state and market.

Three implications can be drawn for inter-regional leadership on policy, politics, practice and knowledge production. First, relatively well-regulated private housing sectors can inspire other housing systems – now that PRSs and informal solutions are often the only possibilities to house people elsewhere – to seek innovative crisis responses incorporating private for-profit interests. Second, vivid alternative tenure innovation such as in the case of limited-profit and municipal housing providers may encourage other countries to creatively explore options for market alternatives. Third, as Kofner (2017) envisaged for the Western Mediterranean countries a new orientation towards social rented housing may take the rich and tested experiences of the DACH countries as mobilizing reference point. In return, the region may benefit from knowledge on how to better incorporate self-help and informal solutions into housing provision in times of disruption and transformation.

Concluding discussion

The central value that this global handbook on housing and welfare receives from three of the conservative continental welfare states - Germany, Austria and Switzerland - lies in a time- and place-sensitive account of how the Bismarckian welfare regimes and the respective national housing systems have co-evolved over time, and with what kinds of mutual effects. In the heartlands of industrialization, welfare systems and housing solutions have been strongly linked with the national economic needs at the time, and – notably in the beginning – with accommodating the potentially threatening working class. It was found that even today, after 150 years, the early societal contracts and the foundational institutions still matter because of pronounced path-dependencies and longitudinal trajectories.

The unique nature of housing as a good and outcome in between market, state, non-profit sphere and household/family differentiates it markedly from other pillars of welfare provision. In the case of DACH countries, relatively power-balanced and outcome-stable private rental systems (PRS) help to keep housing consumption affordable and tenures secure while social, subsidized and cooperative housing has been integrated into housing systems to serve the more vulnerable (urban) households. Welfare relates to housing in contingent, complex ways and nationally distinct fashion via ideology, policy, institutions and investment. Both domains are interdependent but exhibit differing patterns of functional autonomy, institutional entanglement and material intensity. Despite political and institutional contingency DACH-region outcomes regarding housing and welfare are still to this day the envy of many other nations.

Space and time critically matter for analysing welfare-housing relations; highlighting the importance of empirical studies and non-aggregate data synthesis (Esping-Andersen, 1990). In highly urbanizing societies welfare and labour target groups are concentrated in cities, and so is the demand for affordable housing and alternative tenures such as cooperative or municipal housing. German agglomerations and the few bigger cities in Austria and Switzerland thus demand specific analytical attention. A distinctively spatial perspective may also identify new possibilities for housing and labour market interventions, including migrant and refugee integration. Temporal dimensions are also significant. Beyond mere periodization, history matters for understanding the early contracts around welfare and housing, and the initial institutional and financial set-ups that may still be relevant today in predominantly path-dependent systems (Ruonavaara, 2020). Obviously, the few path-breaking historical moments deserve extra attention.

Distinct implications for international, comparative research can be drawn. As the DACH region provides fresh impulses to expand the analytical and explanatory frame of the (re)contextualized housing–welfare question, new bridges could be built between various scholarly communities. As questions of affordability begin to impinge on wider social policy concerns (Hick et al., 2022) we may need to incorporate in our studies more centrally the latest work in *comparative political economy* and national growth model analysis, *political science* that explores how political agendas and reforms are being (re)made in this intervention field, and *urban-centred literature* as cities constitute key sites for contingent and contentious welfare, housing and economy transformations. In addition, close-to-practitioner studies should help to explore good practice development and transfer potential across space and time.

Against emerging priorities for people, place and planet we need to intellectually and politically confront the nationally mediated interdependencies and potentialities of the welfare – housing – economy triangle in order to help shape new impulses for agendas and interventions. One practical proposal may be cited from the literature where for Germany's social housing resurrection rent control, subsidies and investor profit margins can be tailored exactly to specific projects and spatially targeted precisely at the focal points of housing (Kofner, 2017). Are we seeing the advent of New and Progressive Social Housing modelled on universal principles (Wetzstein, 2022), or a continuation of a hybrid, cross-subsidization public housing model with limited benefits to selected groups (Grander, 2019)? Another urgent challenge internationally constitutes the pressing urban land question (focus: ownership, management, lease models and so forth) as a trigger for cross-cutting societal reform. The promise lies in a better shaping of public urban form, re-balancing of housing systems, facilitating re-industrialization, enhancing liveability and community cohesion as well and boosting economic participation.

Yet, in the current poly-crisis historical moment, a profound systemic challenge emerges with respect to national welfare–housing systems. Three processes surrounding housing intersect and create compounding effects. First, the *profitability* of private housing provision is being undermined. In rural areas, there is often a lack of demand and structural public infrastructure constraints may exist. In urban areas, affordability pressures currently turning into a 'cost-of-living'-crisis mean extra costs for labour, material and finance that cannot be fully passed on to housing consumers, and therefore reduce profit margins. Second, *affordability* is being weakened. In urban areas, we find a lack of (retained) affordable housing because less de-commodified housing stock exists and the supply-demand arithmetic for housing has been undermined. And third, a potential *investment freeze* is looming as business and finance models are currently inadequate for sustained investment under inflation/stagflation conditions. In addition, stakeholders face competing investment objectives in response to demographic changes, planetary needs and housing-specific goals such as quality, size and location. Consequently, entire housing systems are at risk of further unbalancing and destabilizing, and of straining housing-welfare relations in the process. The proposed Housing-Welfare Interdependency model may help to identify some deepening and emerging inconsistencies but also potential solutions in this regard.

This intersecting systemic pressure poses very tangible risks for the DACH region. Countries that base housing outcomes strongly on private sector provision are currently challenged not just because landlord-tenant relations may become more tension-ridden and contentious. There is also less capacity for household-based residential self-help in this environment of dominating commercial contractual relations. There are nevertheless opportunities for the region as the traditional corporate 'high-consensus'-orientation may help stakeholders to work through the difficult transformations without serious risks to social cohesion and societal harmony. Furthermore, we may see new innovative housing models emerge – for example an expansion of urban leasehold homeownership, new forms of limited profit provision and the expansion of cooperative and shared living formats – that hold the promise of better reconciling housing processes and outcomes in between welfare, economy and planet for generations to come.

The current intersection of several dynamics – demographic pressures, potential housing construction freeze, the affordability/cost-of-living crisis, the challenge to existing growth models, a potential overburdening of welfare commitments as well as revanchist political pressure in times of widespread prosperity loss – could easily create a fertile ground for path-breaking societal trajectories that would be of concern to many. Yet, on the contrary, we may also face moments where 'boundaries of possibility' may actually expand again (Stephens, 2020). I propose to focus more on the latter. If then, in a context of more contentious international political economies, the value of regional leadership will be assessed against the combined performance of welfare, housing and economy for people, communities and places, then Europe should be reminded of its positive 'third way'-legacy. The DACH region offers some mobilizing reference points in this regard, not least due to the 100 years of exemplary Vienna housing that was borne in times of widespread poverty, acute housing need and near-absence of effective welfare.

References

Amann, W., & Mundt, A. (2005). *The Austrian System of Social Housing Finance*. Institut für Immobilien, Bauen und Wohnen GmbH, Vienna. https://iibw.at/index.php/de-de/component/sobipro/208-the-austrian--system-of-social-housing-finance?Itemid=0.

Armingeon, K., Bertozzi, F., & Bonoli, G. (2004). Swiss Worlds of Welfare. *West European Politics*, *27*(1), 20–44. https://doi.org/10.1080/01402380412331280793

Aspalter, C. (2006). *Fighting Old-Age, Poverty: Universal Pensions Schemes in the Netherlands, Switzerland, New Zealand, and Canada*. National Pension Institute/NPC, Government of Korea.

Balmer, I., & Gerber, J. D. (2018). Why Are Housing Cooperatives Successful? Insights from Swiss Affordable Housing Policy. *Housing Studies*, *33*(3), 361–385. https://doi.org/10.1080/02673037.2017.1344958

Barenstein, J. D., Koch, P., Sanjines, D., Assandri, C., Matonte, C., Osorio, D., & Sarachu, G. (2022). Struggles for the Decommodification of Housing: The Politics of Housing Cooperatives in Uruguay and Switzerland. *Housing Studies*, *37*(6), 955–974. https://doi.org/10.1080/02673037.2021.1966392

Bengtsson, B., & Jensen, L. (2013). Unitary Housing Regimes in Transition: Comparing Denmark and Sweden in a Perspective of Path Dependence and Change. In *At Home on the Housing Market*, RC43 Conference Book of Proceedings, Universiteit van Amsterdam. http://uu.diva2:portal.org/smash/get/diva2:1463975/FULLTEXT01.pdf

Bernt, M., Colini, L., & Förste, D. (2017). Privatization, Financialization and State Restructuring in Eastern Germany: The Case of Am Südpark. *International Journal of Urban and Regional Research, 41*(4), 555–571. https://doi.org/10.1111/1468-2427.12521

Bernt, M., & Holm, A. (2009). Is It, Or Is Not? The Conceptualisation of Gentrification and Displacement and its Political Implications in the Case of Berlin-Prenzlauer Berg. *City*, *13*(2–3), 312–324. https://doi.org/10.1080/13604810902982268

Blackwell, T., & Kohl, S. (2019). Historicizing Housing Typologies: Beyond Welfare State Regimes and Varieties of Residential Capitalism. *Housing Studies*, *34*(2), 298–318. https://doi.org/10.1080/02673037.2018.1487037

Blank, F. (2020). The German Welfare System. The Calm after the Storm. In S. Blum, J. Kuhlmann, & K. Schubert (Eds.), *Routledge Handbook of European Welfare Systems* (pp. 110–128). Routledge.

Bourassa, S. C., & Hoesli, M. (2010). Why Do the Swiss Rent? *The Journal of Real Estate Finance and Economics*, *40*(3), 286–309. https://doi.org/10.2139/ssrn.964637

Cholodilin, K. A., Mense, A., & Michelsen, C. (2016). *Market Break or Simply Fake? Empirics on the Causal Effects of Rent Controls in Germany*. DIW Discussion Papers, No. 1584, Berlin: Deutsches Institut für Wirtschaftsforschung (DIW).

Co-operative Housing. (2022). *Co-Operative Housing/Housing Cooperatives Worldwide/Europe/Austria. About Austria: History*. https://www.housinginternational.coop/co-ops/austria/

Crouch, C. (2009). Privatised Keynesianism: An Unacknowledged Policy Regime. *The British Journal of Politics and International Relations*, *11*(3), 382–399. https://doi.org/10.1111/j.1467-856X.2009.00377.x

Cuennet, S., Favarger, P., & Thalmann, P. (2002). *La Politique du Logement* (Vol. 1). Collection le savoir suisse.

Daley, C., Gubb, J., Clarke, E., & Bidgood, E. (2012). *Healthcare Systems: Switzerland*. Civitas Health Unit.

Egner, B., Kayser, M., Böhler, H., & Grabietz, K. (2018). Lokale Wohnungspolitik in Deutschland. *Working Paper Forschungsförderung*, No. 100, HansBöckler-Stiftung. https://nbn-resolving.de/urn:nbn:de:101:1-2018112311271207265584

Esping-Andersen, G. (1990). *The Three Worlds of Welfare Capitalism*. Princeton University Press.

Esping-Andersen, G. (2010). Prologue: What Does it Mean to Break with Bismarck? In B. Palier (Ed.), *A Long Goodbye to Bismarck? The Politics of Welfare Reform in Continental Europe* (pp. 11–18). Amsterdam University Press.

Esping-Andersen, G., & Rainwater, L. (Eds.) (1987). *Stagnation and Renewal in Social Policy: The Rise and Fall of Policy Regimes*. Sharpe.

FAZ. (2022, December 2). Lage in Immobilienbranche: Im Wohnungsbau Droht ein Dramatischer Einbruch. *Frankfurter Allgemeine Zeitung/Online*. https://www.msn.com/de-de/finanzen/top-stories/lage-in-immobilienbranche-im-wohnungsbau-droht-ein-dramatischer-einbruch/ar-AA14OZqv?ocid=msedgntp&cvid=0b333d263be3401f8ea21b5281b74aff

Friedrich-Ebert-Stiftung. (2022). https://library.fes.de/fulltext/stabsabteilung/00073.htm

Glaser, M. (2017). The Situation of Social Housing in Switzerland. *Critical Housing Analysis, 4*(1), 72. https://doi.org/10.13060/23362839.2017.4.1.326

Grander, M. (2019). Off the Beaten Track? Selectivity, Discretion and Path-Shaping in Swedish Public Housing. *Housing, Theory and Society*, *36*(4), 385–400. https://doi.org/10.1080/14036096.2018.1513065

Hassel, A., Palier, B., & Avlijaš, S. (2020). The Pursuit of Growth. Growth Regimes, Growth Strategies and Welfare Reforms in Advanced Capitalist Economies. *Stato e Mercato*, *40*(1), 41–77. https://doi.org/10.1425/97509

Häusermann, S. (2009). Reform Opportunities in a Bismarckian Latecomer: Restructuring the Swiss Welfare State. In B. Palier (Ed.), *A Long Goodbye to Bismarck? The Politics of Welfare Reform in Continental Europe* (pp. 207–232). Amsterdam University Press.

Heitzmann, K., & Wukovitsch, F. (2015). Towards Social Investment and Social Innovation in EU Member States? First Observations of Recent Developments in Austria. *ImPRovE Working Papers*, (15/19).

Hemerijck, A., & Eichhorst, W. (2009). *Whatever Happened to the Bismarckian Welfare State? From Labor Shedding to Employment Friendly Reforms* (Vol. 4085). Forschungsinstitut zur Zukunft der Arbeit.

Hick, R., Pomati, M., & Stephens, M. (2022). *Housing and Poverty in Europe: Examining the Interconnections in the Face of Rising House Prices*. Cardiff University.

Hilber, C., & Schöni, O. (2016). Housing Policies in the United Kingdom, Switzerland, and the United States: lessons learned. *Cityscape*, *18*(3), 291–332. https://www.jstor.org/stable/26328289

Huber, M., & Orosz, E. (2003). Health Expenditure Trends in OECD Countries, *Health Care Financing Review*, *25*(1), 1–22. https://pubmed.ncbi.nlm.nih.gov/14997690

Kemeny, J. (1995). *From Public Housing to the Social Market. Rental Policy Strategies in Comparative Perspective*. Routledge.

Kemeny, J. (2005). 'The Really Big Trade-off' between Home Ownership and Welfare: Castles' Evaluation of the 1980 Thesis, and a Reformulation 25 Years on. *Housing, Theory and Society*, *22*(2), 59–75. https://doi.org/10.1080/14036090510032727

Kemeny, J., Kersloot, J., & Thalmann, P. (2005). Non-Profit Housing Influencing, Leading and Dominating the Unitary Rental Market: Three Case Studies. *Housing Studies*, *20*(6), 855–872. https://doi.org/10.1080/02673030500290985

Kemp, P. A. (2015). Private Renting After the Global Financial Crisis. *Housing Studies*, *30*(4), 601–620. https://doi.org/10.1080/02673037.2015.1027671

Koch, P. (2021). Tackling the Global Housing Challenges: Relevance and Replicability of Switzerland's and Uruguay's Housing Cooperatives' Policies and Strategies. *Working Paper 3.1 The Role of Housing Cooperatives in Switzerland*, Institute Urban Landscape, ZHAW.

Koessl, G. (2022). From the Housing Affordability to the Cost-of-Living Crisis: Austria's Housing Market in the Wake of the Covid-19 Pandemic. *Working Paper for ENHR-Working Group 'Policy and Research'*, Barcelona, 2022.

Kofner, S. (2017). Social Housing in Germany: An Inevitably Shrinking Sector? *Critical Housing Analysis*, *4*(1), 61–71. https://doi.org/10.13060/23362839.2017.4.1.325

Lawson, J. (2009). The Transformation of Social Housing Provision in Switzerland Mediated by Federalism, Direct Democracy and the Urban/rural Divide. *European Journal of Housing Policy*, 9(1), 45–67. https://doi.org/10.1080/14616710802693599

Leimgruber, W. (2014). The Scramble for Land—Saving the Remains of Rural Space in Switzerland. *Journal of Earth Science and Engineering*, *4*, 612–626.

Matznetter, W. (2002). Social Housing Policy in a Conservative Welfare State: Austria as an Example. *Urban Studies*, *39*(2), 265–282. https://doi.org/10.1080/00420980120102966

Mayrhuber, C., Glocker, C., Horvath, T., & Rocha-Akis, S. (2015). *Entwicklung und Verteilung der Einkommen in Österreich*. WIFO Studies.

Mundt, A. (2018). Privileged but Challenged: The State of Social Housing in Austria in 2018. *Critical Housing Analysis, 5*(1), 12–25. https://doi.org/10.13060/23362839.2018.5.1.408

Mundt, A., & Amann, W. (2010). Indicators of an Integrated Rental Market in Austria. *Housing Finance International*, *25*(2), 35–44.

Oosterlynck, S., Kazepov, Y., Novy, A., Cools, P., Sarius, T., & Wukovitsch, F. (2015). Welfare Systems, Governance and Social Innovation: Case Study Country Profiles of Austria, Belgium and Italy. *Discussion Paper No. 15/17*, Herman Deleeck Centre for Social Policy, University of Antwerp.

Österle, A. & Heitzmann, K. (2009). Welfare State Development in Austria. Strong Traditions Meet New Challenges. In K. Schubert, S. Hegelich, & U. Bazant (Eds.), *The Handbook of European Welfare Systems* (2nd ed., pp. 31 48). Routledge

Österle, A., & Heitzmann, K. (2020). Austrification in Welfare State Change? An Analysis of Welfare State Developments Between 1998 and 2018 in Austria. In S. Blum, J. Kuhlmann, & K. Schubert (Eds.), *Routledge Handbook of European Welfare Systems* (pp. 21–37). Routledge.

Palier, B. (2006). The Politics of Reforms in Bismarckian Welfare Systems. *Revue Française des Affaires Sociales*, *5*, 47–72.

Reinprecht, C. (2014). Social Housing in Austria. In K. Scanlon, C. Whitehead, & M. F. Arrigoitia (Eds.), *Social Housing in Europe* (pp. 61–73). John Wiley & Sons.

Reisenbichler, A. (2021). Housing Finance between Social Welfare and Growth Strategies. In A. Hassel & B. Palier (Eds.), *Growth and Welfare in Advanced Capitalist Economies* (pp. 320–347). Oxford University Press.

Ruonavaara, H. (2020). Rethinking the Concept of 'Housing Regime'. *Critical Housing Analysis, 7*, 5–14. https://doi.org/10.13060/23362839.2020.7.1.499

Schönig, B. (2020). Paradigm Shifts in Social Housing After Welfare-State Transformation: Learning from the German Experience. *International Journal of Urban and Regional Research*, *44*(6), 1023–1040. https://doi.org/10.1111/1468-2427.12914

Stephens, M. (2020). How Housing Systems are Changing and Why: A Critique of Kemeny's Theory of Housing Regimes. *Housing, Theory and Society*, *37*(5), 521–547. https://doi.org/10.1080/14036096.2020.1814404

Wang, D., & Aspalter, C. (2006). The Austrian and Swiss Welfare State System in International Comparison. *Journal of Societal and Social Policy*, *5*(2), 25–49. https://ssrn.com/abstract=1727031

Werczberger, E. (1997). Home Ownership and Rent Control in Switzerland. *Housing Studies*, *12*(3), 337–353. https://doi.org/10.1080/02673039708720900

Wetzstein, S. (2017). The Global Urban Housing Affordability Crisis. *Urban Studies – Critical Commentaries*, *54*(14), 3159–3177. https://www.jstor.org/stable/26428376

Wetzstein, S. (2021). Assessing Post-GFC Housing Affordability Interventions: A Qualitative Exploration Across Five International Cities. *International Journal of Housing Policy*, *21*(1), 70–102. https://doi.org/10.1080/19491247.2019.1662639

Wetzstein, S. (2022). 'New Social Housing' for the Post-Pandemic World! 'Shared Prosperity' —Guided Normative Framing and Potential Synergies for People and Places. In S. Güntner, J. Hauser, J. Lehner, & C. Reinprecht (Eds.), *The Social Dimension of Social Housing* (pp. 296–313). Spector Books.
Wijburg, G., & Aalbers, M. B. (2017). The Alternative Financialization of the German Housing Market. *Housing Studies*, *32*(7), 968–989. https://doi.org/10.1080/02673037.2017.1291917

9

HOUSING AND WELFARE IN ITALY

Daniela Leonardi

The current state of housing in the region

The pandemic has made poverty worse, exacerbating a situation that was already difficult and is now nothing short of alarming. The Caritas report "Failed Reboot?" (Caritas Roma, 2021) presented in Rome on April 1, 2022, casts an unforgiving light on a very serious situation where it appears that "the only thing that's been rebooted is poverty"[1] (Ciccarelli, 2022). As for inequality in Italy, the 2020 Oxfam Report "Time to Care" and its accompanying focus on Italy – tellingly entitled "DisuguItalia"[2] – paint an equally bleak picture. We read that the last 20 years have eroded social equity: the share of Italy's national wealth in the hands of the richest 10 per cent has increased by 7.6 per cent, while that of the poorest half of the population has continued to shrink, dropping by 36.6 per cent. Women and young people are the most heavily penalized. Today, 30 per cent of young people who work do not earn more than €800 gross per month, and 13 per cent rank among the working poor. Furthermore, 11 per cent of Italian women are still forced to choose between working and looking after their children or other family members.

All this is directly tied in with the housing issue. Housing poverty is a particular form of poverty that is intertwined with income, material and health poverty but has its own specificities. It should not be regarded as a sort of secondary level of social differences but as a primary dimension of the unequal distribution of resources. A growing number of people and social groups – especially women with children, young people and people with an immigrant background – are finding it very difficult to access the private housing market.

On January 7, 2021, the last extension of the eviction ban introduced in 2020 by the so-called "Cura Italia Decree" expired. The ban was introduced to support people who were no longer able to pay rent because of job or income loss resulting from the pandemic. Though the ban was extended several times at the request of the tenants' union, the Constitutional Court recently put a stop to further claims, ruling that "the limitation of property rights has reached the maximum tolerable limit". In what is widely believed to be a conservative estimate, the tenants' union puts the number of families risking eviction at around 150,000. The share of tenant households behind on rent in the private rental sector has jumped from less than 10 per cent to 24 per cent since the start of the pandemic. At the same time, the share of households with arrears on mortgage payments increased from 4.1 per cent to 11.9 per cent during the COVID-19 emergency (Housing Europe, 2021, p. 76). The eviction

 DOI: 10.1201/9781003212690-11

Table 9.1 Housing tenure status of Italian households 1977–2020

	1977	*1981*	*1991*	*2000*	*2010*	*2020*
Outright homeownership	47.0	49.7	57.7	62.5	58.3	60.7
Mortgaged homeownership	3.4	4.1	9.6	8.8	11.4	11.4
Private and public rented	43.9	39.7	23.1	19.6	20.7	20.0
Free users	5.7	6.5	9.6	9.2	9.6	7.9
Total	100.0	100.0	100.0	100.0	100.0	100.0

Sources: Bank of Italy - Household Income and Wealth Survey (1977–2020); Author's elaborations.

moratorium thus put the question of rent arrears on hold, neither addressing it in the long term nor resolving it. The measures introduced nationwide made no additional provisions for those who live in rented apartments, probably reflecting the Kemeny model (2005), which states that there is an inverse relationship between homeownership rates and welfare coverage. And according to ISTAT data (2022c, p. 6), 70.8 per cent of Italians owned their own homes in 2021 (Table 9.1).

More severe forms of housing deprivation are also on the rise. Homelessness, for example, was more visible than ever before during the pandemic, particularly when lockdown emptied the city streets. As the people experiencing homelessness were at greater risk of contagion (Benavides & Nukpezah, 2020; Brodkin, 2021), the link between housing conditions and the right to health became clear to all. The problem is increasing both in small towns and in large cities, and the homeless population is becoming increasingly heterogeneous. This poses a growing challenge for policymakers, as it is not easy to design and implement measures for a population with very diverse needs and profiles (Consoli & Meo, 2020; Tosi, 2017). In this connection, Tosi (2017, p. 107) writes of the "heterogeneity of exclusion" arguing that this is the key to understanding both the debate on homelessness and the policies for tackling it.

The author of this chapter has investigated the local monitoring of a national initiative called Notice 4/2016 which funded measures for combating serious adult marginalization and homelessness, and which specifically provided for Housing First and Housing-Led approaches. The investigation found that homeless people experienced the pandemic in very different ways depending on the housing adequacy of the services hosting them; if they were not hosted by any service and lived on the street, their experience was far worse (Leonardi & Stefani, 2021). This demonstrates that the pandemic has confirmed housing as a central issue in promoting individual and collective well-being, as well as ontological security[3] (ibid., p. 94).

The housing issue in Italy undoubtedly presents numerous challenges for social justice. Significantly, despite the country's declining and ageing population, housing is cast in a neoliberal light as a scarce resource, a commodity which is not within everyone's grasp. Even before the pandemic, Ascoli and Bronzini (2018, p. 12) wrote of a "new housing issue". The housing issue has probably always existed, but after decades in which the spread of homeownership fuelled the illusion that housing problems had been largely overcome, the social and economic turmoil caused by the pandemic has made the idea that crises can be averted simply by promoting housing purchases patently unsustainable.

A brief overview of the historical development of the housing regime

Since the post-war period, responses to housing issues have alternated between increasing construction and fielding economic and social measures. However, Tosi (2017) emphasizes that–historically–the policies implemented to address the housing issue have been "not very social" (p. 9).

It was long believed that building programs would drive social programs (Filandri, Olagnero & Semi, 2020, p. 43) but this proved not to be true. Social housing policies have tended to address demand at the upper end of the scale, concentrating on people whose circumstances are closer to normality, but without meeting the needs of those who are more affected by housing exclusion.

Some of the main characteristics that defined the decades from the 1950s to the present day will be outlined below, though obviously in a somewhat simplified form and makes no claim to being exhaustive. The intent is to sketch a picture that can be useful in gaining an understanding of the recent past, as a more detailed historical reconstruction is beyond our scope here.

The 1950s was a period of intense building, as part of the material and symbolic reconstruction of Italy's cities and life trajectories following the war. Unemployment was rampant, housing was hard to come by, and the entire economy was struggling. An entire nation had to be rebuilt.

In this period, the housing situation was extremely serious, especially in the major cities: war damage had left a shortage that was soon compounded by the impact of migration and urbanization. Living in caves, shacks, cellars, shantytowns and self-built dwellings was as much a feature of the South as it was of the big cities of the North. It was estimated that 9 million housing units needed to be built in the space of five years. However, only 20 per cent of these units were to be provided via government programs, while the private market directed its attention towards higher-value construction (Novaro, 2020, p. 22) to meet the needs of an expanding market rather than those identified by government plans (Minelli, 2004, p. 100). Regarding the role of the national government, housing policies accounted for 26 per cent of public spending, with only a fraction earmarked for building public rental housing for the most disadvantaged social classes, by contrast with the prominent role such housing takes in the national production of countries such as Great Britain, the Netherlands, France and Sweden (Novaro, 2020, p. 22).

This period of intense building, which also provided employment for large segments of the population, gradually gave rise to a society of small homeowners or tenants awaiting access to owner status.

Italy has become a nation of homeowners thanks to subsidized and supported housing programs, tax incentives, public policies and funds. Some scholars use the expression "owners at all costs". (Filandri, Olagnero & Semi, 2020; Gainsforth, 2022). The main policies of the time include the Fanfani Act of 1949, otherwise known as the INA-Casa plan, and Law 167 for public and affordable housing of 1962 which led in the following year to GESCAL (GEStione CAse per i Lavoratori), or the Workers' Housing Agency (see below). The INA-Casa Plan for working-class housing was established to address the employment and housing crises (Pavolini & Deriu, 2015; Pilat, 2014). Keynesian in inspiration, the Plan sought to garner support among the middle classes of the time. Accordingly, it had a twofold scope: as a manoeuvre aimed at revitalizing the economy and employment by building low-cost housing, and as a nationwide "institutionalized charity", encouraging all social groups to join together in meeting the needs of the poorest. For the ruling Christian Democrats, moreover, the home was the cornerstone of society, fundamental for the consolidation of the family, hence the party slogan "Not all proletarians, but all owners"[4] (Baldini, 2010, p. 153).

Fanfani, the Plan's promoter, believed that it would set up a virtuous circle of demand for materials and consumer goods. The Plan was funded by withholdings on workers' wages together with contributions from employers and the government, which were then paid to INA, the National Insurance Institute. The idea was that tenants would make a monthly payment for a certain number of years, after which they would own the property. The Plan was highly praised and has been described as an exemplary experience, one of a kind in terms of the results it achieved (Storto, 2018, p. 17).

The GESCAL fund was intended to construct and assign houses to workers. Its operational principle was to build houses for workers with contributions from the workers themselves, from companies and the government. Its goal was to finance affordable social housing with levies on employees' wages and employers, providing less well-off workers with homes at a time when 60 per cent of Italy's residential buildings had significant structural shortcomings. For a sense of the atmosphere of those years, interviews with poor Italian immigrants who had moved to Milan from rural areas in southern Italy provide valuable insights into life in the shantytowns known as *coree*, clustered on the city's outskirts (Alasia & Montaldi, 1975).

In the 1960s, the building situation was highly unbalanced: the free rental market offered attractive income opportunities for landlords, while low-income owners were unable to cope with the costs of housing maintenance (Filandri, Olagnero & Semi, 2020, p. 45). For those who were shut out from a real estate market that catered to the tastes and pocketbooks of the middle and upper classes, the answer was often self-building. A typical example is the so-called *borgate romane*, large unauthorized neighborhoods in Rome, which were built in the 1960s and 1970s and have since been retrospectively regularized through a long and complex process (Chiodelli et al., 2021). In the 1980s, hundreds of such makeshift settlements sprung up in Rome, with a population estimated at over 800,000 inhabitants (Coppola, 2008). The government's reaction was patchy, with sporadic attempts at repression (Chiodelli et al., 2021). A building amnesty introduced in 1985 attracted over 400,000 applications in Rome. The amnesty allowed residents to end their disputes with the state and legalize their property's status, while the municipal administration collected funds to provide primary services in the areas concerned.

The oil and social crisis of the 1970s brought the expansionary thrust to an end, and policies were needed to counter growing inequalities. The "Housing Reform Act"–Law 865 of October 22, 1971–established the ERP Public Housing program for housing units built directly or indirectly at government expense to meet the needs of those who were unable to access the private market. As part of welfare policies based on applicants' housing needs and socioeconomic condition, the law transferred all building stock funded by the INA-Casa and GESCAL programs to the IACP–the agencies that promote, build and manage public housing–and set out a range of programs and means of coordination.

Public intervention in residential construction has led to a distinction between fully subsidized publicly built housing (*edilizia sovvenzionata*), subsidized privately built housing (*edilizia agevolata*) and agreed housing (*edilizia convenzionata*), i.e., private housing provided under the terms of an agreement with the municipal authorities (Caruso, 2017).

- *Edilizia sovvenzionata* (the so-called *case populari*) benefits from direct government subsidies for housing rented to low-income tenants. Units are reassigned if tenants no longer meet income requirements for eligibility.
- *Edilizia agevolata* is built through an agreement between the municipality and construction companies or cooperatives. The builder commits to sell the homes at a fixed price set by the municipality. Such housing is provided both for rent and for sale.
- *Edilizia convenzionata* is a kind of private housing for rent or for sale whose transfer costs or rents are regulated by a specific agreement drawn up between the Municipality and the housing provider.

The 1980s was a period of transition when investments in public housing and rising income from homeownership created situations that were often at odds. Promoting homeownership was thought to ease social tensions. Gainsforth (2022, p. 200), for example, describes it as a means of reducing class conflict that increasingly threatened the interests of the economic and financial élites. By contrast, Grazioli (2017) describes squatting–the occupation of vacant public or private buildings by families and individuals in a condition of severe housing deprivation without the owner's permission–which she interprets as a collective response to social exclusion: while squatting "can be seen as a social response to the housing emergency, it [...] exposes dramatically the responsibilities of governments and governance agents in mismanaging the cities" (Grazioli, 2017, p. 394).

Since the 1990s, an area-based approach to urban housing policies has gained ground, with interventions focusing on local needs (Frangioni, 2022, p. 542). The 1990s also saw the emergence of a neoliberal model. Law 431/1998, "Provisions governing residential leases and termination", added further types of rental contracts and streamlined eviction and termination procedures. For example, Article 3[5] established that "at the first expiry date, the landlord can exercise the right to refuse to renew the agreement, providing the tenant with at least six months' notice thereof". A crucial step was made with by Law 560/1993, which established the conditions under which public housing can be sold, and inaugurated the government's process of divesting from its housing stock. Yet another major step came with Legislative Decree 112/1998, which assigned legislative power to the regional administration on the basis of the principle of subsidiarity.

The 2000s began with a reaffirmation of the will for public regulation of housing interventions. Law 21/2001 introduced "Provisions for reducing housing deprivation and measures to increase the supply of rental housing" and was designed to make more housing available to municipalities on an emergency basis. It also introduced an experimental program whereby municipalities under what the text refers to as "high housing stress" and in areas subject to urban regeneration can increase the supply of accommodation to be permanently rented at below-market rates. Article 4 provides for an innovative urban program promoted by the Ministry of Public Works aimed primarily at attracting private investment to improve infrastructure in derelict neighborhoods with insufficient housing, and also provides for measures to boost employment, increase social integration and bring substandard housing up to code.

As Grazioli (2017, p. 394) points out, institutions no longer distribute public housing to calm social unrest, as they did after the housing struggles erupted in the 1970s and 1980s. Rather, they have gradually shifted to more individually oriented approaches such as rent subsidies and contracts with private firms and cooperatives.

At present, we can say that the neoliberal turn has not been "simply" a question of cutting back. Certainly, the aim has been to reduce housing protections, and there is no doubt that has contributed to making the housing issue worse (Tosi, 2017, p. 35). While the new generation of social policies in some ways perpetuate the traditional approach, they also break with it in several respects. For example, there has been a shift from supply side assistance to demand-side subsidies, and new forms of social supply are making up for the decline in subsidized public housing. On a more traditional front, access to homeownership continues to receive support, measures such as the *Fondo sociale per l'affitto* provide tenants with housing allowances and subsidies for rental assistance on the private market, and programs are in place to encourage owners of vacant housing to rent (Tosi, 2017, p. 37).

The welfare regime and its development

Esping-Andersen's seminal *The Three Worlds of Welfare Capitalism* places Italy in a regime-type cluster together with nations such as Austria, France, and Germany. Esping-Andersen (1990, p. 27) affirms that:

> [I]n these conservative and strongly "corporatist" welfare states, the liberal obsession with market efficiency and commodification was never preeminent and, as such, the granting of social rights was hardly ever a seriously contested issue. What predominated was the

> preservation of status differentials; rights, therefore, were attached to class and status. [...] On the other hand, the state's emphasis on upholding status differences means that its redistributive impact is negligible.

He also emphasizes the important role of the Church, strongly committed to the preservation of traditional familyhood. In this connection, Malpass and Rowlands (2010) affirm that the Church usually behaves in ways that complement rather than challenge the welfare regime and the church traditions of the countries it operates in. For example, in keeping with the belief in "subsidiarity", the Catholic Church endorses the family as the principal welfare provider in Southern European countries. Andrea Muehlebach's (2012) *The Moral Neoliberal: Welfare and Citizenship in Italy* describes morality as perfectly consistent with, and "indeed indispensable, to market orders". In addition to neoclassical economic fundamentalism, market regulation, economic redistribution in favour of capital, and international free trade principles, neoliberalism "at its very core entails a moral authoritarianism that idealizes the family, the nation, God" (Muehlebach, 2012, p. 6). We are seeing the emergence of a highly moralized kind of citizenship at a time when social services are being cut and privatized.

> The Italian state has in the last three decades sought to mobilize parts of the population into a new voluntary labour regime–a regime that has allowed for the state to conflate voluntary labour with good citizenship, and unwaged work with gifting.
>
> (ibid., p. 6)

In this context, family and relationships have a prominent place, especially in young people's access to housing (Allen et al., 2004). In Italy,

> Social insurance typically excludes non-working wives, and family benefits encourage motherhood. Day care, and similar family services, are conspicuously underdeveloped; the principle of 'subsidiarity' serves to emphasize that the state will only interfere when the family's capacity to service its members is exhausted.
>
> (Esping-Andersen, 1990, p. 27)

In the three decades since *The Three Worlds* was published, scholars have continued to debate, update and criticize its theses. For example, a long-standing feminist critique points out that Esping-Andersen's work does not address the gendered division of labour, thus neglecting an important element of stratification (Lewis, 1992).

In 2000, Esping-Andersen turned his attention to the sustainability of welfare systems in the 21st century, discussing how population ageing affects the pension system. Among the challenges facing welfare regimes–and which are especially relevant to Italy–he also emphasizes low birth rates and "[t]he average weekly 'unpaid domestic worktime' of a woman" which in Italy amounts to 45 hours per week, compared to an estimated 25 hours in Denmark (Esping-Andersen, 2000, p. 5). This clearly affects the female population's employment opportunities. Moreover, the pandemic has once again shown that the social reproduction workload still falls almost exclusively on women. In turn, having a large number of unpaid work hours taking care of children and the home affects job loss during lockdowns, a phenomenon that has affected women significantly. And the question of economic independence and the possibility of maintaining a home is no less pertinent to the line of thought we are pursuing here. Esping-Andersen (2000, p. 6) discusses the "familialism" of Italy's welfare regime, as evinced by "the sustenance of the unemployed: in Italy,

95 per cent of unemployed under age 35 continue to live with their parents". In this connection, Gainsforth (2022, p. 17) notes that young people aged 18–34 are often employed in precarious and poorly paid jobs, which together with the scarcity of affordable rents undermines their opportunities to become independent. In many cases, the job difficulties resulting from the pandemic have obliged young people to live with their parents. This, clearly, has done nothing to defang the narrative, common among the country's politicians, that young Italians are *bamboccioni* – big babies – unable to face the responsibilities of adult life.

In Italy, 62.5 per cent of young people between 18 and 34 years old live with their parents, compared to a European average of 48.1 per cent (Censis, 2016). Income support policies for this age group and measures to assist their independence are almost completely absent. The family plays a key role in satisfying its members' housing needs, as it does in the production of other welfare services. This role has further implications for the intergenerational transmission of wealth–and hence of inequality–and in reproducing a familistic welfare regime (Poggio, 2005).

Since Esping-Andersen's (1990) categorization, the Italian welfare system has traditionally been described as corporatist. However, more recent studies have emphasized that the family can have a central role in the allocation of housing resources, going beyond the state/market binomial (Pavolini & Deriu, 2015). This results in the intergenerational transmission of poverty and thus the persistence of inequalities. Often, families have been supported by social services for several generations. Kurz and Blossfeld (2004) see Italy as part of a Southern European regime characterized by a high degree of familism: while public policies are less developed and access to the private mortgage market is more difficult, the family has a predominant role in providing economic support and facilitating access to housing, for example through self-building (Castles & Ferrera, 1996; Pavolini & Deriu, 2015). Public policies centre less on providing social housing and subsidies to households, while they strongly support access to homeownership, especially through tax measures that reduce the cost of mortgages and lighten the fiscal burden on bequests and gifts between members of the same family.

The Italian welfare system has always offered only a limited level of social protection for the most vulnerable groups, a feature that has its roots in the "Mediterranean model" of welfare (Tosi, 2017, p. 83). Edgar et al. (2002) argue that the ineffectiveness of the Mediterranean model in assisting low-income groups has long been attributed above all to the fact that they have provided very little social housing in the rental market and, as we have seen, to a strong tradition of encouraging homeownership. Moreover, the Italian welfare system has historically been categorical rather than universal, and uneven between regions.

The current economic and social crisis is part of a severe housing crisis caused by a particularly weak system of protection policies.

The interrelation between housing and welfare

Significantly, in a special issue of *la Rivista delle Politiche Sociali* (4/2018) on long-standing problems and fresh perspectives on the housing issue, Ascoli and Bronzini (2018, p. 10) wrote that in the Italian welfare system, as we know, housing policies have always been the "Cinderella" of social policies: the goal of adequate housing in terms of quality and size for all those who cannot access it on their own has never been high on the country's political agenda. When Antonio Tosi chose "Is it still possible to think of housing welfare?" as the subtitle of his 2017 book on the connection between housing and welfare, he was not referring to a long-gone glorious past, but to a painting that has never been completed. He also writes that Italy's social policies are "not very social". Ascoli and Bronzini (2018) describe Italian welfare as relying largely on fiscal measures, the most important concerning the housing sector: tax deductions are essentially aimed at those who can afford to buy or renovate a home. Policies benefiting the more economically deprived groups remain weak, and it is important to point out that although not all tenants live in poverty and not all poor people live in rented housing, tenants are more likely to be poor than homeowners (Filandri & Moiso, 2018, p. 59). As ISTAT data (2022b) shows, the incidence of absolute poverty varies according to housing tenure, and the picture is more critical for households living in rented accommodation. In 2021, the number of rental households living in poverty topped 889,000, accounting for 45.3 per cent of all poor households in the country. We can say that housing and welfare have joined in a marriage that is difficult, to say the least.

Recently, great expectations were raised by the National Recovery and Resilience Plan (PNRR) that Italy, like all other EU Member States, had to prepare to access the funds available from the European Union's NextGenerationEU post-pandemic recovery package. However, according to the National Observatory on Housing and Urban Regeneration Policies (2022, p. 8), the National Recovery and Resilience Plan has not brought about the expected change of course. The Observatory (2022, p. 8) paints a picture where the housing issue continues to be relatively marginal for three reasons: insufficient resources, failure to focus on the issue's true centrality, and, lastly, an approach that has piled new initiatives on top of ongoing ones, with no attempt to link them together or even think of ways to overcome the chronic fragmentation and the lack of coordination. Caritas Italy's report on housing and the country's recovery and resilience[6] (2022, p. 5) opened by stating that the NRRP is a challenge that Italy must face in order to guarantee a future of stability and social harmony and that it was thus natural to expect the Plan to include medium-term measures addressing an issue that is crucial to economic and social cohesion, viz., the lack of affordable housing, which is one of the obstacles to overcoming Italy's deep-seated inequality. The Caritas report examines the portions of the Plan's "Missions" that deal with housing, which are covered in several different sections of the official PNRR document. This is because no single Mission is expressly devoted to the housing issue; indeed, the Plan's first priorities do not include urban quality, much less housing. The Caritas Italy study (2022, p. 5) found that the PNRR has not paid much attention or allocated significant financial resources to increasing housing provision, and the availability of low-cost housing in particular. In analyzing the content of the National Recovery and Resilience Plan and its supplementary fund as well as the first efforts to implement the measures, the study shed light on Italy's difficulty in fielding a housing policy that prioritizes the needs of the poorer segments of society (ibid., p. 5). However, as reported in *The State of Housing in Europe 2021*[7] (Housing Europe, 2021, p. 76), the Plan channelled resources into renovating the public housing stock:

> The draft national recovery and resilience plan includes a 'Safe, green and social' Programme, dedicated to public housing. It's been allocated 2 billion Euros, to be used for requalification of the existing public housing stock, including both energy retrofit (from class G to E) and anti-seismic measures.

Moreover, recent measures show that the government's major emphasis continues to be on encouraging the purchase of a first home for all income brackets (Caritas Italiana, 2022, p. 14). This confirms that Italy still struggles to implement a housing policy that prioritizes the most disadvantaged social groups, those who have specific needs and difficulties in affording housing, or who for various reasons cannot access the private market.

What we are seeing in Italy today is a substantial retreat from the housing issue on the part of public policy: government action is sporadic and largely improvised, relying on calls for proposals that in no way reflect a medium-long-term strategic vision, not least because there is no national urban agenda (National Observatory on Housing and Urban Regeneration Policies, 2022, p. 7).

Regional outlook: A serious housing problem, and a housing welfare system that has yet to take shape

While recognizing that attempting to trace the changing configurations of the welfare state back to a set of static forms carries the risk of reifying the very object of interest (Blackwell & Kohl, 2018), we will nevertheless attempt to identify some of the significant similarities and differences between Italy and other Southern European countries.

According to Guillén et al. (2022, p. 149), who adopt a comparative perspective on Southern European welfare states over four decades, the idea that there is a distinct Southern European welfare model in European social protection systems was developed in the 1990s. Ferrera (1996) argued that Southern European countries shared several common traits:

- An institutional structure that has been evolving from a typical Bismarckian occupational approach to a mixed model with the development of Beveridgean universalistic healthcare systems (or, at least, moving towards universalism) since the 1980s
- A highly fragmented corporatist income maintenance system, displaying marked internal polarization both across social groups and across welfare sectors dimensions. Peaks of generosity in the fields of pensions and unemployment protection for core occupational groups actually coexisted with large protection gaps, especially visible in the field of antipoverty and family policies, where underdeveloped social services and the lack of a minimum income safety net were southern trademarks. The lack of a national minimum income scheme, the rudimentary character of antipoverty measures for working-age individuals as well as the scarcity of social services made reliance on households as welfare providers key to maintaining social cohesion.

In Italy, these common trends have led to:

- A high rate of poverty. As ISTAT data (2022b) show, in 2021 just over 1.9 million households (7.5 per cent of the total, down from 7.7 per cent in 2020) and about 5.6 million individuals (9.4 per cent as in the previous year) were in absolute poverty. Thus, absolute poverty is still essentially at the historic high reached in 2020, the year the pandemic began. In 2021, the incidence of absolute poverty was highest among households with a larger number of members, reaching 22.6 per cent among those with five or more members and 11.6 per cent among those with four. For relative poverty, the incidence was 11.1 per cent of all households (as against 10.1 per cent in 2020).
- Intergenerational transmission of poverty and a high rate of youth poverty. Patterns of participation or non-participation in the labour market are among the most significant determinants of poverty status (ISTAT (2022a, p. 18). The demographic group of 18–34-year-olds is particularly affected. Here, the incidence of poverty almost quadrupled, going from 3.1 per cent in 2005 to 11.1 per cent in 2021 (ISTAT, 2022b, p. 2).
- A media campaign aimed at criminalizing recipients of the *Reddito di cittadinanza* minimum income scheme introduced in 2019, considerably later than in other European countries and widely attacked on the grounds that it disincentivizes people from seeking employment and has failed to bring recipients back into the job market. The scheme also provides for a sum

specifically intended for rental payments. By contrast, supporters of the *Reddito di cittadinanza*–or "citizenship income"–argue that it is a mainly redistributive measure whose objective is to guarantee everyone a minimum level of economic resources, especially in this time of economic and social crisis. According to ISTAT (2022a, p. 19)

> [E]conomic support measures provided in 2020, in particular the citizenship and the emergency income measures, allowed 1 million individuals to escape absolute poverty. [...] Those same measures provided for a decrease in the poverty intensity of some of those who remained in poverty. Without subsidies, poverty intensity would have been as much as 10 percentage points higher.

Italy is a prime example of the Southern European welfare model, as its mixed welfare state architecture combines a universalistic national health care system that started in 1978 with substantial functional and distributional imbalances:

> A hypertrophic pension system attracting a large share of social protection expenditure, underdeveloped policy sectors such as social exclusion, family and labour market/unemployment, extreme regulatory fragmentation with generous rules for some occupational categories coupled with severe protection gaps for others, and most notably the lack of a minimum income safety net against poverty.
>
> (Guillén et al., 2022, p. 150)

Although all Southern European countries shared some common elements in the past, they followed different paths in modernizing their social protection systems after the 2007–2008 crisis. Spain and Portugal introduced vigorous reforms modelled on a Continental European approach, while Italy and Greece had less success in adapting to change. "The 2007–2008 crisis and the following austerity plans have proven to be a 'great equalizer' for Southern European welfare states but following a downwards trajectory" (Guillén et al., 2022, p. 171).

The Italian welfare system is increasingly based on monetary transfers and tax benefits. Far from growing, the resources allocated to the major universalist systems, from healthcare services to education, are in fact dwindling, while a series of measures have benefited retirees (Ascoli & Bronzini, 2018, p. 9). A chronic problem in Italy is the high rate of tax evasion, which costs the government over €100 billion every year (Pavolini & Ascoli, 2019). Though evasion is currently declining, it is still very high: the loss in 2019, the last year for which data are available, amounts to an estimated €80.7 billion.[8]

Such were the difficulties facing Italy's housing situation at the beginning of the new millennium. Not only is there no national policy promoting social housing for vulnerable population groups, but "[t]he rental sector constitutes a limited share of the market, with high prices compared to the level of salaries and pensions" (Caruso, 2017, p. 37). The mismatch between the financial capacity needed for homeownership and the percentage of owners creates market rigidity. Together with precarious forms of employment, these factors have excluded aspiring first-time homeowners from the market. Social housing provision in the rental market has shown itself to be insufficient, and woefully inadequate in meeting social needs (Caruso, 2017, p. 38).

The pandemic, the 2008 financial crisis and the interaction between market tensions and policy inertia have exacerbated housing issues around the world. In Italy, however, this worsening is part of a situation that was already particularly problematic, both because of the seriousness of the housing crisis and because of weak policy responses. The 2008 crisis and the pandemic must thus be viewed against the backdrop of long-term housing distress. Italy has long had more serious problems than most European countries (Tosi, 2017, p. 81). The fact that Italy's weak welfare system takes a categorical rather than universal approach makes policies less effective and less fair since adopting an effectiveness and equity perspective means addressing needs and disadvantages with universalist criteria. While a categorical approach can allow for small adjustments, policies cannot be reoriented towards effectiveness and equity. Moreover, the Italian system's high selectivity and stringent eligibility requirements often discourage individuals from applying for benefits and joining the (few) assistance programs.

Concluding discussion: The residential is political

Emphasis on housing measures has recently increased, mainly because the continuing crisis is adding to the number of Italians who cannot afford to buy or rent a home on the private market. Nevertheless, the housing sector is still the "wobbly pillar" (Torgersen, 1987) of the Italian welfare system. Housing is one of the rights denied to an increasingly sizeable portion of the Italian population. Traditional housing policies have failed to guarantee access to affordable and decent housing and the situation is exacerbated by the lack of an integrated welfare system. (Tosi, 2017, p. 103). In the current debate in Italy, there is much talk of the "housing emergency" (Caritas Italiana, 2022), "housing poverty" (Ascoli & Bronzini, 2018), the "social risk" entailed by housing problems (Olagnero, 2018), the "return of the housing issue" (Semi, 2017) and the "housing emergency" (Grazioli, 2017). This is a sign that housing issues are so widespread, and the starting point of so many other inequalities, that they are considered a social risk from which the population must be protected.

As shown by ISTAT data (2022c, p. 7), in 2021, 18.2 million households owned the home in which they live. Renting is more common among less well-off households: 35.5 per cent of households whose main income earner is unemployed lived in rented accommodation, as did 68.5 per cent of households with at least one non-national (a figure that rises to 73.8 per cent for households consisting entirely of non-nationals).

As mentioned earlier, the incidence of absolute poverty is higher among households in rented accommodation. Among households with minor children, the incidence is 28.2 per cent if the home is rented, compared with 6.4 per cent for homeowners and 13.1 per cent for usufructuaries and households in rent-free accommodation. These data highlight the importance of diversifying housing support measures and promoting assistance for people living in rented accommodation.

Historically, Italian policies have encouraged homeownership. This is seen as a means of devolving responsibilities to the private sector and an example of individual self-entrepreneurship, and as such is an imperative of neoliberal housing policy (Stedman Jones, 2012). From this perspective, it falls to the family to enable the younger generation to access a mortgage. Accordingly, the Italian government's reluctance to provide other types of help for those who rent on the private market or who cannot find housing on the private market is based on a neoliberal model, as it is aimed at reducing protection and shifting risks to individuals. The consequences of this neoliberal model are obvious. We cannot ignore the fact that larger and larger segments of the population are in debt and/or excluded from the private housing market: women with young children, young people in precarious employment, families whose migrant background exposes them to prejudice and racism, people in marginalized conditions, and middle-aged men who have lost their jobs or whose disposable income has dropped following a divorce.

A key issue here is the heterogeneity and breadth of isolated or at-risk groups. There is a need to extend existing measures ensuring the right to housing, which is still essentially out of reach for too many people. But as can be seen from this brief historical overview and Italy's growing inequality and dismantling of the social housing system, it is necessary to provide non-stigmatizing programs for the groups in greatest difficulty, as part of a common framework of protected rights. These groups have historically been sidelined in housing initiatives that have often ended up benefiting a target that was already better off. They thus need redistributive interventions and protection from the private market's prejudices and speculation.

This section's title – "The residential is political" – is a tribute to Madden and Marcuse's *In Defense of Housing* (2016). It was chosen to highlight that what we have written so far concerns the relationship between housing and rights and to underscore the connection between the two. This connection shows the breadth and urgency of the question.

> "The dominant view today is that if the housing system is broken, it is a temporary crisis that can be resolved through targeted, isolated measures. In mainstream debates, housing tends to be understood in narrow terms. The provision of adequate housing is seen as a technical problem and technocratic means are sought to solve it [...]. Certainly, technical improvements in the housing system are possible, and some are much needed. But the crisis is deeper than that

housing has to be considered from a wider perspective: as a political-economic problem (Madden & Marcuse, 2016, p. 8).

Housing and how it is provided is a political question and tells us about a society's priorities, how communities organize themselves to face housing risks, and their collective responses (Grazioli, 2017). Protecting the right to housing means protecting people's health (Leonardi & Stefani, 2021); guaranteeing affordable housing means guaranteeing that women can escape violent family situations even if they have little or no independence. For this reason, we can also say that affordable housing is a feminist issue.

At the same time, this chapter sought to illustrate the peculiarities of the Italian system, where welfare and housing have always had a problematic relationship: a relationship that could almost be described as a difficult marriage. As mentioned above, Ascoli and Bronzini (2018) described housing policies as the Cinderella of social policies, while Tosi (2017, p. 81) wrote of policy deficits, stating that the lack of coherent housing welfare is a weakness that has its roots in the Mediterranean welfare model.

Today it is important to recognize that the problems of rental tenants must be addressed through policies that can bring a change of direction and prevent non-owners – or homeowners who cannot cope with maintenance expenses – from being trapped in a life of poverty (Filandri, 2016). We need measures that increase the supply of housing available on the market, for example by providing rental incentives to discourage large landlords from leaving their properties vacant. Mediation services putting landlords and potential tenants into contact are also needed, as are policies that affect the percentage of income that households spend on housing costs, which can otherwise often become unsustainable.[9]

A further path of change could concern the new family models, providing for different forms of co-habitation. The latter aspect is also linked to population aging: today 30 per cent of Italy's over-65s live alone. Here as elsewhere, the pandemic made many problems impossible to ignore: an old age spent in a care home surrounded only by the elderly is hardly a sustainable prospect. We must think of alternatives. And we must build them. As Gainsforth (2022, p. 25) argues, the housing question does not only concern the least among us: it concerns the vision we have of society.

Taking a "welfare-to-housing" approach instead of "welfare-to-work" would mean recognizing the home as a priority (Oxfam Report, 2022). By "welfare-to-housing", we mean an approach that helps unemployed people and others receiving government benefits to find affordable housing. The home would thus be regarded as an essential starting point for tackling growing inequalities and adequate housing as an essential right for all. The house is a prerequisite for embarking on broader paths of economic and social growth; living in decent housing triggers a whole series of citizenship behaviors that in turn contribute to the country's development. (Ascoli & Sgritta, 2015, pp. 28–29).

Policy change could take many directions, but there is no doubt that tackling the housing crisis must not be postponed. Recognizing the close connection between welfare policies and housing policies is the indispensable key to addressing the challenge.

Acknowledgments

I warmly thank Violetta Tucci for helping me with the data. I am in particular grateful for ongoing discussions with Sandro Busso and Sarah Gainsforth, and I would like to thank the editors for their insightful comments.

Notes

1 See https://ilmanifesto.it/roma-italia-lunica-cosa-che-riparte-e-lapoverta?fbclid=IwAR0DN9cB6Qssxap_eZPDSAeLGSK4sZA8tSyGvcBgGiyK0nFqAVSjgVgXrac.
2 See https://www.oxfamitalia.org/wp-content/uploads/2020/01/Disuguitalia_2020_final.pdf. "Disuguitalia" is a play on words, a portmanteau combining the Italian words for "inequality" and "Italy".
3 Ontological security can be defined as a feeling of well-being resulting from a sense of constancy in one's social and material environment which in turn provides a secure platform for identity development and self-actualization (Padgett, 2007, p. 1926).
4 In Italian, "non tutti proletari ma tutti proprietari".
5 https://web.camera.it/parlam/leggi/98431l.htm.
6 https://www.caritas.it/materiali/Italia/qrrp/qrrp_num1_mar2022.pdf?fbclid=IwAR2SOnvEvTocZXEY3dnzpwKk2bJHLvjylQ0SJ3OvaZXyU0h2OlTJL5uW2iw.
7 https://www.stateofhousing.eu/The_State_of_Housing_in_the_EU_2021.pdf.
8 https://www.agi.it/economia/news/2021-12-12/evasione-fisco-cgia-14864674. For further information:https://www.dt.mef.gov.it/export/sites/sitodt/modules/documenti_it/analisi_progammazione/documenti_programmatici/nadef_2021/Rapporto_evasione_fiscale_e_contributiva.pdf.
9 During the first lockdown, the SICET tenants' association interviewed four hundred rental households. In most cases, housing costs absorbed over 70 per cent of the household's income.

References

Alasia, F., & Montaldi, D. (1975). *Milano, Corea.* Feltrinelli.

Allen, J., Barlow, J., Leal, J., Maloutas, T., & Padovani, L. (2004). *Housing and Welfare in Southern Europe.* Blackwell.

Ascoli, U., & Bronzini, M. (2018). Il Welfare, la Casa, L'abitare: Lo Scenario Nazionale. Nota Introduttiva. *la Rivista delle Politiche Sociali, 4*, 9–23.

Ascoli, U., & Sgritta, G. (2015). Segni di Investimento Sociale in Italia? In U. Ascoli, C. Ranci, & G. Sgritta (Eds.), *Investire nel Sociale. La Difficile Innovazione del Welfare Italiano* (pp. 7–30). il Mulino.

Baldini, M. (2010). *La Casa Degli Italiani.* il Mulino.

Benavides, A. D., & Nukpezah, J. A. (2020). How Local Governments Are Caring for the Homeless During the COVID-19 Pandemic. *The American Review of Public Administration, 50*(6–7), 650–657. https://doi.org/10.1177/0275074020942062

Blackwell, T., & Kohl, S. (2018). Historicizing Housing Typologies: Beyond Welfare State Regimes and Varieties of Residential Capitalism. *Housing Studies, 34*(2), 298–318. https://doi.org/10.1080/02673037.2018.1487037

Brodkin, E. Z. (2021). Street-Level Organizations at the Front Lines of Crises. *Journal of Comparative Policy Analysis: Research and Practice*, *23*(1), 16–29. https://doi.org/10.1080/13876988.2020.1848352

Caritas Italiana (2022). *Casa e Abitare nel PNRR,* Quaderni sulla ripresa e resilienza del paese, 1/2022. https://www.caritas.it/materiali/Italia/qrrp/qrrp_num1_mar2022.pdf?fbclid=IwAR2SOnvEvTocZXEY3dnzpwKk2bJHLvjylQ0SJ3OvaZXyU0h2OlTJL5uW2iw.

Caritas Roma (2021). *La povertà a Roma: un punto di vista. False ripartenze?* http://www.caritasroma.it/wp-content/uploads/2022/04/Caritas_Rapporto_2021.pdf

Caruso, N. (2017). *Policies and Practices in Italian Welfare Housing*. Springer.

Castles, F. G., & Ferrera, M. (1996). Home Ownership and the Welfare State: Is Southern Europe Different? *South European Society and Politics*, *1*(2), 163–185. https://doi.org/10.1080/13608749608539470

Censis. (2016). *I Giovani e la Casa. La Generazione dei Millennials di Fronte Alla Questione Abitativa.* https://www.sidief.it/eventi/verso_la_casa_taxi_la_domanda_abitativa_dei_giovani.

Chiodelli, F., Coppola, A., Belotti, E., Berruti, G., Marinaro, I. C., Curci, F., & Zanfi, F. (2021). The Production of Informal Space: A Critical Atlas of Housing Informalities in Italy between Public Institutions and Political Strategies. *Progress in Planning*, *149*. https://doi.org/10.1016/j.progress.2020.100495.

Ciccarelli, R (2022). *L'unica cosa che riparte è la povertà*, newspaper article, il manifesto, April 2, 2022. https://ilmanifesto.it/roma-italia-lunica-cosa-che-riparte-e-la-poverta

Consoli, T., & Meo, A. (Eds.) (2020). *Homelessness in Italia: Biografie, Territori, Politiche*. FrancoAngeli.

Coppola, A. (2008). Le Borgate Romane tra '45 e '89: Esclusione Sociale, Movimenti Urbani e Poteri Locali. In M. Cremaschi (Ed.), *Tracce di Quartieri: Il Legame Sociale Nella Città che Cambia* (pp. 161–186). Franco Angeli.

Edgar, B., Doherty, J., & Meert, H. (2002). *Access to Housing: Homelessness and Vulnerability in Europe*. The Policy Press.

Esping-Andersen, G. (1990). *The Three Worlds of Welfare Capitalism*. Princeton University Press.

Esping-Andersen, G. (2000). The Sustainability of Welfare States into the Twenty-First Century. *International Journal of Health Services*, *30*(1), 1–12. https://www.jstor.org/stable/45131407

Filandri, M. (2016). *Proprietari a Tutti i Costi. La Disuguaglianza Abitativa in Italia.* Carocci editore.

Filandri, M. A., & Moiso, V. (2018). L'insostenibile peso Dell'abitare. Un'analisi Sulla Relazione tra Povertà e Housing Affordability in Italia. *La Rivista delle Politiche Sociali*, *4*, 59–76.

Filandri, M., Olagnero, M., & Semi, G. (2020). *Casa Dolce Casa? Italia, un Paese di Proprietari.* il Mulino.

Frangioni, T. (2022). Una Comunità di Gente che Passa? Social Housing e Comunità a Torino. *Politiche Sociali. Social Policies*, *3*, 539–556.

Gainsforth, S. (2022). *Abitare Stanca. La Casa: Un Racconto Politico.* effequ.

Grazioli, M. (2017). From Citizens to Citadins? Rethinking Right to the City Inside Housing Squats in Rome, Italy. *Citizenship Studies*, *21*(4), 393–408. https://doi.org/10.1080/13621025.2017.1307607

Guillén, A. M., Jessoula, M., Matsaganis, M., Branco, R., & Pavolini, E. (2022). Southern European Welfare Systems in Transition. In L. Burroni, E. Pavolini, & M. Regini (Eds.), *Mediterranean Capitalism Revisited: One Model, Different Trajectories* (pp. 149–171). Cornell University Press. https://doi.org/10.1515/9781501761096-009

Housing Europe. *The State of Housing in Europe 2021*. https://www.stateofhousing.eu/The_State_of_Housing_in_the_EU_2021.pdf.

Kemeny, J. (2005). "The Really Big Trade-Off" between Home Ownership and Welfare: Castles' Evaluation of the 1980 Thesis, and a Reformulation 25 Years on. *Housing, Theory and Society*, *22*(2), 59–75. https://doi.org/10.1080/14036090510032727

Kurz, K., & Blossfeld, H. P. (2004). Introduction: Social Stratification, Welfare Regimes, and Access to Homeownership. In H. P. Blossfeld & K. Kurz (Eds.), *Home Ownership and Social Inequality in Comparative Perspective* (pp. 1–20). Stanford University Press.

ISTAT. (2022a). *2022 Annual Report*. The state of the Nation. https://www.istat.it/it/files//2022/09/Annual-Report-2022_Summary.pdf.

ISTAT. (2022b). *Poverty in Italy*. Year 2021. https://www.istat.it/it/files//2022/07/REPORT_POVERTA_2021_Eng_def.pdf.

ISTAT. (2022c). *Gruppo di Lavoro Sulle Politiche per la Casa e L'emergnza Abitativa.* https://www.istat.it/it/files//2022/09/Istat-Audizione-Politiche-per-la-Casa_06_09_22.pdf.

Leonardi, D., & Stefani, S. (2021). The Pandemic and Homeless People in the Turin Area: The Level of Housing Adequacy Shapes Experiences and Well-Being. *Housing, Care and Support*, *24*(3/4), 93–104. https://doi.org/10.1108/HCS-03-2021-0006.

Lewis, J. (1992). Gender and the Development of Welfare Regimes. *Journal of European Social Policy*, *2*, 159–173. https://doi.org/10.1177/095892879200200301
Madden, D., & Marcuse, P. (2016). *In Defense of Housing: The Politics of Crisis*. Verso Books.
Malpass, P., & Rowlands, R. (Eds.) (2010). *Housing, Markets and Policy*. Routledge.
Minelli, A. R. (2004). *La Politica per la Casa.* il Mulino.
Muehlebach, A. K. (2012). *The Moral Neoliberal: Welfare and Citizenship in Italy.* The University of Chicago Press.
National Observatory on Housing and Urban Regeneration Policies. (2022). *Rilanciare le Politiche Pubbliche per L'abitare.* https://www.forumdisuguaglianzediversita.org/wp-content/uploads/2022/07/DOCUMENTO_Rilanciare-le-politiche-pubbliche-per-l%E2%80%99abitare.x94549.pdf.
Novaro, G. (2020). *Abitare i Margini: Politiche e Lotte per la Casa Nella Torino Degli anni Settanta.* Associazione Gruppo Abele Onlus-Edizioni Gruppo Abele.
Olagnero, M. (2018). Discorsi Sull'abitare. Come e a chi Parlano le Nuove Politiche Abitative. *La Rivista delle Politiche Sociali*, *4*, 25–42.
Oxfam. (2022). *Inequality Kills. The Unparalled Action Needed to Combat Unprecedented Inequality in the Wake of COVID-19.* https://oxfamilibrary.openrepository.com/bitstream/handle/10546/621341/bp-inequality-kills-170122-en.pdf;jsessionid=F8F19E3107DDD30011C8CF9EBA520A58?sequence=9
Padgett, D. (2007). There's No Place Like (a) Home: Ontological Security Among Persons with Serious Mental Illness in the United States. *Social Science & Medicine*, *64*(9), 1925–1936. https://doi.org/10.1016%2Fj.socscimed.2007.02.011
Pavolini, E., & Ascoli, U. (2019). The Dark Side of the Moon: Il Ruolo del Welfare Fiscale nel Sistema di Protezione Sociale Italiano. *Politiche Sociali. Social Policies*, *6*(1), 23–46. https://hdl.handle.net/11393/252836
Pavolini, E., & Deriu, F. (2015). Le Politiche Abitative. In C. Ranci & E. Pavolini (Eds.), *Le Politiche di Welfare* (pp. 143–168). il Mulino.
Pilat, S. Z. (2014). *Reconstructing Italy: The Ina-Casa Neighborhoods of the Postwar Era.* Routledge.
Poggio, T. (2005). La Casa Come Area di Welfare. *Polis*, *19*(2), 279–308. https://doi.org/10.1424/20299
Semi, G. (2017). Il Ritorno della Questione Abitativa. *Rassegna Italiana di Sociologia*, *3*, 697– 704. https://doi.org/10.1423/88041
Stedman Jones, D. (2012). *Masters of the Universe. Hayek, Friedman, and the Birth of Neoliberal Politics.* Princeton University Press.
Storto, G. (2018). *La Casa Abbandonata, Il Racconto delle Politiche Abitative dal Piano Decennale ai Programmi per le Periferie.* Officina Edizioni.
Torgersen, U. (1987). Housing: The Wobbly Pillar Under the Welfare State. *Scandinavian Housing and Planning Research*, *4*(1), 116–126. https://doi.org/10.1080/02815737.1987.10801428
Tosi, A. (2017). *Le Case dei Poveri. È Ancora Possibile un Welfare Abitativo?* MIMESIS EDIZIONI.

10
HOUSING AND WELFARE IN CATALONIA, SPAIN

Melissa García-Lamarca

The current state of housing in the region

Catalonia, like the rest of Spain, is a homeownership society. The most recent official statistics (2011) state that 74 per cent of the population are homeowners, 20 per cent are tenants and 2 per cent live in social housing. Related to tenure status, the state of housing in Catalonia – the second wealthiest autonomous community in Spain – has undergone noteworthy shifts in the first decades of the 21st century. Housing production has swung from record high to record low levels, and a decade-long housing crisis has generated unprecedented levels of evictions and housing precarity. Pressure from highly organized housing movements has prompted policy change to mitigate the latter. This overview of the current state of housing in Catalonia illustrates a paradox: on the one hand, the region has adopted one of the most progressive tenancy laws in Europe and far-reaching cooperative housing programmes in Spain; on the other hand, the region reflects Spain's long history of supporting homeowners and private housing while systematically undersupplying social housing.

Catalonia, like all 16 other autonomous communities in Spain, oversees housing, urban planning, and social services, while the Spanish state retains a number of broader powers over housing policy, property rights and macro-economic policy. While many powers were devolved to Spanish autonomous communities during the transition to democracy in the late 1970s, competencies are not always clear-cut and vary between autonomous communities, with many policy areas having shared or concurrent – and sometimes conflicting – powers between the central state and regions (Colino, 2008).

Currently, Catalonia's 2007 Right to Housing Act and 2014 Right to Housing Plan outline the regulatory framework for housing policy and implementation respectively for the region. The 2007 Act was adopted by a tripartite leftist coalition government that sought to address growing social inequalities in the region which were manifested in many sectors, including housing. While the Act was the first in Catalonia and Spain to attempt to ensure the right to housing by systematically linking land use law and housing, it failed to ensure this right was enforceable in the courts and thus not made effective (Mialot & Ponce, 2017). Despite promoting large-scale social housing programmes and creating a number of mechanisms to reduce social residential exclusion, the Act

DOI: 10.1201/9781003212690-12

and its connected plan have been unable to meaningfully change the systemic political economic housing relations which favour homeownership.

In the post-1975 democratic period, housing production and housing finance for developers and households has been largely driven by market forces. As was the case in other autonomous communities, housing production in Catalonia – a region approximately the size of Belgium with a population of 7.5 million – reached its peak during the country's 1997–2007 housing boom, during which in 2006 housing starts reached over 127,000 units. After plummeting to 3,036 in 2013, its lowest level during the democratic period, housing production stood at 15,000 units per year in 2021 (Catalan Institute of Statistics, 2022a). Most housing production is concentrated in the province of Barcelona, where nearly three-quarters of the Catalan population lives. Housing production in the past ten years has occurred in a context where 11.6 per cent of Catalonia's housing stock (448,356 units) was recorded in the 2011 Spanish census as being empty (i.e., unoccupied and not available for purchase or rent), and 12 per cent as second homes. In 2020, the empty housing stock included at least 29,479 empty homes owned by financial entities that had benefited from the government bailout of the financial system in 2012 (Garcia, 2020).

Similar to the rest of Spain, Catalonia's predominant social housing model is called officially protected housing (*vivienda de protección oficial*, VPO). VPO was created during the Franco dictatorship and traditionally operates as an owner-occupier model. The state provides free or cheap land, contracts bank loans and provides developers and purchasers with favourable financing conditions (Alberdi, 2014). In recent decades more rental VPO housing has been developed, with Catalonia holding the most VPO rental-based dwellings of all autonomous communities (Dol et al., 2017; Generalitat de Catalunya, 2022). However, as VPO rental and ownership housing makes up only 2 per cent of the total housing stock, social housing does not meet current needs and has long been systemically underproduced. About 5–6 per cent of total housing production in Catalonia during the boom was social housing, with annual starts peaking at 9,000–10,000 between 2007 and 2009. This fell dramatically by 2013, with 678 VPO housing starts, increasing by 2021 to 2,615 units per year (Catalan Institute of Statistics, 2022b). Furthermore, VPO housing excludes the most vulnerable members of society because they do not meet minimum income eligibility requirements (Pareja Eastaway & Sánchez-Martínez, 2017). The chronic lack of social housing has led to increasing numbers of households squatting in empty bank-owned housing or building informal housing on empty plots of land (Obra Social Barcelona, 2017; Rodríguez & Larios, 2022).

Increasing housing precarity deeply connected to the financialization of housing across Spain during the 1997–2007 housing boom and its reconfiguration in the post-2008 crisis period as detailed in Section "A brief overview of the historical development of the housing regime" – has triggered various grassroots struggles for housing justice in the past decade. The 2008 financial housing crisis led to an unprecedented avalanche of mortgage foreclosures and evictions, shifting in the post-2012 period largely to rental and squatting evictions. Between 2008 and 2017 over half a million rental and mortgage evictions were executed across Spain, with Catalonia sitting at the top of all autonomous communities with 95,000 evictions (Observatori DESC, 2018). Mobilization and direct action from housing movements have succeeded in providing grassroots solutions to immediate housing needs and creating new housing rights legislation (García-Lamarca, 2017a; Palomera, 2018; Romanos, 2014). Noteworthy in Catalonia are the 24/2015 Law, addressing energy poverty and obliging large landlords to offer social rent before evicting vulnerable households in case of non-payment, reinforced by the 2022 Anti-eviction Law, and the (now overturned) 2020 Rent Law. The latter was proposed in response to growing speculative investment in the rental sector and modifies the 2007 Right to Housing Law to control rents according to a reference index, applied in 61 municipalities across Catalonia assessed as having pressured housing markets (Sindicat de Llogateres, 2022). Despite securing some dimension of housing rights, the main challenge these laws face is from the Spanish Constitutional Court which often overturns regional housing laws on the basis of claims of territorial inequality – indeed the 2020 Rent Law was overturned in early 2022. Furthermore, despite these laws and a temporary eviction ban decreed by the state due to the COVID-19 pandemic, in 2020 around 30 households were evicted across Catalonia every day (Caro, 2021; Garcia, 2020). Finally, in May 2023 the first ever Spanish Housing Law was approved. Despite some enhanced protection for tenants, major concerns remain regarding conversions of long-term rental properties to short- and medium-term lets to avoid regulation, as well as the uncertain and uneven implementation of rent controls.

Finally, in recent years cooperative housing has become a small albeit noticeable phenomenon in Catalonia and the Barcelona area in particular (Gutzon Larsen, 2019). It has grown as a way to diversify social housing tenure away from the mortgage vs rental VPO binary and generate more democratic and sustainable housing options, with the most common structure being right-of-use housing cooperatives built on municipally owned land. While questions of access and affordability are rife, this model has the potential to create non-speculative, use value-oriented housing, a form of housing commons that can meet collective welfare needs (Buil, 2019).

A brief overview of the historical development of the housing regime

Spain's housing regime underwent a significant transformation in the second half of the 20th century. Over half of the country's population were tenants in the 1950s – rising as high as 95 per cent of households in Madrid and Barcelona – while by the turn of the 21st century over 85 per cent were homeowners. Housing production also changed dramatically during this period. The ravages of the Spanish Civil War (1936–1939) left Spain with poor-quality housing that was in short supply, but housing production between 1960 and 2010 was almost three times greater than the population increase during the same period. This left Spain with the largest quantity of second and empty homes in Europe by 2011. This section – focusing largely on the Spanish state since Catalonia lacked housing-related competencies until the post-1980 democratic period – outlines these key periods of housing regime transformation. It shows how institutional, legal and changes made private market housing production and a homeownership-oriented model flourish.

Building a country of homeowners, not proletariats: Housing as a problem and an opportunity

The Spanish Civil War (1936–1939) unleashed destruction that worsened already poor housing conditions across the country, especially since Housing Acts before this period did not address the massive housing shortages for working-class households (Tatjer, 2005). As the decades passed, working-class households were increasingly leaving southern Spain for industrializing cities in Catalonia and the Basque Country, driven by poverty, hunger, the promise of a better life and political exile or to escape from repression. In Catalonia's growing industrial cities, largely in the Barcelona metropolitan area, overcrowded sublet rooms, self-built shacks or in some cases even caves were often the only housing option for many of the tens of thousands of migrants arriving in the mid-1940s onwards (De Andres Creus, 2011; Masjuan, 2010).

The Franco dictatorship (1939–1975) saw the housing situation as both a problem and an opportunity. The problem was constituted by the so-called "mass of second-class Spaniards" (Maestrojuan, 1997, p. 174) who were seen as generating social dangers to the dictatorship's stability.

Rent freezes were included in the first 1946 Urban Letting Act (*Ley de Arrendamientos Urbanos: LAU*), one of the most important pieces of legislation regulating the rights and duties of landlords and tenants as LAUs dictate the duration and conditions of rental contracts (Leal, 2005). Renewed in 1955, rent freezes were driven by housing scarcity and the dictatorship's desire to please a broader social base including shopkeepers, industrialists and tenants (Artola Blanco, 2012; Leal, 2005). The more systemic solution to the housing situation, however, was the creation of policies that built a nation of Christian, Spanish homeowners, put women in their "place" at home, and established the growth conditions for a professional real estate development sector allied with the dictatorship (García-Lamarca, 2022). The first Ministry of Housing was created in 1957, led by the Falangist José Luis Arrese who declared war on slums and subleasing as particularly rampant hotbeds for immoral, proletarian thought and action. The Ministry's deep-seated belief in homeownership was reflected in Aresse's famous first official speech: "we want a country of homeowners, not proletarians."

During Spain's developmental period in the 1960s, when the economy began to open up to foreign capital investment, residential construction started to become an economic motor and building a country of homeowners became possible. The 1960 Horizontal Property Act was an important piece of legislation that enabled the sale of individual apartments within a building, established with the prime objective of encouraging tenants to purchase property (Betrán Abadía, 2002). Private initiatives to construct state-subsidized housing (VPO), legislated in 1963 and regulated in 1968, also grew as the state enabled access to mortgage credit, especially through the Spanish Mortgage Bank (BHE), facilitated legal processes for developers to obtain land, and provided financial help for construction. These public interventions attempting to deal with rapid urbanization, provide public housing, and meet demands from a rapidly expanding tourist industry spurred Spain's first housing boom from 1969 to 1974 (Charnock et al., 2014). While working-class housing demand was made more profitable for private developers during this period, the scale of production was unable to provide housing for the poorest of the working class (Tatjer, 2005).

From state to market rule: Housing, land, and finance in democratic Spain

Franco's death in 1975 led to the end of the nearly 40-year dictatorship, one of the longest in Europe, and inaugurated a return to democracy in Spain. Policies addressing residential housing, land use and mortgage finance were the three axes that reconfigured the housing regime in this period (López & Rodríguez, 2010), following a neoliberal logic that was growing across Europe and much of the Global North at the time.

State-level policies adopted to regulate housing across Spain in the 1980s and 1990s reinforced homeownership as a dominant tenure. The 1985 Boyer Decree eliminated rent control and tenant protection clauses in all new rental lease contracts, but this did not spur a fall in rental prices nor an increase in rental housing supply as intended (Pareja Eastaway & San Marín Varo, 2002). House purchase was encouraged by making both mortgage principal and interest payments tax-deductible. From 1985 to 1989 this included both main *and* second residences, contributing to Spain's second short but intense housing boom (1985–1992) upon joining the European Economic Community (EEC) in 1986 (García Montalvo, 2003). Thus it was largely the socially excluded and lowest-income earners who rented, mainly because they were unable to purchase property (Trilla, 2014a). Housing plans in the 1980s left out the lowest income segments of the population as VPO housing was captured by higher income groups due to a combination of economic crisis, policy to reduce inflation and high interest rates (Rodríguez Alonso, 2004). In sum, housing policies failed to provide accessible and affordable housing for all, as rental legislation, tax incentives, and Housing Plans encouraged homeownership.

In terms of land use, ideas around economic liberalization began to penetrate Spain and its urban planning and land use institutions after joining the EEC in 1986. The original 1956 Land Act, which regulated urban expansion, was changed in 1998 to redefine concepts of "developable" and "non-developable" land, with the latter opened up to extend the potential of urban development to all land across the country (Puig Gómez, 2011). The notion of the social function of ownership – present in previous iterations of the Land Act, where the public determination of different land classifications controlled landowners' right to build – was thus removed from planning obligations, and private land ownership came to operate at the core of the urban development process (Roca Cladera & Burns, 2000). Through their devolved competencies, Catalonia adopted the 2/2002 Act in lieu of the 1998 Land Act to control urban planning and regulate housing development. While local governments were enabled to make building permission contingent on land owners ceding a proportion of their land for public use, extremely attractive profit margins during the housing boom meant that most municipalities sold this land to developers or leased it for private use. Despite having a majority in Parliament, attempts from the Catalan leftist coalition government in power during the 2000s to reduce land speculation, foster the rental market, and curb skyrocketing housing prices during the boom were ultimately unsuccessful due to the magnitude of the profits made by the private sector and taxes collected by the public sector. In terms of the latter, in 2004 alone local land and building tax revenues generated over €9 billion for municipalities across Spain (Pou, 2007).

Finally, liberalizing Spanish housing finance systems in the 1980s was critical in enabling the credit explosion in the 2000s. Until 1980, channels to finance housing production in Spain were publicly controlled. Consumer mortgages were provided either by the public BHE, an entity absorbed into a larger public bank called Argentaria in 1991 then progressively privatized until it was absorbed into the private bank BBVA in 1999, or local savings banks. The 1981 Mortgage Market Regulation Act deregulated access to finance by enabling the entrance of private operators, privatizing public financing avenues, and building a deeper integration of financial and mortgage markets (Alberdi, 1997). It also established the financial infrastructure to securitize debt through covered bonds, with the 1992 Securitization Vehicles Act creating the legal basis for Spanish residential mortgage-backed securities (RMBS) (Nasarre-Aznar, 2002). Finally, the geographical deregulation of local savings banks in the 1990s led to their spectacular growth in offices and personnel, with the best way to make quick money being mortgage and development loans.

Spain's 1997–2008 financialized housing boom and post-2008 crisis

As Spain became more integrated into the EEC, falling interest rates and the adoption of the euro combined with neoliberalized housing, land and finance-related legislation unleashed the country's "urbanization tsunami" (Fernández Durán, 2006). In a decade, one-quarter of Spain's land area was paved over and built upon, with the country proudly becoming the European leader in cement consumption and fifth in the world after China, India, the US, and Japan (Naredo et al., 2008). House prices skyrocketed from €720 per square metre in 1997 to a peak of €2,100 per square metre in 2008 and the country's total housing stock increased by almost 6 million units. At its peak, this represented more than the housing production in Italy, France, and Germany combined (López & Rodríguez, 2011). Spanish financial entities granted at least 1 million mortgages to vulnerable segments of society between 2003 and 2007, with a securitized volume reaching €49.9 billion or 20.7 per cent of the total volume of RMBS issued in the country from 2003 to 2008 (Ezcurra Pérez,

2012). Housing production ultimately exceeded population growth by almost three times, and the dictatorship's desire to create a nation of homeowners and for construction to act as a growth axis was finally realized (see Table 10.1).

The burst of the housing boom in 2008 generated mass unemployment, plummeting housing prices and a socio-economic crisis unseen in democratic Spain. As foreclosures and evictions grew, the state prioritized saving the financial sector through tens of billions of euros in public funds and reconfiguring the real estate sector as a place for foreign investment in rental properties (García-Lamarca, 2021). Recent housing policies and plans have not been able to stem the over half a million registered evictions that have taken place across Spain from 2008 to 2017 and housing precarity being experienced especially in larger cities in Catalonia and across Spain.

The welfare regime and its development

Similar to the development of the housing regime, the incremental development of Spain and Catalonia's welfare regime also reflected the shift from dictatorship to democracy. A corporatist (Esping-Andersen, 1990), authoritarian welfare regime best characterizes the Franco dictatorship, with primary roles for the Fascist-Catholic state-church alliance and the family, especially the patriarchal vision that charged women with social reproduction chores in the home. Responding to internal social demands and European political and socio-economic pressures, Spain's welfare regime attempted to move closer to European standards during the transition to democracy (Gallego et al., 2005). During the democratic transition welfare policy competencies like health care, education, social services, unemployment benefits and pensions were largely decentralized to autonomous communities, with this process being asymmetrical and complex between regions (ibid.). The subsequent sections outline these changes focusing on the role of different actors in welfare provision, the marketization of individual welfare and questions of social stratification.

Authoritarian, paternalist and corporatist welfare regime: The Franco dictatorship

Although there is a part of the Spaniards' collective imaginary that identifies the Franco dictatorship as providing the first great push to Spain's modern welfare regime, a newer generation of historians point to the exaggerated nature of these claims and suggest a much more complex reality (González Madrid & Ortiz Heras, 2018). In the first decade of the dictatorship, state-driven welfare actions were focused on creating a "productive" society with men and women in their "correct" places. For example, the Family Subsidy (1938) supplemented paltry incomes and disincentivized women from working outside the home, while the introduction of Statutory Sickness Insurance (1942) aimed to enable workers to return to work as soon as possible to support the

Table 10.1 Tenure in Spain, by household (%), 1945–2017

	SPAIN				
	1950	*1960*	*1980*	*2000*	*2017*
Homeownership	45.9	51.9	73.1	80.7	76.7
Private rented	51.2	43.0	20.8	11.0	19
Public/social rented	2.9	1.9	3.0	2.0	2.0

Source: Rodríguez Alonso (2004). 2017 data from INE Life Conditions Survey.

national economy (Arriba & Moreno, 2005; Moreno, 2020). Public support for those not engaged in paid labour was delivered through charity in a paternalistic fashion, with private organizations affiliated with the Catholic Church delivering poverty alleviation programmes (ibid.). Beyond propaganda, of which there was much, González Madrid and Ortiz Heras (2018) argue that the Franco regime barely developed a comprehensive social security system. The nascent system was primarily supported by the individual savings of the weakest workers and some token contributions by the state, with subsequent iterations in the 1960s expanding and consolidating the existing model. Thus, welfare was reliant mostly on the family with some minor support from the state and the church which actively sought to reinstall a patriarchal, family based social order.

The first truly welfarist-oriented interventions in housing under the dictatorship happened when capital began to flow into Spain in the late 1950s and early 1960s, during Spain's developmentalist period. The Ministry of Housing adopted three Social Urgency Plans for Madrid (1957), Barcelona (1958) and the Basque Country (1959) to encourage private capital to collaborate in providing state-subsidized, privately owned housing for the increasing influx of working-class migrants from southern Spain to these cities (Betrán Abadía, 2002; López Simón, 2018). By 1970, the state-financed 180 public working-class housing estates across 15,000 hectares as per the 1961–1976 National Housing Plan, with the number of housing units per estate ranging from 500 to 10,000 (Capel, 1975). However, most of these housing estates were of poor quality, located on the periphery and lacking infrastructure and services, leading some to dub them "vertical shanties" (De Andres Creus, 2011). The most public VPO housing ever produced in Spain was built in the 1960s-1970s: peaking at close to 250,000 units in 1965 and from 1971 to 1975 reaching nearly 200,000 units per year (Trilla, 2014a).

Welfare expansion and deregulation in democratic Spain and Catalonia post-1975

As the dictatorship (1939–1975) coincided with the maximum expansion of the welfare state across Western Europe (Moreno, 2020), Spain had a fair bit of "catching up" to do in the transition to democracy. The latter developed under the notion of modernization and Europeanization, which, under social democrats, not only led to a growth of social support services but also the liberalization of welfare provision (Lessenich, 1996). Spain's integration into the EEC in 1986 pushed the country to follow a pattern of welfare convergence with its European counterparts as social entitlements became more generalized, welfare expenditure increased and social service provision was diversified to private and third-sector organizations (Arriba & Moreno, 2005). A system was established where social protection schemes are accessed either through participation in the formal labour market – a form of social insurance whereby a worker's contribution to social security makes them eligible for unemployment payments, sickness benefits, pensions and other types of support – or people's territorial residence status, upon which universalist schemes like education and health care are based. The state plays a minor and insufficient role in social assistance, especially in comparison to other European countries, meaning that familial-based systems of assistance prevail (Gallego et al., 2005).

Despite post-1977 policies aiming for greater universalization of social security, social spending only rose to 8.6 per cent of GDP, in contrast to countries such as France and Germany which spent almost one-quarter of their GDP on welfare provision (Moreno, 2020). Overall, the weak corporatist welfare arrangements characterizing the Franco dictatorship period eased reforms towards universalism in health care and education in the 1980s and recalibration in the 1990s, but welfare's inherently conservative and family based character has meant an insufficient expansion of care policies and social assistance (Guillén & León, 2016). Ultimately, some strong universal elements were built at the same time that social assistance was underdeveloped and successive waves of labour deregulation took place, normalizing precarious labour conditions and familial dependence (Gallego et al., 2005).

Since the 1980s, Catalonia has overseen many devolved welfare-related competencies including health care, education and housing. The state has gradually reduced its role in housing policy during the democratic period providing the market with a greater role through what has been dubbed "debtfare" (Soederberg, 2014), referring to individual welfare provision through indebtedness. The two Spanish government social housing programmes in the 1990s continued to encourage low- to middle-income households to become owner-occupiers through VPO as well as a new scheme called Housing under Controlled Prices (*Vivienda de Precio Tasado*, VPT). The latter provided public loans according to the weighted family income to purchase a new or secondhand main dwelling as long as met certain size, use and price requirements (Pareja Eastaway & San Martin, 1999). Mortgage indebtedness became especially acute during the 1997–2007 housing boom, where outstanding mortgage debt in Spain increased over fourfold from €154.5 billion in 1999 to €674 billion in 2008 (European Mortgage Federation, 2011).

Starting in the 1990s, Spain saw a historically unprecedented growth in immigration: immigrants represented 2 per cent of Spain's population in 1999 and 12 per cent in the 2011 census. The fact that many of the non-European immigrants were employed in temporary and poorly paid construction and service sectors meant that the duration and quantity of social security support they received was limited when they experienced unemployment because they often fell outside the non-comprehensive social insurance system (Bruquetas Callejo & Moreno Fuentes, 2015). Work in the informal economy is widespread compared to other European countries, and is the only option for undocumented immigrants, meaning that they are necessarily excluded from employment benefits. Due to the general lack of welfare provision in terms of housing, non-European immigrants in particular are heavily impacted by unemployment due to racism in the rental sector and face high rental or mortgage payments, among other disadvantages (Provivienda 2020; ibid.).

The bursting of the 1997–2007 real estate bubble and the subsequent rapid rise of public debt – largely to rescue the financial system – as well as massive unemployment and increased poverty has put great strain on welfare systems across Spain, and has generated calls for the improvement of welfare coverage and provision. As Dol et al. (2017) note, the absence of support mechanisms for troubled households unable to pay for their housing forced them to turn to family, friends, social movements, or non-profit organizations for assistance.

The interrelation between housing and welfare

Despite the intent and range of proposals in Spanish and Catalan housing policy to ameliorate the conditions of households, housing remains an extremely weak component of the welfare state in general (Gallego et al., 2005). Housing has been characterized as the weakest and most vulnerable dimension of the Spanish welfare state (Gomà, 2003), and indeed it could be argued that housing does not even truly form part of the welfare state. Various governments have had aspirations to provide public housing for populations in need, and thus establish a support system and expand the welfare regime, but both Catalonia and Spain have largely failed in turning this aspiration into practice. This reality is reflected today not only in the chronic lack of social housing but also in total public expenditure on social housing: just 0.1 per cent in Catalonia and Spain (Bosch & Trilla, 2021). Aside from reduced eligibility, the limited housing allowance programmes for low-income tenants in market rental housing and other government support services do not address existing social stratification or other structural housing access problems but rather provide insufficient sticking plaster solutions.

The relationship between housing and Spain's wider welfare regime has not changed dramatically over time. Perhaps ironically, it was during the latter part of the Franco dictatorship (the 1960s–1970s) that the highest levels of state housing support existed. This however took the form of subsidized mortgaged homeownership, which used public subsidies to support privately owned housing, meaning that dwellings did not become part of long-term collective welfare provision. State intervention in housing during this period arguably had the largest historic impact in reducing social stratification with regard to housing – in the sense of bringing swaths of working-class people into an ostensibly "middle-class" homeowner category – but since the lowest income populations were not eligible for VPO housing they remained in the market-based rental sector, unable to "build wealth" and achieve housing stability through housing investment like their wealthier counterparts. This became a serious problem post-1985 when rent controls and tenant rights were eliminated, leaving tenants with little or no legal protection against exploitation by landlords. The state-sanctioned 1990–2000s debt-driven model of housing provision appeared to reduce social stratification as increasing numbers of people accessed homeownership and ostensibly saw their wealth rise as their homes increased in value, but the crisis has illustrated how such a "debtfare" approach (Soederberg, 2014) to welfare is fundamentally unsustainable and has ultimately deepened social inequality (García-Lamarca, 2022).

Housing plans and policies up until the 2008 crisis have further reinforced private homeownership, and recent attempts to address social housing provision are insufficient. In this light the market has long been the most important player in housing provision for a range of social groups, with low-income groups being relegated to the rental sector – and since the 2008 crisis in more extreme cases forced to squat in bank-owned housing with or without the support of social movements (Obra Social Barcelona, 2017). This is often households' only option because so-called "emergency" housing mechanisms in most cities are unable to temporarily rehouse evicted households in a timely fashion: in Barcelona, the waiting time to be rehoused is over two years, with more than 600 families on the waiting list (Habitatge, 2021).

Thus, despite the fact that Article 47 of the Spanish constitution enshrines the right to decent and adequate housing, the latter exists as a market commodity rather than a social right. This is also true in Catalonia, where it is telling that an article excluded from the Right to Housing Act due to lack of consensus was one that would have changed the notion of housing as a good for exclusive use and individual enjoyment to one of collective interest (Trilla, 2009). In the Spanish and Catalan cases, it is clear that housing only becomes a social right through the active struggles of housing movements pushing for legal changes and taking direct action to block evictions, occupy banks for debt forgiveness or occupy empty housing owned by financial entities. New Catalan legal frameworks controlling rent and obliging large property owners to rehouse households facing eviction, among other elements that seek to ensure housing rights, have provided a little more protection for the most vulnerable households, but these hard-won legal gains are unevenly implemented and are constantly at risk of being overturned by the Spanish Constitutional Court.

The emerging interest in cooperative housing in Catalonia within the VPO framework is in this light quite promising, as it can greatly contribute to long-term, transferrable and collective welfare provision. During the 2010s, the La Borda cooperative housing project successfully pressured the city of Barcelona to cede municipally owned land slated for rental- or mortgage-based VPO housing production to the cooperative. This required a right-of-use agreement as well as complex and arduous financing arrangements from the social and solidarity economy, which are formal and informal socio-economic arrangements that seek to place the satisfaction of people's needs above profit-making, but funding was finally secured to develop the building (Buil, 2019). Under the leftist municipal government in power since 2015 that emerged in part from housing movements, the right-of-use cooperative housing model has been embraced through several rounds of VPO cooperative housing competitions in Barcelona and increasingly across Catalonia (Avilla-Royo et al., 2021). At the same time, work needs to be done to increase the accessibility of this model as low-income people don't have the initial capital downpayment required by cooperatives; participation in coop development also requires significant time which is often in short supply.

Regional outlook

Spain, and Catalonia, are often grouped into what is termed a Mediterranean welfare model, proposed as a fourth model outside of Esping-Andersen's (1990) tripartite characterization of the welfare state. This characterization emerged from a range of scholars who felt that Southern European countries had distinct dimensions that made them not quite "fit" in the Esping-Andersen model (Guillén & León, 2016). Considering the welfare realities of Spain, Portugal, Italy and Greece, this model involves the combination of insurance schemes and universal programmes with a strong role for the family. There are many historical parallels between Southern European countries that explain why their welfare regimes have developed in a similar pattern: later industrialization patterns in comparison to Northern Europe; the existence of authoritarian dictatorships until the 1970s in Portugal, Greece and Spain; and the creation of welfare structures that predated democracies and thus had no civil society participation, to name a few (Andreotti et al., 2001). Furthermore, the Southern European modus operandi when addressing immigration and housing tends to focus on area-based urban programmes and emergency policies rather than comprehensive forward-thinking policies that reshape broader systems and redistributive welfare arrangements (Arbaci, 2008). Spain and Catalonia echo several realities of the regional characterization of housing in relation to the welfare state such as marginal social expenditures on housing, a very small social rental housing stock, relatively high levels of homeownership and a poorly protected rental sector.

In terms of the last two tenure categories, today's Spain arguably does stand slightly apart in its predominance of homeownership and extremely poor protection of tenants, the 2020 Catalan Rental Law accepted. The Spanish state's ideological and material drive for homeownership was also remarkably strong compared to other Southern European countries. Similarly, the restructuring and expansion of mortgage financing in the 1980s was much deeper than in other southern European countries, perhaps due to Spain's keenness to meet the neoliberal demands of entry into the EEC. This process enabled greater levels of indebtedness and thus "individualized" forms of market- and asset-based welfare provision. This is reflected in Spain's notably higher level of residential mortgage debt per capita during the housing boom: according to the European Mortgage Federation, in 2008 Spain had rates nearly three times greater than Greece and Italy, and 50 per cent more than Portugal. Another difference between countries in the region is that self-built housing was relatively more significant in Italy, Greece and Portugal than in Spain. The Spanish promotion of private real estate development has been attributed to stronger institutionalized connections between developers and systems of urban land release and strategic restructuring of the mortgage system in the 1980s (Allen et al., 2004), building from historically strong relationships between private developers and the dictatorial regime.

A one-size-fits-all Southern European model, however, is also complicated by marked internal regional differences. Within Spain, Gallego et al. (2003) argue that the Basque Country deviates most from the Spanish "familial" welfare model through a more public-communitarian focus due to the active role of government and a strong tradition of corporatism; it is also the autonomous community with the highest social spending and lowest unemployment levels. Catalonia, in turn, is described as a market-communitarian model as the private market plays a larger role in welfare provision compared to other regions (ibid.). Dol et al. (2017) consider the intersection between the regionalization of housing and welfare, unpacking the differences between the responses to the 2008 financial housing crisis in the Basque Country, Andalusia and Catalonia. While the authors underline that regional housing crisis interventions do not fit seamlessly into regional welfare variations described by Gallego et al. (2003), the Basque Country is the region that "has committed itself most strongly to building a somewhat less risky housing system" (p. 594). They also underline the need to account for organized housing movements that have pushed different autonomous communities to take action, which has provided some, albeit limited, respite.

Concluding discussion

This chapter has underlined the insufficient and distinctive role of housing in the Catalan, and Spanish welfare regimes and the impacts this reality has had on housing affordability and security. The supposed constitutional and regional right to decent and adequate housing does not exist in practice, and the strength and predominance of market-based private real estate development ensures that housing exists primarily as a tool for economic growth and investment rather than a process that supports vulnerable households and meets collective welfare needs by providing affordable homes.

The foundation for much of Catalonia's and Spain's missing housing dimensions within welfare regimes is connected to the historic fostering of owner-occupation and its implications. From past to present, this has meant that the rental sector has long been insufficiently protected and supported and that social housing has been systematically underproduced. Furthermore, and ironically, the predominant homeownership VPO social housing provision model has served to generate individual rather than collective wealth-building, where the latter is critical in its ability to contribute to and strengthen the welfare state. Perhaps because such a large proportion of the Catalan/Spanish population are homeowners, households focus on the investment value of housing over its use value, since – once paid – housing brings household wealth and security in the medium to long term (Trilla, 2009), a reality backed by the long political economic push to create a nation of homeowners. Yet, as Trilla (2014b) notes, since housing currently is a source of inequality and social exclusion within the Spanish welfare regime, how can this reality be addressed in a system of free-market housing production, free-market financing and free-market pricing? Inspired by Arbaci (2008), the important question here is: how can forward-thinking approaches to housing in the context of collective welfare provision confront and attempt to transform this market-based housing provision chain?

Extending and mainstreaming some promising initiatives underway is a good place to start. First, is for the state to take decisive action to win a battle long demanded by housing movements: to turn empty housing owned by banks benefitting from the publicly funded rescue of the financial system into social housing (García-Lamarca, 2017b). The most logical place to intervene now is in Spain's "bad bank", the SAREB (Byrne, 2016), a 45 per cent publicly owned asset management company created in 2012 to manage the so-called toxic assets (failed debt and housing) owned by banks who were rescued with €52 billion. While tasked with a goal of selling its assets at a profit by 2027, in early 2022 it emerged that the SAREB would become 100 per cent publicly owned and its €35 billion debt would become public debt (Vargas Martín, 2022). This is the moment to turn the remaining existing housing held by the SAREB and publicly rescued banks into social housing.

Another action could be to extend public support for right-of-using housing cooperatives in order to expand a non-speculative model of housing provision, especially for VPO housing, and for middle-class housing alternatives. At the moment, one of the most significant barriers in the limited Catalan experience to date has been the financial deposits required by households to access a cooperative: this can be up to 30,000 euros, an amount that many working-class households cannot afford. Public subsidies could help cover part of this cost; indeed it is a far superior option to subsidizing mortgage-based VPOs where the property ultimately becomes private rather than public property. More training and capacity-building support are also required to diversify participation in cooperative housing projects beyond white Spanish populations with medium-high levels of formal education. Further exploration of turning bank-owned housing squatted by households with no housing alternatives in collaboration with housing movements into right-of-use housing cooperatives is another emerging avenue to explore, providing security of tenure and shifting more housing into the social sphere. Finally, the development of middle-income cooperative housing projects, which could take place on privately owned land, could be an important step to consider shifting the housing development model more broadly away from a speculative investment-based approach to one grounded in a vision of collective and transferrable welfare provision. All of these options require the state to step up and take decisive action to ensure a transformation in housing production and provision that prioritizes collective societal needs, especially for marginalized populations over those of individuals.

References

Alberdi, B. (1997). Recent Mortgage Market Development in Spain. *Housing Finance International, 11*, 23–30.

Alberdi, B. (2014). Social Housing in Spain. In K. Scanlon, C. Whitehead, & M. F. Arrigoitia (Eds.), *Social Housing in Europe* (pp. 223–237). John Wiley & Sons, Ltd. https://doi.org/10.1002/9781118412367.ch13

Allen, J., Barlow, J., Leal, J., Maloutas, T., & Padovani, L. (2004). *Housing and Welfare in Southern Europe*. Blackwell.

Andreotti, A., Garcia, S. M., Gomez, A., Hespanha, P., Kazepo, Y., & Mingione, E. (2001). Does a Southern European Model Exist? *Journal of European Area Studies, 9*(1), 43–62. https://doi.org/10.1080/14608460120061948

Arbaci, S. (2008). (Re)Viewing Ethnic Residential Segregation in Southern European Cities: Housing and Urban Regimes as Mechanisms of Marginalisation. *Housing Studies, 23*(4), 589–613. https://doi.org/10.1080/02673030802117050

Arriba, A., & Moreno, L. (2005). Spain – Poverty, Social Exclusion and "Safety Nets". In M. Ferrera (Ed.), *Welfare State Reform in Southern Europe: Fighting Poverty and Social Exclusion in Italy, Spain, Portugal and Greece* (pp. 110–162). Routledge.

Artola Blanco, M. (2012). La Transformación del Mercado de Alquiler de Fincas Urbanas en España (1920–1960). *Biblio 3W: Revista Bibliográfica de Geografía y Ciencias Sociales, XVII*(988), 1–15.

Avilla-Royo, R., Jacoby, S., & Bilbao, I. (2021). The Building as a Home: Housing Cooperatives in Barcelona. *Buildings, 11*(4), 137. https://doi.org/10.3390/buildings11040137

Betrán Abadía, R. (2002). De Aquellos Barros, Estos Lodos. La Política de Vivienda y Postfranquista. *Acciones e Investigaciones Sociales, 16*, 25–67.

Bosch, J., & Trilla, C. (2021). *Housing System and Welfare State. The Spanish Case Within the European Context.* The Social Observatory, La Caixa Foundation. https://oobservatoriosocial.fundacaolacaixa.pt/en/-/housing-system-and-welfare-state.-the-spanish-case-within-the-european-context

Bruquetas Callejo, M., & Moreno Fuentes, F. J. (2015). Precarización y Vulnerabilidad de la Población Inmigrante en la España en Crisis. El papel del Estado de Bienestar. *Panorama Social, 22*, 139–151. https://doi.org/10.13039/501100003329

Buil, D. M. (2019). Housing Cooperatives in Barcelona, Between the Commons and Social Production of Habitat: A Step Towards Housing Communalization in Barcelona? *ACME: An International Journal for Critical Geographies, 19*(3), 665–683.

Byrne, M. (2016). Bad Banks and the Urban Political Economy of Financialization: The Resolution of Financial–real Estate Crises and the Co-constitution of Urban Space and Finance. *City*, *20*(5), 685–699. https://doi.org/10.1080/13604813.2016.1224480

Capel, H. (1975). *Capitalism y Morfología Urbana en España*. Los Libros De La Frontera.

Caro, P. (2021). Cada día se Producen 19 Desahucios en la Provincia de Barcelona, *20 Minutos*, 23 September. https://www.20minutos.es/noticia/4831169/0/cada-dia-se-producen-19-desahucios-en-la-provincia-de-barcelona/.

Catalan Institute of Statistics. (2022a). *Habitatges Iniciats. Per Tipus D'edificació. Comarques i Aran,* Àmbits *i Províncies*. https://www.idescat.cat/indicadors/?id=aec&n=15693

Catalan Institute of Statistics. (2022b). *Habitatges Iniciats amb Protecció Oficial. Per Règim de Protecció. Comarques i Aran,* Àmbits *i Províncies*. https://www.idescat.cat/indicadors/?id=aec&n=15695

Charnock, G., Purcell, T., & Ribera-Fumaz, R. (2014). *The Limits to Capital in Spain: Crisis and Revolt in the European South*. Palgrave Macmillan.

Colino, C. (2008). The Spanish Model of Devolution and Regional Governance: Evolution, Motivations and Effects on Policy Making. *Policy and Politics*, *36*(4), 573–586.

De Andres Creus, L. (2011). *Barraques: La Lluita dels Invisibles*. Ara Llibres.

Dol, K., Mazo, E. C., Llop, N. L., Hoekstra, J., Fuentes, G. C., & Etxarri, A. E. (2017). Regionalization of Housing Policies? An Exploratory Study of Andalusia, Catalonia and the Basque Country. *Journal of Housing and the Built Environment*, *32*(3), 581–598. https://doi.org/10.1007/s10901-016-9528-z

Esping-Andersen, G. (1990). *The Three Worlds of Welfare Capitalism*. Polity Press.

European Mortgage Federation. (2011). *Hypostat 2010: A Review of Europe's Mortgage and Housing Markets*. Brussels.

Ezcurra Pérez, M. M. (2012). *Análisis del Impacto de la Titulización Hipotecaria Española en la Concesión de Préstamos Subprime y en la Estabilidad Financiera Bancaria*. [Doctoral dissertation, Universidad de Santiago de Compostela].

Fernández Durán, R. (2006). El TSUNAMI Urbanizador Español y Mundial. *Boletín CF+ S*, *38/39*. http://habitat.aq.upm.es/boletin/n38/arfer_3.html

Gallego, R., Gomà, R., & Subirats, J. (2003). *Estado de Bienestar y Comunidades Autónomas: La Descentralización de las Políticas Sociales en España*. Tecnos. https://dialnet.unirioja.es/servlet/libro?codigo=4248

Gallego, R., Gomà, R., & Subirats, Joan. (2005). Spain: From State Welfare to Regional Welfare? In N. McEwen & L. Moreno (Eds.), *The Territorial Politics of Welfare* (pp. 103–126). Routledge.

Garcia, G. 2020. "Bancs, Sareb i Fons D'inversió Acaparen Milers de Pisos Buits". *La Directa*, 29 September 29. https://directa.cat/bancs-sareb-i-fons-dinversio-acaparen-milers-de-pisos-buits/García-Lamarca, M. (2017a). From Occupying Plazas to Recuperating Housing: Insurgent Practices in Spain: From Occupying Plazas to Recuperating Housing. *International Journal of Urban and Regional Research*, *41*(1), 37–53. https://doi.org/10.1111/1468-2427.12386

García-Lamarca, M. (2017b). Reconfiguring the Public through Housing Rights Struggles in Spain. In J. Hou & S. Knierbein (Eds.), *City Unsilenced: Urban Resistance and Public Space in the Age of Shrinking Democracy* (pp. 44–55). Routledge.

García-Lamarca, M. (2021). Real Estate Crisis Resolution Regimes and Residential REITs: Emerging Socio-Spatial Impacts in Barcelona. *Housing Studies*, *36*(9), 1407–1426. https://doi.org/10.1080/02673037.2020.1769034

García-Lamarca, M. (2022). *Non-Performing Loans, Non-Performing People: Life and Struggle with Mortgage Debt in Spain*. University of Georgia Press.

García Montalvo, J. (2003). La Vivienda en España: Desgravaciones, Burbujas y Otras Historias. *Perspectivas Del Sistema Financiero*, *78*, 1–43.

Generalitat de Catalunya. (2022). *Habitatges Protegits Amb Protecció Vigent*. https://habitatge.gencat.cat/ca/dades/estadistiques_publicacions/indicadors_estadistiques/estadistiques-de-la-politica-dhabitatge-/habitatges-proteccio-vigent/

Gomà, R. (2003). Las Políticas de Vivienda de las Comunidades Autónomas. In R. Gallego, R. Gomà, & J. Subirats (Eds.), *Estado de Bienestar y Comunidades Autónomas* (pp. 142–163). Madrid: Tecnos.

González Madrid, D. A., & Ortiz Heras, M. (2018). El Franquismo y la Construcción del Estado de Bienestar en España: La Protección Social del Estado (1939–1986). *Pasado y Memoria Revista de Historia Contemporánea*, *17*, 361–388. https://doi.org/10.14198/PASADO2018.17.13

Guillén, A. M., & León, M. (2016). *The Spanish Welfare State in European Context*. Routledge. https://doi.org/*10*.4324/9781315552552

Gutzon Larsen, H. (2019). Barcelona: Housing Crisis and Urban Activism. In P. Hagbert, H. Gutzon Larsen, H. Thörn, C.Wasshede (Eds.) *Contemporary Co-Housing in Europe* (pp. 74-93). Routledge.

Habitatge. (2021). *El Consejo de la Vivienda Social de Barcelona Presenta los Resultados de los Cinco Primeros años del Plan por el Derecho a la Vivienda*, 11 November. https://www.habitatge.barcelona/es/noticia/el-consejo-de-la-vivienda-social-de-barcelona-presenta-los-resultados-de-los-cinco-primeros-anos-del-plan-por-el-derecho-a-la-vivienda_1119016.

INE. (2017). *Encuesta de Condiciones de Vida, Base 2004*. Instituto Nacional de Estadística. https://www.ine.es/jaxiT3/Tabla.htm?t=4566

Leal, J. (2005). La Política de Vivienda en España. *Documentación Social, 138*, 63–80.

Lessenich, S. (1996). España y los Regímenes de Estado del Bienestar. *Revista Internacional de Sociología, 13*, 147–161.

López, I., & Rodríguez, E. (2010). *Fin de Ciclo: Financiarización, Territorio y Sociedad de Propietarios en la Onda Larga del Capitalismo Hispano (1959–2010)*. Traficantes de Sueños.

López, I., & Rodríguez, E. (2011). The Spanish Model. *New Left Review*, *69*(May–June), 5–29.

López Simón, I. (2018). El Chabolismo Vertical. Los Movimientos Migratorios y la Política de Vivienda Franquista (1955–1975). *Huarte de San Juan, Geografía e Historia, 25*, 173–192.

Maestrojuan, F. J. (1997). «Ni un Hogar sin Lumbre ni un Español sin Hogar» José Luis de Arrese y el Simbolismo Ideológico en la Política del Ministerio de la Vivienda. *Príncipe de Viana Pamplona, 210*, 171–187.

Masjuan, E. (2010). *El problema de l'habitatge obrer i la creació de les coves*. Sabadell City Council. http://ca.sabadell.cat/Coves/d/problemHabObrer.pdf.

Mialot, C., & Ponce, J. (2017). Ten Years of the French DALO and the Catalan Right to Housing Act: European Innovation in the Fields of Land Use Planning and Housing. *Journal of Comparative Urban Law and Policy*, *2*(1), 1–28.

Moreno, L. (2020). Reforma Social, Franquismo y Estado del Bienestar Democrático. In Damián González Madrid & Manuel Ortiz Heras (Eds.), *El Estado del Bienestar Entre el Franquismo y la Transición* (pp. 21–48). Sílex ediciones. https://digital.csic.es/handle/10261/219319

Naredo, M., Carpintero, Ó., & Marcos, C. (2008). *Patrimonio Inmobiliario y Balance National de la Economía Española (1995 2007)*. Fundación de las Cajas de Ahorros.

Nasarre-Aznar, S. (2002). The Funding of the Mortgage Loans in Spain by the Issue of Mortgage Securities. Their Legal Structure. *Housing Construction: An Interdisciplinary Task*, Vol. I, XXX IAHS World Congress on Housing, pp. 347–354.

Obra Social Barcelona. (2017). ¡*La Vivienda Para Quien la Habita! Informe Sobre Okcupación de Vivienda Vacía en Catalunya*. https://obrasocialbcn.net/presentacio-informe-ockupacio-obra-social-bcn/

Observatori DESC. (2018). *Los Desahucios del 2008–2017: Una Vulneración Grave de los Derechos Humanos que no se Detiene*. https://observatoridesc.org/es/node/4320.

Palomera, J. (2018). Els Sindicats de Llogaters i la Lluita per L'habitatge en el nou Cicle de Financiarització. *Papers: Regió Metropolitana de Barcelona: Territori, Estratègies, Planejament, 60*, 156–163. https://raco.cat/index.php/PapersIERMB/article/view/339248

Pareja Eastaway, M., & San Martin, I. (1999). General Trends in Financing Social Housing in Spain. *Urban Studies, 36*(4), 699–714. https://doi.org/10.1080/0042098993411

Pareja Eastaway, M., & San Marín Varo, I. (2002). The Tenure Imbalance in Spain: The Need for Social Housing Policy. *Urban Studies, 39*(2), 283–295. https://doi.org/10.1080/0042098012010297

Pareja Eastaway, M., & Sánchez-Martínez, T. (2017). More Social Housing? A Critical Analysis of Social Housing Provision in Spain. *Critical Housing Analysis*, *4*(1), 124–131. https://doi.org/10.13060/23362839.2017.4.1.331

Pou, V. 2007. *En Profundidad: El Impacto del Mercado Inmobiliario en las Finanzas Públicas*. Madrid: BBVA https://www.bbvaresearch.com/KETD/fbin/mult/SIES_0706_SituacionInmobiliaria_21_tcm346-176831.pdf.

Provivienda. (2020). ¿Se *Alquila? Racismo y Xenofobia en el Mercado del Alquiler*. https://www.provivienda.org/informe-se-alquila-racismo-xenofobia-mercado-alquiler/

Puig Gómez, A. (2011). El Modelo Productivo Español en el Período Expansivo de 1997–2007: Insostenibilidad y Ausencia de Politicas de Cambio. *Revista de Economía Crítica, 12*, 64–81.

Roca Cladera, J., & Burns, M. C. (2000). The Liberalization of the Land Market in Spain: The 1998 Reform of Urban Planning Legislation. *European Planning Studies*, *8*(5), 547–564. https://doi.org/10.1080/713666428

Rodríguez Alonso, R. (2004). Infrautilización del Parque de Viviendas en España: Aparición de Viviendas Vacías y Secundarias. *Boletín CF+S, 29/30*, 131–177.

Rodríguez, J., & Larios, G. (2022). Les Barraques que mai van Desaparèixer. *La Directa*, *538*, 4–7.
Romanos, E. (2014). Evictions, Petitions and Escraches: Contentious Housing in Austerity Spain, *Social Movement Studies*, *13*(2), 296–302,
Sindicat de Llogateres. (2022). *La Regulació dels Lloguers, Punt per Punt*. https://sindicatdellogateres.org/guia-regulacio/
Soederberg, S. (2014). *Debtfare States and the Poverty Industry: Money, Discipline and the Surplus Population*. Routledge.
Tatjer, M. (2005). La Vivienda Obrera en España de los Siglos XIX y XX: De la Promoción Privada a la Promoción Pública (1853–1975). *Scripta Nova*, *IX*(194 (23)). http://www.ub.edu/geocrit/sn/sn-194-23.htm
Trilla, C. (2009). La Llei del Dret a L'habitatge: Aspectes més Destacats. *Revista Catalana de Dret Públic*, *38*, 97–117.
Trilla, C. (2014a). Desigualdad y Vivienda. *ACE: Architecture, City and Environment = Arquitectura, Ciudad y Entorno*, *9*(26), 95–126. https://doi.org/10.5821/ace.9.26.3685.
Trilla, C. (2014b). Habitatge i Estat del Benestar a Catalunya. *Revista Econòmica de Catalunya*, *69*, 109–124.
Vargas Martín, J. (2022). El Banco malo, o la Oportunidad Perdida de Tener un Parque de Vivienda Pública. *Público.es*, 19 January.

11
HOUSING AND WELFARE IN RUSSIA AND THE FORMER SOVIET UNION

Alexander Kalyukin and Alexander Puzanov

The current state of housing in the region

The housing question has always been topical in Russia. Fyodor Dostoevsky depicted the claustrophobic, overcrowded rental apartments in pre-Revolutionary St. Petersburg in his *Crime and Punishment* and *Poor Folk*; Mikhail Bulgakov critiqued the way of life and manners of the residents in the early Soviet shared communal flats in his *The Master and Margarita* and *Heart of a Dog*; Dmitri Shostakovich's operetta *Moscow, Cheryomushki* satirically portrayed housewarming in a newly commissioned prefab post-war apartment house. For most of the 20th century, Soviet Russia had struggled to solve the housing issue under the auspices of a centrally planned economy. Until 1991, the state (municipalities and large enterprises) took on the prominent role in financing, building, and allocating housing as private property was largely nationalized. Despite the Soviet postwar mass housing programme towering above European competition in both absolute and relative numbers at the end of the 1950s and the beginning of the 1960s (see Smith, 2010, pp. 103–104), it still fell short of universally improving the housing conditions of citizens. After the fall of state socialism and the systemic switch to neoliberal capitalism, the role of the state has been significantly diminished. The large-scale giveaway privatization of public housing engendered a significant class of asset-rich (yet cash-poor) households who were nominally delegated responsibility for housing provision on their own. While the state now takes on the role of promoting market-based solutions to the issue, the housing question still stands.

Over the past 30 years since the dissolution of the Soviet Union, Russia has developed a functioning real estate and construction sector in a market economy. According to Rosstat,[1] the housing stock of the Russian Federation at the end of 2021 was 4 billion square metres of total floor area,[2] which corresponds to an average of 27.8 square metres per person (27.6 and 28.3 square metres in urban and rural areas, respectively). The number of housing units has been greater than the number of households since the mid-2000s; this ratio is currently 1.04. The number of housing units per 1000 people was about 460 in 2021, which matches the European average. Such a level of housing provision in Russia is at a historical high and is associated with a relatively high volume of housing construction against the backdrop of low population growth and protracted retirement of dilapidated and obsolete housing during the last 60 years. Previously, the average floor area per person was 19.2 square metres in 2000, 13.4 in 1980, and only 8.8 in urban areas in 1960.[3] It should be

DOI: 10.1201/9781003212690-13

noted, however, that the Rosstat annual statistics accounts for all housing units including vacant or abandoned housing stock which determines somewhat high numbers of housing provision. The 2020 Census[4] data demonstrate that the average floor area in occupied dwellings was only 19 square metres per person while the ratio of the number of occupied housing units to the number of private households was only 0.92.[5] Conducted using a different methodology, the annual Rosstat survey of household budgets shows that the average floor area in occupied dwellings was 23.4 square metres per person in 2021.

Despite the quantitative improvements over the past decades the housing problem, and particularly the housing shortage, remains. The annual Rosstat survey of household budgets records that in recent years the number of rooms in the occupied housing stock is only approaching the population size – in 2021 there were 1,024 people per 1,000 rooms.[6] Second, the quality of the housing stock (the level of wear and tear) and the provision of basic utilities remains a serious problem. The share of dilapidated and obsolete housing (buildings with at least two-thirds of wear and tear) was 11.3 per cent of total housing stock in 2021 and even higher at almost 13 per cent considering apartment houses. Third, housing consumption is very socially differentiated, and these differences tend to grow. In Russia, the accounting norm of floor area per person (that is, the minimum level at which an individual qualifies as in need of housing) is, on average, 13 square metres, although this is set by local authorities and hence may vary. In 2021, 10.9 per cent of all households had less than 13 square metres per person; in the lower quintile of the distribution by income the share of needy households was 25.9 per cent, among the households with children under 16 years old the share was 28 per cent, among young families – 17.1 per cent, and among families with more than three children – 54.2 per cent. These categories of citizens were the most affected by the housing shortage. Those who enjoy a relatively high level of housing consumption are not only the well-off households (in the upper quintile of the income distribution, the average floor area per person is more than 30 square metres and only 2.6 per cent of households have less than 13 square metres per person) but also the families of non-working pensioners (39.4 square metres and 0.3 per cent, respectively).

The differences in housing provision in urban and rural areas are insignificant and have a levelling tendency: the rural housing provision exceeded the urban by 8.6 per cent in 2010 and only by 2.5 per cent in 2021. This is a consequence of the continued concentration of housing construction in urban areas. It should be noted that the quality and availability of housing utilities in urban areas are significantly higher than in the countryside.

Along with an objective assessment, it is important to also consider the subjective evaluation of the housing shortage. According to the 2020 Rosstat comprehensive survey of living standards, 23.8 per cent of Russians were not satisfied with their housing conditions (due to cramped living and/or poorly maintained housing stock), while only 13.9 per cent intended to improve their housing conditions in the foreseeable future.

The tenure structure of the Russian housing stock should be analyzed in relation to the ownership structure. According to Rosstat, as of 2021, the vast majority (91.5 per cent) of the Russian housing stock (floor area) was owned by private individuals, of which nearly 10 per cent are estimated to be rented out mainly on an informal basis (that is, without paying the required taxes and fees) by individual 'non-professional' landlords (see Puzanov, 2018, p. 299). The balance of 2.3 per cent of private housing stock is owned by legal entities, mostly commercial enterprises and corporations. Most of them let their housing assets on non-market terms to attract workers, and as such this can be regarded as quasi-social rental housing (despite legally using commercial rental agreements). Noteworthy is the absence of large international corporations operating on the Russian rental housing market. Housing cooperatives were legally converted into condominiums during the market transition, and the new 'cooperative ownership' is both negligible and provisional.[7] The state and municipalities retain ownership of 6.2 per cent of all housing stock, most of it being social and "special purpose" (dormitories and so on) rental housing, although public authorities may let it under commercial terms. Therefore, the total share of owner-occupied housing may be assessed at 81.5 per cent, while the rental housing stock makes up the remaining 18.5 per cent (see Table 11.1).

The contemporary tenure structure in Russia has been largely determined by the mass giveaway privatization of public (state and municipal) housing following the demise of the Soviet Union.[8] While the prevalence of homeownership may imply a high degree of household responsibility, in reality, the homeowners do not bear the full burden of property maintenance. For example, capital repairs of apartment houses are heavily subsidized even though homeowners are required to contribute to the regional repairs' funds (no less than 5 per cent of total costs are to be covered by the owners), whereas those living in dilapidated and obsolete housing subject to demolition are entitled to new residential property free of charge which is provided and paid for by the regions supported with subsidies from a specialized federal fund. Consequently, the differences in housing-related expenses between an owner and a tenant are insignificant.

The active expansion of housing commissioning and mortgage lending since 2005 significantly increased the affordability of housing in the market: the share of families who can buy housing with their own and borrowed funds has gone up from 16.8 per cent in 2005 to more than 55 per cent in 2021 (The Institute for Urban Economics, 2022). However, a significant proportion of households do not have and in the foreseeable future will not have the opportunity to improve their housing conditions without state support.

The social housing sector in Russia is rather small and based on the non-privatized portion of Soviet-era public housing. The share of state-owned stock is 2.3 per cent and municipal stock is 3.9 per cent as of 2021. Rents are heavily subsidized and are well below market levels while utility bills, although paid for by tenants, are set at levels below the economic feasibility which means that any investment activity in this sector is possible only at the expense of budget funds. In Russia, the local authorities are responsible for keeping the list of families in need of improving their housing conditions, either through the provision of social housing or through assisted procurement of ownership. A queue for social housing dating back to the Soviet times includes today about 2 million families, or 4 per cent of their total number, according to Rosstat. Since 2005, the Housing Code of Russia stipulates that only low-income citizens (that is, with incomes lower than the regional threshold, usually 1.5–2 subsistence levels) may join this queue (previously and during Soviet times, all those in need of improving their housing conditions could be qualified), although due to the numerous exceptions to this rule, this category makes up less than a half of those newly registered. Annually, only 4 per cent of families on the waiting list are

Table 11.1 Historical shares of housing tenures in Russia, per cent of total floor area (*urban)

	*1926**	*1940**	*1960**	*1980*	*1990*	*2000*	*2010*	*2021*
Home/personal ownership	48.7	31	32	30.6	22	54.2	73.2	81.5
Cooperative ownership	0.9	0	n/d	3.4	6	n/d	n/d	n/d
Private (non-state) rented. Including subletting	n/d	n/d	n/d	3	5	11.1	12.4	12.3
Public/social rented	50.4	69	68	63	67	34.7	14.4	6.2

Source: Authors' calculations based on the 1926 Census, Goskomstat RSFSR, and Rosstat data.[15]

provided with accommodation and their housing conditions are improved (during the Soviet times the share was over 10 per cent). While 92 thousand families were provided with accommodation and their housing conditions were improved in 2021, not all of them were indigent and obtained accommodation on the terms of social tenancy. The average waiting period in the social housing queue is about 25 years.

The volume of new social housing stock construction is small and tends to decrease, according to Rosstat: in 2010, it amounted to only 1.4 million of 4.1 million square metres (34.1 per cent) of newly built housing provided to families that were listed as in need of improving their housing conditions, and in 2021 to only 0.3 millions of 1.4 million square metres (21.4 per cent) of such housing.[9] Therefore, most of the households on the waiting list improve their housing conditions by obtaining ownership on the market with the help of budget subsidies, preferential housing lending programmes, the purchase of housing in instalments, and other programmes, including the use of their own savings.

New housing construction is mainly designed for sale into the ownership of individuals and is financed by households themselves, including the construction of apartment buildings by means of equity participation. This imposes significant commercial risks on prospective homebuyers and is a source of serious tensions in the real estate market (there also have been cases of fraudulent developers) requiring periodic government intervention. Currently, there is a transition to the system when the funds of future homebuyers are blocked in escrow accounts in banks that carry out financing of construction projects.

A brief overview of the historical development of the housing regime

The historical evolution of the housing regime in contemporary Russia has been determined by the legacy of the 70-year period of state socialism, a decade of socio-economic turbulence following the demise of the Soviet Union in 1991 and the transition to a market economy, as well as the economic rebound in the 2000s. The following presents a brief account of the key historical periods in the development of Russian housing.

Despite the acute housing shortage that the Soviet Union inherited from tsarist Russia (and that was only aggravated by two world wars and a civil war), under state socialism and its centrally planned economy housing was given a relatively low priority compared to productive sectors such as industry. The nationalization/municipalization and redistribution of housing after the 1917 October Revolution only slightly improved the living conditions (private ownership of urban land and homeownership in cities larger than 10,000 inhabitants was abolished by decree in 1918), but the unfolding interwar industrialization and urbanization exacerbated them further. According to the 1926 Census, the state and municipalities owned more than half of urban housing stock (which included nationalized formerly private housing – see Table 11.1) while the average urban floor area per person was only 5.9 square metres. The share of personal housing[10] in urban areas was still relatively high in the inter- and post-war period: 48.7 per cent of total housing stock (floor area) in 1926, 30.5 per cent in 1940, and 29 per cent in 1950 (Kalyukin & Kohl, 2020, p. 1775). This was associated with large volumes of self-building: the share of individual construction was around 40 per cent of the total in the 1930s and over 50 per cent in the 1940s (see Figure 11.1). In fact, after the Second World War, individual housing construction was encouraged by the fourth five-year plan and by accessible state loans and allocation of plots of land, particularly in medium-sized and small cities, in order to contribute to post-war reconstruction and to alleviate the housing shortage (Andrusz, 1984). The overall housing commissioning volumes remained low until the mid-1950s as housing provision was mainly the concern of households themselves. State housing construction

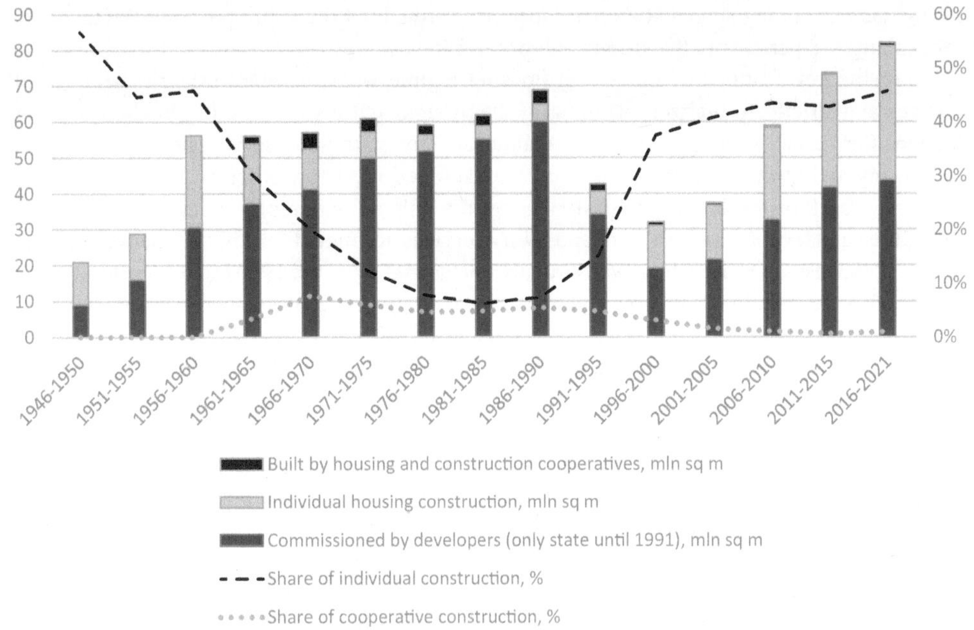

Figure 11.1 Commissioning of housing in Russia by developer type, year averages (left scale), and the share of individual and cooperative construction (right scale).

Source: Authors' calculations based on Rosstat data.

served primarily ministerial and industrial interests (DiMaio, 1974), which focused on accommodating the elites (such as political nomenklatura, cultural intelligentsia, and specialists – as a part of the Soviet reward system), law enforcement agencies, as well as attracting workforce, inter alia, to new cities created for industrial production or mineral extraction.

It was not until 1956 that housing became a political issue when Khrushchev famously proclaimed an ambition to procure every family with a separate apartment by 1980. After that, housing construction surged: yearly averages of total housing commissioning almost doubled in 1956–1960 as compared to the early 1950s (see Figure 11.1). In this, the state (represented by large state enterprises, ministries and local soviets all of which managed their own housing provision and allocation lists) played a key role becoming not only the main actor within the housing regime but also the main investor. The share of state and municipal enterprises and organizations in total housing commissioning exceeded 85 per cent by the end of the Soviet period. The post-war Soviet housing system pursued an egalitarian goal of raising the living standards of the entire population by means of mass housing construction in the form of apartment houses (Harris, 2013; Varga-Harris, 2015).[11] This large-scale programme marginalized self-building: it was banned in cities with populations over 100,000 in 1963, and its share in all housing construction gradually fell to about 6 per cent by the end of the Soviet period. Consequently, the share of personal housing in total stock decreased to approximately 30 per cent by 1980 and to an all-time minimum of 22 per cent by 1990 (see Table 11.1). Meanwhile, new actors such as housing and construction cooperatives and collective farms appeared. Cooperative housing, first permitted in 1924 and banned in 1937, was reintroduced in 1958 and tended to cater for the white-collar middle class who were poorly placed in the social housing queue (Andrusz, 1992). The share of cooperatives in total housing commissioning stayed at around 5 per cent throughout the entire post-war Soviet period after peaking at 8 per cent in the second half of the 1960s.

One of the key features of the Soviet housing regime was the centralized allocation of all resources and strictly formalized planning of the volume and distribution of new housing construction. Furthermore, management and maintenance of state housing was heavily subsidized (the housing and utility services sector, however, suffered from low efficiency and poor service quality as a result of residual funding). Rents covered only a fraction of total operating costs and none of the capital costs. In 1990, rent and utility services accounted for only 2.5 per cent of total household expenses, a share which was virtually unchanged since the 1920s (Goskomstat RSFSR, 1991, p. 133). Tenants in public housing enjoyed extremely strong occupancy rights which could be inherited, bequeathed to relatives or even exchanged with other holders of user rights. Such security connected with tenancy made it a de facto form of individual ownership (Smith, 2010), or quasi-homeownership (Stephens et al., 2015). The distribution of housing through waiting lists kept by municipal authorities and state enterprises was the predominant means of housing provision; in 1990, 26 per cent of all urban households were on waiting lists (Goskomstat RSFSR, 1991, p. 213). In addition to a large public housing sector, private rental housing (subletting) was regarded as a necessary evil performing a useful function of mitigating housing shortage and supporting labour mobility. Subletting was permitted both in public rental units and personally owned housing, although the authorities rarely had the capacity or the political will (for ideological reasons) to regulate and institutionalize this submarket. As a result, subletting was largely informal and no official data are available for this tenure form, although some estimates suggest that it housed 5–10 per cent of the urban population in the Soviet Union (Hegedüs & Puzanov, 2018).

At the end of the Soviet period, Russia's housing stock was characterized by the dominance of the state. Nationally, 67 per cent of the stock was state-owned (that is, managed by either the municipalities or state enterprises and by government agencies) – see Table 11.1. In urban areas, 79 per cent, and in larger cities as much as 90 per cent of housing was owned by the state. The dominance of state ownership at the advent of the market economy in Russia's housing sector dwarfed the corresponding role of the state in Eastern European countries which began housing reform with a substantial private sector (Kosareva & Struyk, 1993, pp. 82–83). Approximately 6 per cent of the housing stock (and up to 10 per cent in larger cities) was made up of housing cooperatives that received considerable state support, including access to cheap bank loans.[12]

After the fall of the Soviet Union in 1991, Russia transitioned from the state-socialist plan-driven system of housing relations to market relations premised on private homeownership rights, in the capitalist sense. The privatization of housing began during perestroika in 1988 with the transfer of fully paid-off cooperative units into ownership of their residents and was extended in 1992 to state and municipal units.[13] This free-of-charge and unconditional privatization to the benefit of sitting tenants (albeit voluntary and reversible), in principle, formalized the housing relations that were de facto previously available to tenants in socialist public housing with its strong and transferrable/exchangeable occupancy rights. Also, since tenants were able to decide whether they wanted to privatize their flats, this resulted in a situation where almost every apartment building in Russia contains units in both private and public ownership (Puzanov, 2013, p. 227). Among those who became owners through privatization many were low-income households unable to bear the full cost of homeownership. Moreover, the privatization in favour of sitting tenants has entrenched and exacerbated Soviet-era inequalities favouring better-off households (Yemtsov, 2007). As a result of the mass giveaway privatization, the share of private housing stock has radically increased while the share of the state and municipal housing stock has plummeted. Similarly, in terms of tenure structure, the share of owner-occupied housing has dramatically increased since 1991 as the share of public/social rental housing decreased tenfold (see Table 11.1).

The volume of total housing commissioning declined from 49 million square metres in 1991 to just 30.3 million in 2000; it surpassed the peak Soviet output of 1987 (72.8 million) in the year 2014 (more than 84 million) and currently peaks at almost 93 million in 2021 (see Figure 11.1). The structure of housing commissioning has also undergone drastic changes: according to Rosstat, the share of housing production by state and municipal enterprises and organizations fell from 80.5 per cent in 1990 down to 20.8 per cent in 2000, to 9.4 per cent in 2010, and to less than 3 per cent in 2021. The share of private organizations (developers) expanded from 22.1 per cent in 2000 to 41.7 per cent in 2021. Simultaneously, the share of individual housing construction (that is, financed by personal and borrowed funds) increased from 9.1 per cent in 1990 to 41.6 per cent in 2000, to 43.7 per cent in 2010, and to 53 per cent in 2021.[14]

After 1991, the Russian government has considered rental housing as a residual segment of housing policy; most of its efforts have been directed at supporting the homeownership. Ultimately, public (state and municipal) rental housing in modern Russia emerged out of the leftovers from non-privatized state or municipal housing stock and is meant to supply accommodation only to the neediest households under social tenancy contracts (Puzanov, 2013). Its share is steadily diminishing, dropping to 6.2 per cent in 2021 (see Table 11.1). The private rental sector survived the transition and increased in size but continues to provide only temporary housing solutions to tenants. As was mentioned above, apart from homeowners subletting their housing, this sector includes also the quasi-social rental housing of private legal entities such as enterprises and corporations, which together amount to 12.3 per cent of total housing stock in 2021 (see Table 11.1). As a result, Russia experiences a high demand for rental apartments, coinciding with a pronounced scarcity of legal rental housing supply. Both legalizing and institutionalizing informal renting and subletting (while creating favourable conditions for the market-based provision of rental housing) are being considered, but with limited results so far (Peppercorn & Taffin, 2013).

During the first post-Soviet decade, the state capital investments in housing construction and subsidies for housing and utilities were drastically reduced, while simultaneously a programme for housing and utility subsidies for low-income families was implemented. Since 1999, there has been an increase in government spending on housing support for citizens, including programmes for the resettlement from the dilapidated and obsolete housing stock, provision of housing at the expense of the federal budget to military personnel and, to a limited extent, some other categories of citizens. The 2005 Housing Code categorized the public obligations by different levels of public authority (federal, regional, municipal) and defined categories of citizens eligible for housing support. In the same year, fundamental changes occurred in the system of housing finance: an increasing portion of construction has been financed by private developers while the state budget no longer plays an important role. Construction of multi-family dwellings involves today a broad use of private investment, including household savings and mortgage loans. The Russian residential mortgage debt as a share of GDP increased from 0.1 in 2005 to 10.1 in 2021 (EMF-ECBC, 2022, p. 156). These relatively low numbers of indebtedness could be explained by cultural resistance to debt and the established perception of housing as a fundamental social right that should be guaranteed by the state, hence creating a unique system of "property without markets" in which most housing has been privatized but never fully commodified (Zavisca, 2012). The share of transactions in the primary and secondary real estate market (their ratio being 20/80) with the use of housing mortgage loans has increased from only 3.6 per cent in 2005 to more than 50 per cent nowadays. The expansion of mortgage lending since 2005 significantly increased the affordability of housing: if in 2004, only 9 per cent of households could buy housing with their own and borrowed funds, in 2005, their share jumped to 16.8 and exceeded 55 per cent by 2020 (The Institute for Urban Economics, 2022). Since 2014, an affordable rental housing market and a non-profit housing fund (in parallel with public/social housing) for low-income citizens have been developed. The preferential mortgage programmes have aimed to support certain categories of citizens who need to improve their housing conditions but do not have an opportunity to accumulate sufficient funds to purchase housing. Since 2019, support measures for individual housing construction have been also introduced.

The welfare regime and its development

The welfare regime in Russia has undergone a series of dramatic transformations over the course of the 20th century. During tsarist times, it was state- and privately funded charities, churches and private organizations that took on the role of welfare institutions in helping the destitute populations. The main sources of social security for workers during that time were mutual assistance funds, and the first social insurance laws were passed in 1912. After the 1917 October Revolution, the establishment of the modern social welfare system began. Welfare was seen as a comprehensive social institution premised on the state monopoly on social services and provision. After the demise of the Soviet Union in 1991, the state struggled to fulfil its welfare commitments as much of the population experienced a drastic decline in living standards. At the same time, the welfare system was liberalized with the admission of private actors (although the state retained a key position) which decreased the role of social provision and elevated the role of social security.

The Soviet Union was an ostensibly communist state and as such ideologically opposite to a capitalist one. While Esping-Andersen's (1990) seminal typology of welfare regimes concerns capitalist states, some researchers (for example, Sidorina, 2005; Vinogradova, 2004) argued that the Soviet Union's comprehensive institutional system based upon universalist and egalitarian principles resembled the social democratic model yet also exhibited some features of the corporatist model regarding the reliance on employment-based welfare provision (via enterprises/firms) and the status- and loyalty-based distribution of social benefits. Others (for example, Lapina, 2007) maintained that the Soviet welfare institutes and practices were entirely different from the capitalist system. Indeed, if under capitalism welfare systems seek to reduce inequalities in the labour market, under state socialism – which aims at full employment, maintains low and undifferentiated wages, and subsequently stimulates high female employment to achieve acceptable family living standards (Kornai, 1992) – the welfare regime rather rewarded individual labour contributions, discipline and political loyalty through privileges and material recognition, including plots of land, access to better quality apartments and summer houses *(dachas)*.[16] Therefore, within the Soviet welfare regime, the role of employment was paramount while the reward system from employment was paradoxically a source of certain social differentiation, even though the overall equalization tendencies were still stronger.

The Soviet state sought to institutionalize welfare, rejecting the pre-Revolutionary view of it as a "residual" sporadic activity only brought up to fix social problems once the market and the family failed to do so (Madison, 1964, p. 191). The welfare provision was universal and comprehensive in its scope, managed mainly by the central state and, since 1957, also by the local trade union committees allowing for a certain decentralization of decision-making. As a result of the state monopoly on social services provision, private and/or voluntary agencies were absent.

In 1919, contribution-free insurance was introduced with the establishment of a nationwide social security fund. However, under Stalin, social security benefits were considered rather a reward that was bestowed on the most productive workers. Under Khrushchev, social security coverage was expanded aiming to eliminate the inequalities in benefits, financed by designated government funds and employer contributions. By the 1970s, the Soviet Union achieved the development of free healthcare and education, introduced disability and war veterans' benefits (who particularly enjoyed priority for housing provision), as well as a gradually expanding pension provision (reforms of 1956 and 1964) and maternity and child welfare instruments (maternity leave, daycare services). Importantly, the provision of healthcare, education, and daycare was tied with residency registration (*propiska*), which elevated the role of housing (Smith, 2010), although healthcare could also be linked with the workplace.

Despite the universalist character of the Soviet welfare system, in terms of allowances and benefits, it was industrial employees who enjoyed broad social security coverage (with the exception of unemployment benefits), while collective farmers until the mid-1960s had to rely on mutual aid funds (Madison, 1964, p. 193). Therefore, although the Soviet regime was committed to promoting distributive equality, justice, and solidarity – through guaranteed full and secure employment, state-controlled and heavily subsidized prices for essential goods (including housing-related expenditures), free healthcare and education, and egalitarian wage policies – the welfare provision was nevertheless biased towards industrial workers (Cook, 1993). In this, the role of employment and affiliation with specific enterprises/firms was paramount. Those unwilling or unable to work had to rely on help from their families.

The backbone of the Soviet welfare system had been the social consumption funds established in the 1960s. Their main function was to ensure state control over expenditures and consumption of the population aimed at levelling the socio-economic inequalities by combining both the socialist principle of distribution (according to labour, such as pensions and unemployment benefits) and the communist principle (according to needs, such as healthcare and education). These funds ensured free and equal access to education and healthcare, financed pensions, social benefits, and holiday pay, as well as covered maintenance of the housing stock. Approximately 40 per cent of social consumption funds were financed directly from the state budget (through indirect and revenue taxes), this share gradually decreasing since the 1970s, while the share of state social insurance budget (funded by payroll deductions) gradually increased to more than 28 per cent, and the share of funds of enterprises and organizations had been stable at around 20 per cent (Ivanova, 2011, p. 194). While the state-financed funds covered the entire population, the latter two funds were financed from the enterprises' profits and distributed to the employees, therefore, stimulating a collective material interest in the development of production. The bulk of payments and services was legally fixed and subject to annual increases and as such independent of changes in the economic situation (Ivanova, 2011, p. 227). Social consumption funds were meant to promote an ideology of equitably regulated 'rational consumption' since the formation of individual and social needs of the population outperformed the economic possibilities of their satisfaction (Ivanova, 2011, p. 188). These funds were thus a crucial part of a systematic policy of consumer price subsidies that held down the cost of living to an extraordinary degree.

The social consumption funds were distributed, in equal parts, through cash payments (including pensions, benefits, vacation pays, and stipends) and free benefits and services (education and healthcare, as well as virtually free housing and utility services). At the end of the Soviet period, educational, medical, housing and other social benefits constituted around 30 per cent of the average income of workers and employees, the rest being wages (Goskomstat RSFSR, 1991, p. 123). Throughout the 1980s, total expenditures on social welfare programmes increased at a rate of 7–8 per cent a year. In 1989, pensions accounted for more than 70 per cent of all social security and insurance payments which, in turn, comprised about 43 per cent of total expenditures of all social consumption funds combined. Education accounted for 23 per cent, healthcare for more than 15 per cent, and housing stock maintenance for 7.2 per cent of all welfare expenditures in 1989. Total welfare payments from all these funds accounted for 27 per cent of the national income in 1989, of which social security and insurance payments and benefits accounted for more than 11 per cent (Goskomstat RSFSR, 1990, pp. 10, 139).

The fall of the Soviet Union in 1991 launched a series of negative socioeconomic processes such as a decrease in real incomes, surging inflation, and shortage of consumer goods. After the relatively stable late Soviet period, industrial workers and middle classes experienced impoverishment and a decline in living standards. Therefore, there was an urgent need to introduce new emergency instruments of social protection such as a voucher system, compensation payments, and in-kind assistance to families and children. The 1993 Constitution stipulated that Russia was a social state aimed at creating conditions that ensured a decent life for all its citizens. Accordingly, the government pursued the indexation of salaries, pensions, and benefits – however, unsuccessfully both because it was incommensurable with the rates of inflation and contradicted the real economic capabilities of the state as well as the IMF's demands to cut social spending. As a result, the government was faced with a budgetary need to roll back welfare expenditures, although the social need was rather for their expansion (Szelényi & Mihályi, 2019, p. 92). Importantly, the notions of subsistence level and minimum wage introduced already in the 1960s (and enshrined in federal laws in 1997 and 2000, respectively) still defined welfare provision levels; the former serves as the poverty threshold, while the latter determines the unemployment, maternity and other social insurance benefits. Holding down the minimum wage (in fact, well below the official subsistence level of income) became a means of holding down social expenditures (Standing, 1996, p. 234). In Russia, the minimum wage equalled subsistence level only in 2018, and was constitutionally fixed to stay higher than subsistence level in 2020.[17]

The ideological switch from state socialism to neoliberal capitalism coupled with the economic crisis of the early 1990s paved the way for reforming the system of welfare provision. Welfare reforms have been devised to make access to benefits based on individual citizenship rather than group/collective membership (such as industrial enterprise affiliation). Simultaneously, state commitments were reduced to entitlements as the priority had switched from a universal provision to an individual-targeted categorical and – to a lesser extent – means-tested social assistance. The evolution of the welfare system under the market economy involved a further responsibilization of citizens[18] and the development of social insurance which included compulsory and voluntary services concerning pensions and medical care managed by dedicated funds. The state social insurance fund finances temporary working disability benefits, maternity and child allowances, as well as occupational disease and accident benefits for workers and members of their families. Therefore, the welfare reforms have involved a shift from direct funding of welfare by the government to financing through a system of insurance funds (which nevertheless assumes a certain degree of continuity with the Soviet system). This was meant to standardize and neutralize the welfare system, as well as ensure the transparency of budgetary flows. Even the giveaway privatization of public housing could be regarded as a welfare policy in which the population was endowed in kind with real estate assets.

Political reforms of the first post-Soviet decade allowed for supplementary medical insurance (1991 and 1993), charities (1995), and provision of social services by NGOs (1996) which may apply for public support and participate in public procurements, as well as supplementary private pension insurance (1998) to complement the safety net provided by the state (although currently less than 1 per cent of all pension provision). For instance, the state-funded compulsory healthcare insurance, covering most of the basic services and even acute dental care, could be complemented with private insurance in order to receive quicker and specialized medical help. The existence of private alternatives as well as the reduced promises by the state contributed to the shortening of queues for social goods and services, although they remained prominent regarding public housing and municipal daycare places. Simultaneously, the social welfare system continued to give high priority to old-age pensioners and veterans, retaining the focus on appraising the past patriotic (military and working) service and considering their active participation in electoral campaigns. In 1995, a federal law guaranteeing extensive social benefits to veterans was passed; it involved free-of-charge transport benefits, supplementary medical services, and housing and utility discounts.

Despite the strong economic recovery after the turbulent 1990s and the financial crisis of 1998, the restructuring of the welfare state continued as only the broadest and most visible benefit cuts were avoided. The reduction of social and other subsidies was regarded as being key to fiscal stabilization, completing market transformation, and modernization of the economy (Cook, 2007). For instance, a series of pension reforms followed. The reform of 2001 regulated pensions based on employee contributions into a pension plan (22 per cent of salary up until a specific limit, after which the amount is reduced to 10 per cent) while also guaranteeing benefits for those unable to work. The contemporary pension in Russia comprises three parts: social part (basic pension paid to unemployable citizens), insurance part (wage-based), and funded part (formed with insurance contributions of employers and return on their investment, temporarily frozen since 2015). In 2018, the government decided to gradually increase the normal retirement age from age 60 to 65 for men and from age 55 to 60 for women. In case a retired non-working individual's total income is lower than the subsistence level of a pensioner, supplementary social assistance in cash is paid up to that level.[19]

In 2005, seeking to increase the transparency of budgetary flows, the Russian government carried out the benefit monetization reform which replaced some in-kind social benefits (such as the free supply of medicines, free public transport, and free health resort services for veterans) with in-cash payments. Initially, cash compensations did not account for the inflation which triggered mass protests across the country and rendered the reform widely unpopular. Also, since 2007, the state introduced several policies in response to demographic challenges (declining birth rates and high death rates). Increases in state-funded benefits to mothers and children were meant to compensate for women's lost earnings. Moreover, an expansive pronatalist policy was passed which included a large one-off payment – the maternity (family) capital – for the birth of a child (until 2020 only for the second and subsequent ones), as well as subsidies for daycare costs, increases in maternity leave benefits, and programmes to aid prenatal and neonatal health, orphans and adoptions.

Just like during the Soviet period, the family remains an institution of social welfare in Russia, although its role in welfare provision can only be approximated based on secondary sources. At the beginning of the transition in the 1990s, when the direct role of the state in welfare provision was reduced, the role of kinship became more important, especially in relation to childcare and elderly care (Szelényi & Mihályi, 2019, p. 36). Also, in the case of Russia, *dachas* and plots of land in the countryside have served not only recreational purposes but have been crucial for households' self-provisioning of fruits and vegetables, particularly in the 1990s (Lovell, 2003). Like in many other post-socialist countries in Central and Eastern Europe, the household in Russia has been a source of intergenerational support and self-help. One indicator that approximates the level of these transfers is the number of people partially or fully dependent on support from others: their share declined somewhat from 26.4 per cent according to the 1989 Census to 25.3 per cent according to the 2020 Census. Another indicator highlighting the role of the family is the share of multi-generational households (married couples and/or single parent(s) living with their parents and/or relatives) which assumes intergenerational dependence. Based on data from the Censuses, the share of such households was 13.5 per cent in 1989 and 19.5 per cent in 2010; according to the Rosstat comprehensive survey of living standards, in 2018, the share of such households was 18.6 per cent (see calculations by Eliseeva, 2020, pp. 18–19). According to our estimates based on the 2020 Census data, the share of multi-generational households declined to 13.8 per cent which is comparable with the 1989 level, although it should be noted that the categorization of family types is rather ambiguous.

Let us demonstrate the scope of the Russian welfare regime. One indicator that shows the load on the social welfare system is the share of the population with wages below the official subsistence level. Since 1991, this share has fallen from more than 30 per cent to 11 per cent in 2021. The share of all welfare benefits in the income of the population in 2020 was 21.5 per cent, according to Rosstat. The share of social assistance and allowances to GDP was 3.9 per cent; the share of all welfare expenditures to GDP was 12.7 per cent. Subsidies and benefits for housing and utility payments comprised 2.8 per cent of all welfare expenditures (down from 3.6 per cent in 2015). Of all welfare payments in 2020, pensions accounted for more than 66 per cent which is comparable with the late Soviet levels (although current welfare payments, unlike the Soviet social consumption funds, account only for incomes and do not reflect, for instance, medical insurance contributions).

The contemporary Russian welfare system can be said to continue delivering a broad range of minimum services and benefits for the whole population, even though the state now shares some of its responsibilities with private actors. The contemporary welfare system in Russia, therefore, combines elements of Soviet-era legacies and market-based mechanisms of means-tested social security, and as such, represents a peculiar combination of elements of social-democratic, corporatist, and liberal models according to Esping-Andersen's (1990) typology.

The interrelation between housing and welfare

Housing and welfare formed a unique nexus in the Soviet Union: strong occupancy rights in public housing and the persistence of personal homeownership were reinforced by the welfare rights tethered to residency (Smith, 2010). Since the mid-1950s under the centrally planned economy, the role of the state in housing provision for the population was decisive. The 1977 Soviet Constitution stipulated the citizens' right to housing and the public control over its fair distribution. Citizens whose living space was under the sanitary norm of 9 square metres (the de facto threshold was even lower at 5 square metres) could join a municipal or enterprise housing waiting list, the latter offering less than half the waiting time compared to the former (Morton, 1980, p. 239). Housing provision was considered an important element of the social consumption funds expenditures, primarily the fund of socio-cultural activities and housing construction formed at the enterprises. Although under the centrally planned economy social consumption funds were considered an instrument of equalization of the level of citizens' well-being, contradictory trends in terms of housing provision were observed: on the one hand, the criterion of housing need was fundamental in determining the right to state housing support; on the other hand, access to additional volumes of housing consumption often "replaced" the possibility of obtaining money income, the differentiation of which was significantly limited under the planned economy. Such additional volumes of consumption were provided by the increased quality of housing provided to certain elite categories (first of all, the nomenklatura), and the right to so-called additional floor area which was taken into account when determining the size of housing provided to certain categories of citizens (employees of state, professional, party, cooperative and other public organizations, officers with the rank of colonel or higher, certain categories of scientists, artists and so on). Thus, the housing provision played an ambivalent role within the welfare regime. At the same time, quasi-market relations were also present in the sphere of housing in the form of subletting an apartment exchange with a shadow side payment through the so-called second or informal economy (Perevedentsev, 1990).

Consequently, housing consumption was stratified even during the Soviet period. According to Alexeev's estimates (1988), at the end of the 1980s in the USSR, the income elasticity of consumption of state-owned housing (considering informal shadow incomes) was estimated at 0.15 for Soviet Russia with other northern regions and 0.30 for southern Soviet republics. With the transition to a market economy, the level of stratification of housing consumption has increased. Already by 2004, the average value of the per capita income elasticity of housing consumption between quintiles increased to 0.19 by 2004 and reached 0.21 in 2021 (the highest value of the income elasticity of housing consumption was recorded in 2015 at 0.26). At the same time, the highest values of elasticity were recorded in the transition from the bottom quintile to the second one (0.35) and from the fourth quintile to the top one (0.27). In general, the differentiation in terms of housing consumption remains at a significantly lower level in comparison with the differentiation in terms of income. Based on Rosstat data, in 2021, while the income of households in the top quintile of the distribution in terms of average per capita income exceeds the income of households in the bottom quintile by 6.2 times, the difference in the level of housing consumption of such households is only 1.8 times.

The commodification of housing during the Soviet regime was low. According to some estimates (Kosareva et al., 2011, pp. 30–31), in 1989, 1.8 million families improved their living conditions (4.5 per cent of all families), while 59.1 per cent of all improvements were associated with the allocation of rental housing built by state enterprises, organizations, and collective farms. One-third of housing improvements, however, were associated with quasi-market relations in the sphere of housing exchange (see Table 11.2).

Table 11.2 Distribution of households who improved their living conditions

Indicator	*1989*		*2009*		*2021*	
	Number of families, 000s	*Share of families who improved housing conditions, %*	*Number of families, 000s*	*Share of families who improved housing conditions, %*	*Number of families, 000s*	*Share of families who improved housing conditions, %*
Total, of which:	1,812	100	1,991	100	3,743	100
Individual housing construction	84	4.6	207	10.4	374	10.0
Housing cooperative construction	66	3.6	7	0.4	13	0.4
State or municipal rental housing, kolkhoz rental housing	1,071	59.1	9	0.5	33	0.9
Market housing purchase, quasi-market housing exchanges	591	32.6	591	64.7	3,323	88.8
of which with public financial support	0	0	n/d	n/d	330[20]	8.8

Source: Authors' calculations based on Kosareva et al. (2011, pp. 30–31) and Rosstat data (2021).

With the transition to the market economy, housing has largely lost the characteristics of a public good and acquired the characteristics of a market commodity: in 2021, 89 per cent of all housing improvements were associated with the acquisition of housing ownership on the primary and secondary markets (see Table 11.2). The role of the state is to support the purchase of housing into private property for certain groups of citizens. Another important element of state policy has been the support of the real estate market, which led to the skewing of state support towards relatively wealthy groups of citizens by means of providing a property tax deduction for income tax on the purchase of housing and payment of interest on housing mortgage loans, as well as preferential mortgage programmes.

Even though it is the low-income families that are the focus of state housing support according to the Russian Constitution, the provision of social housing to this social group has in reality not become a real priority of housing policy. The provision of such families with social housing is attributed to the obligations of local budgets that are constantly decreasing. Besides that, the municipalities are not interested in investing in the construction of social housing which could be then privatized by the sitting tenants. The current situation also means that municipalities are unable to construct and maintain social rental housing as separate buildings, and instead social housing is represented by individual flats in mixed-ownership apartment houses (Puzanov, 2013, p. 228). In practice, welfare support is provided to citizens according to the categorical principle. Out of the approximately 150 categories entitled to housing support, some are distinguished according to the principle of social vulnerability and need for housing support (such as young families, including those who recently had a child, orphans, and the disabled), while others based on belonging to certain professions and having merits and distinctions for the patriotic (military and labour) service to the state. It should be noted that housing support for some categories of citizens considers their income and financial situation, but the share of such citizens is small, no more than 15 per cent.

Table 11.3 provides an estimate of the structure of public expenditures (and lost revenues) on housing support for citizens in 2017–2021, and as such reflects the real priorities of housing policy. The average annual amount of such expenditures, according to estimates, is about 1 trillion rubles (approximately EUR 12.5 billion), or 0.8 per cent of GDP. Some 42 per cent of public expenditures relate to housing support of separate categories of citizens allocated on various grounds, and only 2 per cent – to housing support based on the recognition of citizens as low-income. More than a quarter (28 per cent) of public expenses are related to using the maternity (family) capital for the purpose of improvement of housing conditions. This measure lies within the welfare framework and is not directly connected with the housing support of citizens, yet in practice, in 2021, 58 per cent of beneficiaries used this capital to improve their housing conditions, according to Rosstat (in the 2010s this share reached 90–95 per cent). Finally, 28 per cent of public expenditures are related to the support of the real estate market, which benefits families who can afford to buy housing without state support.

In terms of the role of the family in housing provision, the case of Russia is similar to other post-socialist countries of Central and Eastern Europe (Lux et al., 2022; Stephens et al., 2015). Here, the household has remained an important source of housing welfare through intergenerational support which is evidenced by the relatively high number of extended families sharing housing, as was noted above. Another indicator is the high level of individual housing construction (53 per cent in 2021, according to Rosstat) comprising self-provision where the work is contracted out and, to a smaller degree, self-building. Such high levels of individual construction suggest that housing welfare is created from within the family, although the preferential mortgage lending for individual housing construction since 2019 has also contributed to this.

Table 11.3 Structure of public expenditures on housing support in 2017–2021, per cent

Total expenditures	*100*
Federal expenditures, of which:	57
Net expenditures on selected categories	20
Provision of maternity (family) capital used for improvement of housing conditions	28
Mortgage subsidies	9
Regional expenditures, of which:	36
Expenditures on selected categories (including federal transfers)	17
Foregone revenues due to income tax deductions	19
Municipal expenditures, of which:	7
Housing provision to low-income households	2
Expenditures on other selected categories	5

Source: Authors' calculations based on Rosstat data.

Drawing on Esping-Andersen's (1990) typology of welfare regimes, Hoekstra (2003) has emphasized three interrelated parameters of welfare state models in relation to state interventions in the housing market: the level of housing decommodification, the level of housing stratification, as well as the proportions of joint participation of the state, the market and households in the provision of housing and in influencing housing policy. In this regard, it can be argued that the Soviet housing regime entailed the elements of the social-democratic and corporatist models. The elements of the latter model were a high level of public expenditures in the housing sphere, housing provision for needy households irrespective of their income level at the expense of state budget funds, and a high share of rental (public/social) housing in the housing tenure structure. In turn, the elements of the corporatist model were a high share of resources of state enterprises in housing provision (including social development and housing construction funds in the 1970s–1980s), the correlation of the level of housing consumption with the membership in a particular organization/corporation (nomenklatura, military personnel, employees of strategic enterprises, and so on), and the predominance of individual housing construction in rural areas.

During the transition to a market economy, there was an obvious and expected phasing out of the elements of the social-democratic model and the emergence of elements of the liberal model as a backbone. However, a more detailed analysis shows that it was not so straightforward. The most important element of the *liberal model* in the organization of the housing sector in modern Russia is the presence of market-based mechanisms of means-tested state housing support. At the same time, the specificities of the income distribution of the population (with an overall high level of differentiation, most of the population has incomes below and significantly below the average/median), as well as the current high level of concentration of housing construction at local markets in the hands of several developers which determines the monopoly rent and low elasticity of supply – all this limits the development possibilities within the liberal model. In the foreseeable future, at least half of households will not be able to purchase housing on the market without some form of state support.

In addition, due to inertia within the housing sector, the correlation between the level of income and the level of housing consumption, although increasing, remains at a fairly low level in comparison with other countries with market economies. This is because many households who were provided with housing during the Soviet period continue to use the same residential premises (having changed their status from a tenant to owner-occupier). Simultaneously, the predominance of multi-family dwellings in the structure of the housing stock determines the cohabitation of households with different incomes in one apartment building as a typical situation. The associated political constraints determine the preservation of essential elements of the *social democratic model*. First, the state is still obliged to subsidize capital repairs of apartment houses and the resettlement of owners of dilapidated and obsolete housing irrespective of its tenure status. Second, the level of direct and indirect subsidization of housing and utility services remains high, which determines the overall high level of public expenditures in the housing sphere, the high proportion of them aimed at subsidizing current consumption.

Under these conditions, the importance of elements of the *corporatist model* has endured and even strengthened. This is manifested in the increase in the share of individual housing construction in the total commissioning of new housing and the significant role of intergenerational transfers in the housing provision of the population. Additionally, a significant share of the budget funds allocated to the housing sector is formed on the principle of support of individual 'corporations' represented by the relevant categories of citizens, for example, military personnel, civil servants, and so on. These costs constitute the bulk of expenditures to support the housing consumption of citizens. Thus, the current state of the housing regime in Russia represents a peculiar combination of the three welfare models (Kosareva et al., 2015, p. 110). This combination was not the result of a social contract reflecting the particularities of established institutions but was formed largely spontaneously suggesting a highly path-dependent configuration of the Russian housing welfare regime.

Regional outlook

The countries of the former Soviet Union (FSU) share many characteristics in the housing sphere and housing policy. The common past, the presence of ongoing post-Soviet economic cooperation (except for the Baltic republics that joined the European Union but followed the post-socialist trend of super-homeownership) as well as the shared socio-cultural background determine the commonality of the models of housing legislation and housing policy, which is often associated with borrowing and hence repetition of others' mistakes. There are five common trends in the development of the housing sector of the region.

First, mass privatization of housing and marginalization of social housing stock swept across the entire FSU. As a result, in most countries of the region, the share of privately owned housing exceeded the indicators of the Russian Federation: for example, the share of privately owned housing is about 98 per cent in Lithuania, 94.2 per cent in Belarus, and 99.8 per cent in Armenia. Housing privatizations decimated the stock of public housing, while the growing market-based housing production has been almost entirely focused on homeownership. The mass privatizations, however, mostly failed to produce either genuine homeowners or professional rental investors willing and able to undertake maintenance and capital repairs of common property in apartment houses (Peppercorn & Taffin, 2013).

Second, due to the economic crisis of the first decade of the market transition, total housing commissioning collapsed while there was also a sharp reduction in housing construction at the expense of budget funds. For example, in Armenia, the volume of housing construction in 2000 amounted to only 62,000 square metres in contrast to 849,000 square metres in 1991. Third, all countries attempted to adapt market-based housing provision mechanisms by developing mortgage systems (to date, the most attractive conditions of mortgages and the largest volume of mortgage lending, in addition to Russia, have been formed in Kazakhstan). Fourth, there is an absence of an institutional private rental sector. Rental supply is smaller than expected in mature market economies with similar patterns of demography, incomes, mobility needs, and building inventory. The formal private rental sector is residual and there is mostly informal subletting market in individual apartments.

Fifth, the Soviet-era obligations for state housing support to certain categories of citizens (that is, those on municipal housing waiting lists) still linger in the absence of transparent mechanisms for the fulfilment of these obligations. Still, not all families who were listed as in need of improving their housing conditions managed to get housing support. At the same time, the elements of the Soviet and the market housing policy continue to coexist. For example, in Armenia, Belarus, Russia, and Ukraine, two types of social housing are institutionalized: an inherited and difficult-to-reform Soviet-era public housing and a new one, not subject to privatization and aimed at housing support for certain categories of citizens (the lower middle class in Russia, low-income families in Ukraine, families of refugees and victims of natural disasters in Armenia, and low income and some other social categories of citizens in Belarus). The stock of this "new" social housing is very small and does not have a significant impact on the housing provision.

Moreover, the FSU countries exhibit several national and regional characteristics. First, there is a problem of informal (spontaneous) construction. The issues of regulating the growth of informal settlements and prevention of spontaneous construction are directly dependent on the degree of development of spatial planning systems in the FSU countries. Also, the lack of documents confirming the right of ownership and permission to build new dwellings or upgrade the existing ones may promote informal developments. The problems of informal housing are also associated with the use of precarious constructions and living in areas unsuitable and unfavourable for life. Addressing the issues of legalization or reconstruction of informal settlements is an essential element of housing policy in Central Asian countries. The scale of informal building in the European part of the region has been small. However, even here several violations of current legislation become a serious problem, such as the construction of apartment buildings in areas designated for individual housing construction or using buildings for permanent residence that are not intended for this purpose (for instance, not provided with basic engineering infrastructure). Second, natural disasters and armed conflicts have forced people to change their place of residence, having made the provision of housing for refugees, displaced people, and others who have lost their homes a long-term priority of housing policy in countries such as Armenia, Azerbaijan, Georgia, Tajikistan, and Ukraine. The restoration of the destructed and damaged housing stock is an acute issue. Third, in the Baltic countries, as well as in Armenia, restitution has become a significant element of housing policy.

Presently, many governments in the FSU region are attempting to pull diverse housing-related reforms into comprehensive and cohesive strategies focused on the provision of housing to segmented household groups as opposed to the traditional Soviet-era focus on the quantity and quality of the housing stock. The housing codes and legislative acts of FSU countries determine the categories of citizens who may be allocated social housing, and the principles for determining the categories are similar to those used in Russia. Usually, these are socially vulnerable (such as orphans, large families, single elderly citizens, veterans, and the disabled) low-income populations in need of better housing conditions, as well as certain categories of civil servants. However, the volume and pace of construction of social housing and the use of other tools of housing support are still insufficient due to the limited financial and organizational capacity of central and local authorities. There is still a shortage of affordable housing for socially vulnerable segments of the population or people with low incomes. On average in the region, since 2000, only 3–4 per cent of needy households improve their housing conditions every year. In the poorest countries of the region (Tajikistan, Kyrgyzstan), this number is just over 2 per cent. This means that the average waiting time for housing support in the countries of the FSU region is between 25 and 50 years.

At the same time, the FSU countries differ significantly in both the severity of the housing problem and the number of resources that can be used to provide state housing support to citizens and to support housing construction. As of 2021, the average floor area per person in the region varies from over 33 square metres in the Baltic countries and Armenia,[21] from 23 to 29 square metres in Belarus, Kazakhstan, Russia and Ukraine, and from 13 to 16 square metres in Uzbekistan and Kyrgyzstan. The commissioning of housing per person in 2021 ranged from 0.9 square metres in Kazakhstan to 0.2 square metres in Kyrgyzstan[22] (in Russia and Belarus it was 0.5–0.6 square metres per person).

To sum up, the active housing support is currently being implemented in Russia, Belarus and Kazakhstan, and the housing regimes that are emerging in these countries have many similar features. In other FSU countries, apart from the Baltic states, the housing regime as such is still developing. While legal and policy documents establish general priorities for housing policy, in practice, there is a regime of 'minimum state intervention,' reserved for the most acute problems emerging in the housing sector.

Conclusions

In this chapter, we examined the historical development and the current state of housing and welfare regimes in Russia. The specificities of this development are intricately linked with the fundamental changes in the political and economic regime since 1991 following the demise of state socialism and the transition to neoliberal capitalism. Although the transition to a market economy entailed a certain retreat of the state, the socialist-era legacies and the associated path dependencies (ranging from the concrete Soviet-built environment to the organization of social housing queue to the citizens' expectations of the role of the authorities in housing provision and its maintenance) suggest that the role of the state has been redefined; it still remains pivotal yet seeks to delegate some of its responsibilities to the market and citizens themselves.

The current level of housing provision (measured as the average floor space of total housing stock per capita) and the total volume of housing commissioning (both in absolute and relative numbers) are at a historical high. However, housing provision is more socially differentiated than during the Soviet period and these differences have a tendency to increase. Fewer than a quarter of Russians are dissatisfied with their living conditions, although only about 14 per cent intend to improve that. At the same time, about half of households nowadays are able to purchase housing with their own and borrowed funds. This is due to the general increase in the level of well-being associated with the economic rebound in the 2000s, but it can also suggest that the cultural resistance to debt and mortgage loans observed in the 2000s (Zavisca, 2012) may be gradually fading away.

The tenure structure of Russian housing has shifted drastically since 1991: during the Soviet period, it was dominated by public tenancy, after the introduction of the market economy it has been dominated by owner-occupancy. Although the share of owner-occupation exceeds 80 per cent according to our estimates, it is still slightly lower than in some other super-homeownership post-socialist societies. The expansion of homeownership is attributed to the mass giveaway privatization since the beginning of the transition which left behind a form of state legacy welfare (Stephens et al., 2015). Apart from the responsibilization and budget cost-cutting imperatives, the privatization of housing was also a means to endow citizens with property to compensate for the falling living standards incurred by the transition to a market economy. The social rental sector is rather small at 6.2 per cent and emerged from the non-privatized portion of Soviet-era public housing. Any investment activity in this heavily subsidized sector is only possible at the expense of budget funds. Private renting is largely informal and hard to measure, but our estimates suggest its enduring residualism irrespective of political and economic regimes.

The Soviet-era nexus of housing and welfare where the state played a key role (Smith, 2010) is no longer relevant. Housing has largely lost its status as a public good and has become commodified when it comes to the privatized former public housing stock and especially the new-built stock. Today, almost 90 per cent of all housing improvements are associated with the acquisition of ownership on the primary and secondary markets. The role of the state has been to support the purchase of housing into private ownership for certain groups of citizens on the market whilst still subsidizing the housing and utility services as well as capital repairs of apartment houses. The new housing construction has been characterized by the confluence of self-financing and state-supported market mechanisms such as preferential mortgage programmes (which leads to increasing inflation risks). The public/social housing sector has inherited the queuing system from the Soviet era but has been gradually shifting from universalist to market-based means-tested support. The waiting time in the queue is incredibly long and only 4 per cent of families are provided with accommodation annually, while only 2 per cent of public housing-related expenditures go to the housing provision for low-income households. As a result, the ever-topical housing question in Russia is becoming particularly acute for the least well-off households. Intergenerational transfers remain the primary means of improving housing conditions for such households, while the depopulation trends in the majority of Russian regions only emphasize the role of such transfers.

Notes

1 Rosstat stands for the Federal State Statistics Service of Russia. The data this chapter refers to can be found at https://rosstat.gov.ru/folder/10705.

2 In Russia, the total floor area of a housing unit is calculated as living space plus the kitchen, bathroom, corridor, balcony, and wardrobe space.

3 The minimum "sanitary norm" of 9 square metres of living space per person (equivalent to about 12 square metres of total floor area) stipulated in 1922 was only, on average, achieved in cities by the 1970s (Kalyukin & Kohl, 2020, p. 1,774).

4 The 2020 All-Russian Population Census was delayed due to the COVID-19 pandemic and was conducted during autumn 2021.

5 The 2020 Census data accounted only for individual (separate) and communal flats and individual houses; dormitories were not included.

6 About 40,000 households participated in the survey. It should be noted that the housing provision may be underestimated due to the non-sampling of rich households with an extremely high level of housing consumption.

7 Cooperative ownership is provisional until members pay up their shares and thereby acquire ownership. After repayment of all installments by members, the operations of such cooperatives are regulated by the laws governing homeowners' associations.

8 Importantly, homeowners in Russia enjoy a legal provision to both privatize and de-privatize a state or municipal property free-of-charge once; in the latter case, the authorities are obliged to buy back.

9 The construction of new social housing is concentrated in a few relatively economically well-off regions of Russia (for example, Moscow and the Republic of Tatarstan). In the rest of the country, those in a queue are mainly provided with vacated secondary social housing stock. In Russia as a whole, no more than half of all social housing is provided in newly constructed buildings.

10 Personal property was a distinctive legal tenure under state socialism different from the notion of private property. Mark Smith characterizes personal property as "something akin to a medium-term leasehold largely stripped of any cash-generating potential" (Smith, 2010, p. 19). Personal property "derived from a citizen's own labor, and as it was only for personal use … by definition it could not actively exploit another citizen" (Smith, 2008, p. 285). Therefore, despite a de jure abolition of private property in 1918, individual ownership effectively persisted in Soviet cities as a form of decommodified personal property.

11 According to the 1989 Census, only 11.7 per cent of the population in Soviet Russia were housed in buildings constructed before 1950, while 39.4 per cent lived in housing stock constructed during the 1950–1960s and almost 49 per cent in houses built in the 1970–1980s. For comparison, according to the 2020 Census, 8.8 per cent of the population lived in houses built before 1957, while more than 22 per cent lived in houses built during 1957–1970, and almost 43 per cent in houses built in 1971–1995.

12 Access to cooperatives was regulated by the state authorities. In practice, this opportunity was available only to residents in large cities. The right to use a housing unit in a cooperative was granted only based on membership in that cooperative.

13 Giveaway privatization of state and municipal housing stock was set to end by March 2007 but was extended five times since then and indefinitely extended in February 2017.

14 Additionally, in 2021, the share of organizations (developers) in mixed ownership in total housing commissioning was 0.3 per cent, the share of cooperatives was 0.8 per cent while the balance 1.3 per cent was unspecified by Rosstat.

15 The relative size of housing tenures during the period under review has changed significantly, but unfortunately the available data only allows us to provide an approximation of the tenure development. For instance, there is no statistics on private rented housing and subletting. Also, since 1993, cooperative housing ownership is not distinguished from private ownership of legal entities.

16 While departmental *dachas* were indeed a part of the Soviet reward system targeting the nomenklatura, higher-ranked officers, employees of government agencies, the intelligentsia (for instance, writers) and so on, the rest of the population was entitled to plots of land in the countryside (also referred to as *dachas*) for allotment gardening as a means of subsistence with only sheds and shacks to store tools permitted to be built and dwelling activity severely restricted. The third *dacha* type, garden-plot associations and cooperatives, were formed under the auspices of local trade unions and enterprises/firms and allowed standard design buildings (see Lovell, 2003).

17 In 2021, the subsistence level was RUB 11,908, or approximately EUR 149; the federal minimum wage was RUB 12,792, or approximately EUR 160.

18 Guy Standing (1996, p. 231) noted that policy-makers, politicians and national and foreign "advisers" shared a worrying sentiment that under the state socialist regime "the populations were 'over-protected' so that they need a period of re-education to alter their consciousness." The outcome of the transition had been an enormous increase in inequalities as the poor have lost both benefits and wage income. So, if in 1989 the Gini coefficient in Russia was 0.29, by 1996 it more than doubled to almost 0.64 and somewhat decreased to just under 0.55 in 2015 (Novokmet et al., 2018, fig. B24 in Appendix B).

19 According to Rosstat, more than 20 per cent of all pensioners kept on working after reaching retirement age in 2021, the main reason for this being a low level of pension provision as well as insufficient incomes (salaries) of their closest relatives.

20 Excluding beneficiaries of homebuyers' income tax deductions and those who used maternity capital to improve housing conditions.

21 Emigration in the post-Soviet era contributed to the high levels of floor area provision.

22 Self build account for 80 per cent of housing commissioned in Kyrgyzstan.

References

Alexeev, M. (1988). Market vs. Rationing: The Case of Soviet Housing. *The Review of Economics and Statistics*, *70*(3), 414–420. https://doi.org/10.2307/1926779

Andrusz, G. D. (1984). *Housing and Urban Development in the USSR*. State University of New York Press.

Andrusz, G. D. (1992). Housing Co-operatives in the Soviet Union. *Housing Studies*, 7(2), 138–153. https://doi.org/10.1080/02673039208720730

Cook, L. J. (1993). *The Soviet Social Contract and Why It Failed: Welfare Policy and Workers' Politics from Brezhnev to Yeltsin*. Harvard University Press.

Cook, L. J. (2007). *Postcommunist Welfare States: Reform Politics in Russia and Eastern Europe*. Cornell University Press. https://doi.org/10.7591/9780801460098

DiMaio, A. J. (1974). *Soviet Urban Housing: Problems and Policies*. Praeger Publishers.

Eliseeva, I. (2020). Osnovnaya sem'ya: Predystoriya i Perspektivy [The main family: Background and Prospects]. *Demograficheskoye Obozreniye*, 7(4), 6–35. https://doi.org/10.17323/demreview.v7i4.12042

EMF-ECBC. (2022). *HYPOSTAT 2022: A Review of Europe's Mortgage and Housing Markets* (pp. 1–188). European Mortgage Federation. https://hypo.org/app/uploads/sites/3/2022/11/HYPOSTAT-2022-FOR-DISTRIBUTION.pdf

Esping-Andersen, G. (1990). *The Three Worlds of Welfare Capitalism*. Polity Press.

Goskomstat RSFSR. (1990). *Narodnoye Khozyaystvo RSFSR v 1989 g. [National Economy of the RSFSR in 1989]*. Respublikanskiy informatsionno-izdatel'skiy tsentr.

Goskomstat RSFSR. (1991). *Narodnoye Khozyaystvo RSFSR v 1990 g. [National Economy of the RSFSR in 1990]*. Respublikanskiy informatsionno-izdatel'skiy tsentr.

Harris, S. E. (2013). *Communism on Tomorrow Street: Mass Housing and Everyday Life after Stalin*. Woodrow Wilson Center Press.

Hegedüs, J., & Puzanov, A. (2018). The Private Rental Sector Under Socialism. In J. Hegedüs, M. Lux, & V. Horváth (Eds.), *Private Rental Housing in Transition Countries* (pp. 41–69). Palgrave Macmillan. https://doi.org/10.1057/978-1-137-50710-5_2

Hoekstra, J. (2003). Housing and the Welfare State in the Netherlands: An Application of Esping-Andersen's Typology. *Housing, Theory and Society*, *20*(2), 58–71. https://doi.org/10.1080/14036090310000634

Ivanova, G. (2011). *Na poroge 'Gosudarstva Vseobshchego Blagosostoyaniya': SOTSIAL'NAYA Politika v SSSR (seredina 1950-kh – nachalo 1970-kh godov) [On the brink of the 'welfare state': Social policy in the USSR (mid-1950s - early 1970s)]*. Institut rossiyskoy istorii RAN.

Kalyukin, A., & Kohl, S. (2020). Continuities and Discontinuities of Russian Urban Housing: The Soviet Housing Experiment in Historical Long-Term Perspective. *Urban Studies*, *57*(8), 1768–1785. https://doi.org/10.1177/0042098019852326

Kornai, J. (1992). *The Socialist System: The Political Economy of Communism*. Clarendon Press. https://doi.org/10.1093/0198287763.001.0001

Kosareva, N., Polidi, T., Puzanov, A., & Tumanov, A. (2011). *Sravnitel'nyy Analiz Potrebleniya i Raskhodov v Zhilishchnoy Sfere [Comparative Analysis of Consumption and Expenditures in the Housing Sector]*. Higher School of Economics. https://www.hse.ru/data/2011/04/05/1211687705/K_P_P_T.pdf

Kosareva, N., Polidi, T., & Puzanov, A. (2015). *Zhilishchnaya Politika I Ekonomika v Rossii: Rezul'taty I Strategiya Razvitiya [Housing Policy and Economy in Russia: Results and Development Strategy]*. Higher School of Economics. https://urbaneconomics.ru/sites/default/files/ieg_zhilishchnaya_politika_i_ekonomika_v_rossii.pdf

Kosareva, N., & Struyk, R. (1993). Housing Privatization in the Russian Federation. *Housing Policy Debate*, *4*(1), 81–100. https://doi.org/10.1080/10511482.1993.9521125

Lapina, N. (2007). *Uroki Sotsial'nykh Reform v Rossii: Regional'nyy Aspekt. Analiticheskiy obzor [Lessons of Social Reforms in Russia: Regional Aspect. Analytical review]*. Institut Nauchnoy Informatsii po Obshchestvennym Naukam RAN.

Lovell, S. (2003). *Summerfolk: A History of the Dacha, 1710–2000*. Cornell University Press. https://doi.org/10.7591/9781501704574

Lux, M., Sunega, P., & Kažmér, L. (2022). Families and the Reproduction of Homeownership in a Post-Socialist Context. In R. Ronald & R. Arundel (Eds.), *Families, Housing and Property Wealth in a Neoliberal World* (pp. 72–91). Routledge. https://doi.org/10.4324/9781003092117-4

Madison, B. (1964). Social Welfare: Soviet Model. *Social Service Review*, *38*(2), 191–205.

Morton, H. W. (1980). Who Gets What, When and How? Housing in the Soviet Union. *Soviet Studies*, *32*(2), 235–259. https://doi.org/10.1080/09668138008411295

Novokmet, F., Piketty, T., & Zucman, G. (2018). From Soviets to Oligarchs: Inequality and property in Russia 1905-2016. *The Journal of Economic Inequality*, *16*(2), 189–223. https://doi.org/10.1007/s10888-018-9383-0

Peppercorn, I. G., & Taffin, C. (2013). *Rental Housing: Lessons from International Experience and Policies for Emerging Markets*. The World Bank. https://openknowledge.worldbank.org/handle/10986/13117

Perevedentsev, V. (1990). Kvartirnyy Vopros I Perestroyka [The housing problem and perestroika]. *Voprosy Ekonomiki*, *8*, 74–79.

Puzanov, A. (2013). Russia: The Persistence of the Socialist Legacy? In J. Hegedüs, M. Lux, & N. Teller (Eds.), *Social Housing in Transition Countries* (pp. 225–243). Routledge. https://doi.org/10.4324/9780203095904-25

Puzanov, A. (2018). Russia: A Long Road to Institutionalisation. In J. Hegedüs, M. Lux, & V. Horváth (Eds.), *Private Rental Housing in Transition Countries* (pp. 287–309). Palgrave Macmillan UK. https://doi.org/10.1057/978-1-137-50710-5_12

Sidorina, T. (2005). *Dva veka Sotsial'noy Politiki [Two Centuries of Social Policy].* Rossiyskiy gosudarstvennyy gumanitarnyy universitet.
Smith, M. B. (2008). Individual Forms of Ownership in the Urban Housing Fund of the USSR, 1944-64. *The Slavonic and East European Review*, *86*(2), 283–305.
Smith, M. B. (2010). *Property of Communists: The Urban Housing Program from Stalin to Khrushchev*. Northern Illinois University Press.
Standing, G. (1996). Social Protection in Central and Eastern Europe: A Tale of Slipping Anchors and Torn Safety Nets. In G. Esping-Andersen (Ed.), *Welfare States in Transition: National Adaptations in Global Economies* (pp. 225–255). SAGE Publications. https://doi.org/10.4135/9781446216941.n8
Stephens, M., Lux, M., & Sunega, P. (2015). Post-Socialist Housing Systems in Europe: Housing Welfare Regimes by Default? *Housing Studies*, *30*(8), 1210–1234. https://doi.org/10.1080/02673037.2015.1013090
Szelényi, I., & Mihályi, P. (2019). *Varieties of Post-communist Capitalism: A Comparative Analysis of Russia, Eastern Europe and China* (Vol. 157). Brill Academic Publishers.
The Institute for Urban Economics. (2022). *Dinamika Dostupnosti Zhil'ya v Rossii v 2006–2021 gg. [Housing Affordability Dynamics in Russia in 2006–2021].* The Institute for Urban Economics. https://www.urbaneconomics.ru/sites/default/files/dinamika_dostupnosti_zhilya_v_rossii_v_2006-2021_gg_0.pdf
Varga-Harris, C. (2015). *Stories of House and Home: Soviet Apartment Life during the Khrushchev Years*. Cornell University Press. https://doi.org/10.7591/9781501701849
Vinogradova, E. (2004). Sotsial'naya Politika: Istoricheskiy, Teoreticheskiy, Prakticheskiy Aspekty [Social policy: Historical, Theoretical, Practical Aspects]. *Ekonomicheskiye i Sotsial'nyye Problemy Rossii*, *1*, 6–52.
Yemtsov, R. (2007). *Housing Privatization and Household Wealth in Transition* (Research Paper No. 2007/02; pp. 1–32). United Nations University UNU-WIDER. https://www.wider.unu.edu/publication/housing-privatization-and-household-wealth-transition
Zavisca, J. R. (2012). *Housing the New Russia*. Cornell University Press. https://doi.org/10.7591/9780801464300

12

HOUSING AND WELFARE IN EAST-CENTRAL EUROPE – THE CASE OF HUNGARY

József Hegedüs

The present chapter focuses on Hungary, while also pointing out some basic similarities and differences with the most recent Central and Eastern European (CEE) members of the European Union.[1] These countries share several common elements in their history, especially between 1945 and 1990, but also have had important similarities after transitioning from planned economies to various forms of liberal market economies. There is an ongoing debate in the housing and welfare literature as to whether their historical similarities have led to the emergence of more or less the same welfare and housing regimes in these countries. Their development was dominated by Soviet political and military rule until 1990, and despite their differences, the common features of their housing and welfare regimes were predominant in this period. This chapter focuses on the post-1990 period. We argue that until the financial crisis of 2008, post-socialist CEE countries experienced very similar outcomes, and therefore, the transformation of their housing regimes followed a similar logic; but after the crisis, differences in economic and political situations intensified, which then amplified the likelihood of divergence between the region's housing and welfare regimes.

The chapter follows the method of interpretive sociology and adopts a historical institutional ist approach. Changes to the elements of the housing regimes or the introduction of social policy measures are the result of the interplay and aspirations of political and economic actors. These stakeholders can be national actors (political parties, trade unions, business groups, public authorities, and non-profit organizations) or international organizations. The emergence of, and changes in, different regimes are fundamentally brought about by the interactions of the dominant political actors determined by the internal power relations in a given country, so their different policy measures and their expected or unexpected outcomes will not necessarily be in line with the expectations of political ideology.

Housing in Hungary in the context of the new member states and EU

The housing situation in Hungary is presented in comparison with the other countries of the European Union, where the countries of the EU are divided into three groups by housing indicators.[2] Country groups used in this comparison are based on the presumed (dis)similarities of their welfare and housing regimes (Hegedüs et al., 2017; Hegedüs, 2018).

 DOI: 10.1201/9781003212690-14

In overall comparison, the CEE EU Member States (EU-11) differ markedly from the older Member States of Western, Northern, and Southern Europe (See Table 12.1).

- GDP per capita in the EU-11 countries lag significantly behind old Member State averages, despite the slowly narrowing socio-economic gap;
- CEE countries face ongoing population decline, which is not counterbalanced by immigration;
- The role of the informal economy is significantly larger in the CEE countries than in the most developed European economies;
- Housing market processes are less developed: cash-based transactions and intergenerational and interfamily (monetary and in-kind) transfers remain significant, which also modifies class relations;
- The states are too "weak" to integrate the society on basic liberal democratic principles; new elites have strong incentives to hold on to power, and in response to this some governments might take extreme measures to reorganize societies.

Table 12.1 Basic housing indicators of CEE EU Member States (Periphery II EU-11) in comparison to Northern and Western Europe and Southern Europe (Periphery I) – latest available data[3]

		Northern and Western European EU MSs	*Periphery I Southern European EU MSs*	*Periphery II EU–11*	*Hungary*	*EU avg.*
I1	GDP per capita (PPP) USD 20021	56,387	42,536	36,959	36,850	50,551
I2	Population change, 2001–2021(2001=100)	108.0	109.2	95.9	95.6	105.7
I3	Share of foreign-born population, 2021	16	13	4	6	13
I4	Population (million), 2021	287.3	129.0	101.8	9.7	518.1
I5	Population as share of EU population (%), 2021	55	25	20	2	100
I6	Average house price (EUR/m2), 2020	3,727	2,294	1,629	1,657	3,033
I7	Average market rent (EUR/m2/ month), 2021	18.3	14.5	10.3	15.1	15
I8	Total Outstanding Residential Loans to GDP Ratio (%), 2021	60.8	31.0	17.3	9.5	45.2
I9	Change of residence in past five years (%), 2012	27	11	7	7	19
I10	Annual housing construction between 2010 and 2021 per 1000 inhabitants (average population of 2001 and 2021)	3.8	4.3	3.0	2.4	3.8
I11	Rental (subsidized) (%), 2021	9.9	8.7	7.6	3.9	9.1
I12	Owner-occupied housing (%), 2021	60.6	74.8	88.0	91.7	69.3

(*Continued*)

Table 12.1 (Continued)

		Northern and Western European EU MSs	*Periphery I Southern European EU MSs*	*Periphery II EU–11*	*Hungary*	*EU avg.*
I13	Dwelling per 1,000 inhabitants (2021 or the latest available)	496	531	464	462	498
I14	Rooms per capita, 2021	1.7	1.8	1.5	1.6	1.7
I15	Average dwelling size (m2), 2012	99	97	69	76	92
I16	Share of overcrowded housing, 2021 or latest (%)	8.3	18.4	31.3	13.5	15.2
I17	Share of the population facing severe housing deprivation, 2019 or latest	2.1	3.8	8.2	7.8	3.9
I18	Share of housing expenditure in disposable household income (%), 2021 or the latest	21.8	18.1	17.1	12.5	20.0
I19	Share of households in housing arrears, 2021	7.5	12.5	8.7	11.2	8.9
I20	Share of the informal economy (% of GDP), 2012	11	20	23	22	14

Economic development and demography

EU-11 countries lag significantly behind the most developed European economies (of Northern and Western Europe) in numerous socio-economic measures. Per capita GDP PPP in the EU-11 – including Hungary – is, on average, 27 per cent lower than in the old EU Member States, and 34 per cent lower than in the group of the most developed countries.

Apart from a brief period of stagnation between 2009 and 2016, the population of the European Union has shown consistent growth; by contrast, the population on the periphery has decreased steadily. Between 2001 and 2021, the EU-11 lost 4 per cent of its population, while the old EU Member States saw an average of 5.7 per cent population growth. Mitra (2001) characterized the transitional societies as "grey and poor", referring to their ageing populations and low birth rates. These countries generally exhibit net emigration.

The housing sector: Quality, tenure structure, and financing

The dwelling size (floor area) in the EU-11 is 80 per cent of that in the old EU Member States, and the number of rooms per resident is 0.4 lower than the Western and Northern European average. On the other hand, the share of overcrowded dwellings in the CEE countries (including Hungary) is five times greater than the average of the old Member States. House prices and market rents are, on average, half the equivalent rates in Western Europe (although the reliability of this data is questionable).

The mass privatization of formerly public rental housing reduced the public social rental housing sector to 3 per cent or less of national housing stocks in most transition countries – as opposed to 10–12 per cent in the most developed countries in Northern and Western Europe.

Housing market transactions and residential mobility indicators also show a systematically lower rate of housing mobility: on average, 7 per cent of people in the EU-11 have moved house in the past five years, compared to 18 per cent in Northern and Western Europe. This is both a cause and a consequence of the dominance of home ownership in CEE countries and is linked to housing market rigidity as well as demographic processes. In the CEE's housing finance systems, outstanding mortgages account for about 17 per cent of GDP, in contrast to 45 per cent in the old EU Member States.[4]

Housing deprivation, income inequality, and affordability

The discrepancies in the quality of the housing stock are manifested in the share of households experiencing severe housing deprivation: in the CEE region, this indicator is five times greater than in Western and Northern European countries. In Hungary, 17 per cent of households suffer severe deprivation. Due to the shortage of affordable rental housing, low-income households must make do with the lowest segment of the private rental market or with poor-quality municipal social housing.

In the transition countries, the emerging predominance of free markets was accompanied by a radical increase in income inequality: the share of people who struggle to make ends meet is significantly higher than in the most developed countries. EU-SILC surveys showed in 2011 that in the post-crisis period, some 40 per cent of CEE households had trouble making ends meet, compared to 15 per cent in Northern and Western European countries. Five years later, in 2016, the average figure was 4 per cent in the most developed European regions; 16 per cent in the Southern European Member States; and 13 per cent in the transition countries. In Hungary, 17 per cent of households struggled to make ends meet in 2016. The regional component of inequalities within the European Union is therefore significant. In the new EU Member States, one-fifth of households are in arrears with their housing payments, even though housing-related costs claim a similar share of household budgets across the European regions (20–23 per cent).

A brief overview of the historical development of the housing regime

Before 1945: The regulated private market

Housing policy in the inter-war period was shaped by two factors: rent control in the urban rental sector and the housing crisis caused by refugees following the Trianon Peace Treaty. The urban housing situation was already very serious in the period before 1914, and state and municipal housing construction failed to alleviate it to any significant extent. The introduction of rent controls during World War I led to a decline in housing construction, which only aggravated the housing crisis due to the influx of refugees. The government's declared aim was to phase out rent control (returning to pre-war rent levels by 1928) and to boost housing construction. However, the economic crisis of 1928 forced the government to postpone the abolition of rent control. In 1935, there were 2 million dwellings in Hungary, 70 per cent of which were owner-occupied and 30 per cent rented; 84 per cent of housing in Budapest was rented, 58 per cent in other towns and 13 per cent in villages (Ruisz, 1941).

From the 1920s, to promote housing construction, the government granted a 10–20-year tax exemption and a moratorium on rent controls for new homes. Moreover, direct government programmes were launched in the late 1920s, partly by legally obliging the two insurance companies (OTI[5] and MABI[6]) to keep 30 per cent of their assets as housing loans, and later as bonds issued by the National Housing Loan Bank. This helped to kick-start urban rental housing construction, mainly in Budapest. In the countryside, small housing schemes were launched, also with state-subsidized loans. The ONCSA (National People and Family Protection Fund) scheme, which has helped many families with children to get their own homes, is a notable example.

1948–1959: Regime change and the impact of class struggle on the housing market

After the end of the Second World War, housing policy was concentrated on the renovation of war-damaged properties. After the communist takeover, the government used only housing policy instruments that did not divert resources away from industrial restructuring.

At the same time, the reorganization of the economy and the establishment of new power structures generated a significant demand for housing in the capital and in the main economic centres (county seats). The relocation of the former elite to the countryside in 1951 freed up a significant housing stock for the new working class and party elite. In addition to the Real Estate Management Company of the Capital (FIK), more than 30 companies had the right to "acquire" apartments and provide them to their subordinates, to be used as service apartments. Before the state mergers, the FIK played an important role in housing provision (by using market instruments to acquire depreciated housing, mainly from landlords).

Housing nationalization took place in 1952–1953 when the state expropriated dwellings with more than six rooms and set up state-owned housing management organizations, run by the local councils (the so-called IKVs). In the 1950s, the practice of renting out flats led to the creation of shared tenancies, where a dwelling was divided into several flats with the same legal title, and typically with shared kitchen, toilet, and bathroom facilities.[7] The Nagy[8] government of 1953, which built some new housing in the 1950s and aimed to appease extreme left-wing groups, announced a more intensive housing policy, which it was largely unable to deliver. The 1950s were a chaotic period (ad-hoc interventions, often variable orders, opaque housing allocation) for housing policy, when the emphasis was on redistributing existing housing wealth, and hardly any new housing was built. (Hegedüs & Tosics, 1983)

1960–1970: Early socialist model – housing emerging as a political priority

Hungary represented one of the most liberal models of the centrally planned economy, which could partly be explained by the political consolidation of the Kádár administration[9] following the 1956 Revolution. The real change was the 15-year housing plan drawn up in 1958, which undertook to build 1 million dwellings in 15 years (an average of 6.6 dwellings per thousand people per year). The introduction of the "new economic mechanism" in 1968 illustrated the move towards a more decentralized and more efficient economic system, which was vastly curtailed by the collapse of the Prague Spring in 1968. The housing system had a dual "market" structure. In the *state sector*, there were different directly controlled allocation channels (typically affecting not only the council housing stock, but the new, owner-occupied, state-built pre-fabricated housing stock as well), where income and need factors played an important role in allocation, while in the *private sector* household incomes and informal social networks played the decisive role. Both sectors worked within the limits of the central planning system, which regulated the supply of building materials, the accessibility of labour for private construction, the construction industry, and access to finance. (Moreover, privately-owned property was also controlled: one family was legally allowed to own only one housing unit and one secondary unit.)

Table 12.2 Tenure structure in Hungary, 1970–2011

	1935	*1970*	*1980*	*1990*	*2001*	*2011*
Owner occupied	75.0%	66.5%	71.5%	72.3%	90.0%	88.0%
Municipal housing	2.0%	33.3%	28.3%	19.0%	3.7%	2.7%
Corporate housing	2.0%	–	–	3.7%	1.0%	1.0%
Private rental and other	26.0%	0.3%	0.2%	5.0%	5.3%	8.3%
Total	100.0%	100.0%	100.0%	100.0%	100.0%	100.0%

Sources: Ruisz, 1941; CSO.

The housing finance system in the socialist housing model was based on the centrally controlled state-owned bank (National Saving Banks – referred to as "OTP" in this article, based on its Hungarian acronym), state budget allocation, and as a supplement, the state-owned companies could contribute as well. From an economic and sociological standpoint, the typical loan product (35-year, low fixed interest rate loans, at around 1.5–3 per cent) was effectively a form of subsidy. There was no proper underwriting procedure in place; instead, eligible households were assigned as potential buyers and enjoyed automatic rights to subsidy and subsidized loans. In 1973, for example, 35 per cent of the total investment in the state-controlled sector was financed through loans, 53 per cent through budget subsidies, and 12 per cent from household savings, while in the private sector, the respective contributions were 24, 1, and 75 per cent (Hegedüs & Tosics, 1996, p. 249). The official salary provided a safeguard against the risk of non-payment or default, and thanks to the wage garnishment[10] and full employment system, lending did not entail any significant risk to OTP (Table 12.2).

1970–1980: The state-regulated socialist model

By the late 1960s, the political leadership realized that the 15-year plan could only be completed by amending the financial scheme. To balance the budget, the proportion of public investment in housing had to be reduced. A more important change was the 1971 introduction of a new, modern system of housing subsidies (ultimately in force until 1993). The previous system was characterized by ad hoc solutions. There was no explicit codified system for the allocation of council housing, and the new (vacated) housing units were allocated by the bodies that built or managed them (e.g., state-owned companies). According to the new Housing Law of 1971, the allocation of both public rental housing units and owner-occupied units (developed by cooperatives and OTP) was based on a points system for applicants on the housing waiting list, grouped according to their financial situation. In the case of council flats, a fee was payable; for owner-occupied flats, a cash down payment was expected with the possibility of a 35-year preferential loan with a fixed 1 per cent interest rate. State-controlled public and owner-occupied housing was deeply subsidized. The political elite, however, had a way to retain their advantage through other channels of the housing distribution system: state-owned firms and organizations (like the police or military) could buy housing allocations, as could political organizations such as the Young Communist League (KISZ). County[11] Party Committees could also act as developers when allocating housing to their employees. OTP remained the only bank involved in housing finance for a long period. Private housing was subsidized to a much lesser extent, typically through capped state-subsidized loans.

This sector typically consisted of self-build (with self-help elements) and second-home construction with workers and materials.

1981–1989: Moving away from the autocratic, one-party controlled market model

As the macroeconomic conditions of Hungary worsened at the turn of the 1980s, housing policy opened towards the market, in the hope that increasing private actors' incentives to invest in housing could ease the budgetary burden while maintaining the volume of the output. The financial disadvantages of the private sector diminished by making private single-family house building (self-help construction) eligible for construction subsidies, increasing the upper ceiling of the OTP loan, permitting the use of employers' credit, and even offering local council housing grants. Other measures to promote private investments included the increase of the land supply (Hegedüs, 1992). From 1985, small savings cooperatives were granted the right to issue housing loans, which in principle introduced competition within the banking sector. Despite this, OTP remained in a near monopoly position in the housing finance system (in which it controlled 85 per cent of the loan portfolio in 1989).

Housing output decreased in the late 1980s despite the shift in the subsidy system (see Figure 12.1). House prices increased (by 150 per cent between 1980 and 1989), and OTP had difficulty selling new subsidized apartments in multi-unit buildings in some cities, due to their high cost and the population's shift of preferences towards houses (supported by the changes in the subsidy system).

The other – and more important – sign of the crisis was the increase of interest rates and the technical bankruptcy of the state bank OTP. The outstanding housing loan portfolio increased rapidly from 3 to 14 per cent of the GDP between 1980 and 1989. The housing loan portfolio of long maturity and fixed low rate (1–3 per cent) caused a huge budgetary outlay on the books of OTP and other saving cooperatives as the nominal interest rate increased.

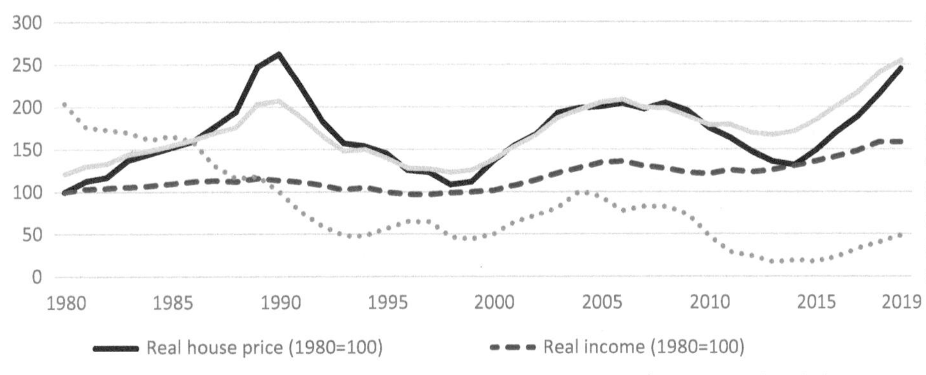

Figure 12.1 House price, income, and new construction in Hungary 1980–2019.

Sources: CSO, MRI.

This dual housing system operated until the early 1980s when the difficulties of the socialist economy model appeared. The 1980s were a period of opening-up in housing policy when market elements were introduced in an attempt to increase the efficiency and performance of the economy. The council thus obtained housing much more cheaply than if it had built it. And in 1986, the process of housing privatization was launched, slowly, with several rules, but with great interest. Within four years, 50,000 flats had already been sold to tenants.

1989–1999: Transition to a market-based housing finance system

The housing finance system technically went bankrupt in 1988. Housing subsidies reached 3.3 per cent of the GDP in 1989 (World Bank, 1991), which was unmanageable. To decrease the budget burden, the new government in 1990 offered an option to pay back the loan at a 50 per cent discount. By 1997, 83 per cent of the loans were repaid, while the remainder became a budget burden of the heavily state-subsidized mortgage "old loans" (49 and 29 per cent of the housing subsidies in 1992, 1994 respectively) (Hegedüs et al., 1996, p. 88).

The other major change in the housing sector was the drastic "give-away" privatization program, which had started at the end of the eighties but gained real impetus with the Housing Law of 1993 which gave sitting tenants the Right to Buy (purchase their rental home at a significant discount). From 1990 till 2001, the municipally owned public rental stock decreased by 74 per cent, from 721,000 to 189,000 units; and sitting tenants typically paid heavily discounted rates, at 10–15 per cent of the market price. This resulted in grave consequences for the social housing policy after the transition, but it also meant an immediate relief for the budget, partly in the form of the revenue from selling the housing stock, and partly from the redemption of state subsidy for the rehabilitation of municipal stock.

The housing mortgage market and housing output shrank during high inflation periods (despite attempts to introduce deferred payment mortgages – DPMs – and other housing construction subsidies): mortgage loan portfolio/GDP ratio decreased from 13 to 2 per cent, and new annual construction dropped from 50,000 to 20,000 between 1989 and 1999. The typical loan product was a variable rate loan in HUF with a small interest rate subsidy and limited tax exemption for mortgage repayment (Hegedüs & Várhegyi, 2000). At the end of the 1990s, the positive macroeconomic changes influenced household behaviour: house prices increased by more than 40 per cent in the three years before the mortgage expansion started in 2000.

2000–2008: Emergence of the mortgage market

The period of 2000–2008 saw the emergence and rapid expansion of mortgage lending in Hungary. By 2000 the overall economic performance of the country had improved considerably; the GDP grew by 4 per cent in 2000, and the fiscal deficit decreased to 2.9 per cent. Furthermore, the unemployment rate fell to around 5–7 per cent, and after a long period of decline house prices increased by 42 per cent in real terms between 1998 and 2000. The government[12] started a new housing policy focusing on stimulating new construction, which was considered to be a main driver of economic growth. The most significant element of this policy was the support for housing mortgages introduced in 2000 to make mortgages more affordable (Hegedüs & Somogyi, 2005). An additional element in the subsidy program was the substantial increase in the rate of Personal Income Tax (PIT) mortgage payment allowance on new housing construction which had been introduced in 1994. The interest rate on subsidized loans fell to around

4–6 per cent by 2002, while market interest rates of housing loans remained at around 14 per cent. Moreover, in 2002 the government increased the subsidy for contract savings and the housing construction grant.[13] By 2004, this subsidy system placed an unmanageable burden on the budget, exacerbating irresponsible government spending after 2002. The subsidies were drastically cut in 2004.

The cut in subsidies did not stop the expansion of the market, as new foreign exchange (FX) mortgage loan products were introduced, which seemed more affordable than the HUF-based loans with reduced subsidy. Annual new lending in foreign currency went from almost zero in 2004 to 90 per cent of the mortgage loan portfolio in 2007. The market interest rate of Swiss Franc-denominated loans was around 6 per cent, while for HUF loans it was above 10 per cent. As Hungary was an EU member and most banks operating in the country were in European ownership, concerns about the financing were dismissed (Hegedüs, 2010).

2008–2014: Crisis management

The consequences of the financial crisis on the housing sector in Hungary were very much like those experienced elsewhere. Annual new housing construction decreased from 36,200 to 7,300 dwellings between 2008 and 2013; real house prices were more than 30 per cent lower in late 2013 than in early 2008; housing transactions decreased by 40 per cent in the same period, and the share of non-performing loan increased from 5 to 20 per cent between 2008 and 2014.

The initial response to the crisis by the 2009–2010 interim government focused on managing the fiscal deficit, which was one of the conditions of the IMF loan. An important element of the fiscal adjustment program was the drastic cut of housing subsidies: mortgage interest subsidies and the housing construction grant were abolished.

In 2010, when the Orbán government came to power with a "supermajority", he engineered a confrontation with banks to fit the administration's "unorthodox" goals of social transformation. Housing policies were dominated by the management and mitigation of social conflicts caused by the credit crunch. The key framework was laid out in a 2011 action plan.[14]

One of the most controversial programs was the early repayment scheme. This allowed borrowers to repay their mortgage in a single lump sum at a more favourable exchange rate between September 2011 and February 2012. Early repayment targeted more affluent debtors with significant savings, creating huge losses for the banks while disregarding issues of fair redistribution and social justice. Eighteen per cent of borrowers used this opportunity (ca. 170,000 contracts) generating a total loss of one billion EUR in the banking sector, or a 5700 EUR relief per debtor. The government acted to incentivize early repayment: savings banks were ordered to offer loans, the tax office did not examine the origins of repayments, state organizations and companies were encouraged to lend to their employees, and banks were threatened with repercussions if they sabotaged the process (Csizmady et al., 2019).

The least controversial, most successful program was offered by the National Asset Management Ltd. (NET). The social housing enterprise was supported by an alliance of banks. Through NET, the state bought mortgages, turning mortgaged housing units into rental accommodation, whilst allowing the former owners to remain as tenants, paying a fixed and very low rent (around 10–15 per cent of the market rent). Thirty-six thousand units were included in the program, seven times more than expected in the initial plan. Eligibility criteria and conditions changed repeatedly as the program grew. It is difficult to estimate the overall cost of the program, but it is thought to have been around EUR 500 million, or EUR 1 400 per household.[15]

Those whose status was in between that of the most vulnerable and the most affluent were unable to take advantage of the early repayment program or the NET as they lacked savings for the former and did not meet the strict criteria of the latter. By 2018, the number and proportion of defaulters had decreased, in large part thanks to outsourcing cases to debt-collecting companies. The conversion of FX loans to HUF loans, on the other hand, did make a significant difference, even though leading experts of the Hungarian National Bank found this solution problematic.

Banks survived these years in a weakened political position, although they could compensate for some of their losses by increasing service fees. With restructured loans and the sale of non-performing loans, they could rid themselves of political responsibility and prepare for the shifting circumstances as creditors.

2015 onward: The recovery of the housing market and towards the second crisis

In 2014 the government decided to give a substantial boost to the housing market. The Family Home Allowance (FHA) program – a de facto continuation of the "social policy allowance" discontinued in 2009 – was launched in July 2015 for an indefinite period. VAT on construction was also cut from 27 to 5 per cent in January 2016 (up to HUF 5 million,[16] or ca EUR 15,000 per dwelling[17]), but only for a limited period of three years. FHA was very generous, especially for families with three or more children (offering up to HUF 10 million, or ca. EUR 30,000, in non-repayable grants), but it had very restrictive and pro-middle-class conditions. Boosting housing construction was an obvious goal of the FHA (as well as the VAT allowance), and it embedded the pro-ownership housing policy as an integral part of the government's family policy. This injection of subsidies and low interest rates (in line with the broader European trends) also helped to re-launch mortgage lending, despite stricter post-crisis lending regulations. However, FHA did not seem to significantly increase the number of childbirths. Between 2015 and 2019 the conditions of the scheme were modified nine times, to loosen conditions and include more middle and upper-middle-income young families with (planned) children.

After 2010 the Hungarian government followed an anti-migration policy, which highlighted the importance of demographic processes. In the CEE EU Member States, including Hungary, the population has long been stagnating or – more often – decreasing, the fertility rate is lower than the EU average, and net migration is negative. The housing and family policy measures did not affect measurable change in this direction.

The welfare regime and its development

The evolution of the welfare regime in Hungary is analyzed in three major periods: the pre-WWII era, the communist period, and the post-transition period. The latter cannot be clearly classified into the existing welfare regime typologies (despite numerous attempts). State control over prices and wages represent the quasi-decommodification of important segments of the market and have become a key characteristic of Hungary's welfare regime. However, the decommodification is only partial (hence "quasi") because full control cannot technically be exerted in a modern economy with a high share of foreign trade. At the same time, the second or informal economy resulting from state failures has also played a major direct or indirect welfare role in many areas (services, education, health system, housing, recreation, etc.). The rapid liberalization of prices and wages in the 1990s and the change in the functions of the informal economy led to novel challenges, to which the countries of the region sought similar solutions (primarily through instruments aimed at reducing market imbalances), but their approaches began to diverge after the 2008 crisis. One likely reason for this is the different political ideologies and economic policies. The chapter traces these stages through the example of Hungary.

At the turn of the 20th century, Hungary attempted to adopt a welfare system similar to Western European examples, although with some delay and less extensive intervention (Tomka, 2006; Szikra & Tomka, 2009), with social insurance and some more comprehensive social programs for state employees. By the end of the 1920s, a compulsory insurance system (similar to its Czech and Polish counterparts) was established, and a relatively generous old-age and sickness pension scheme was introduced at relatively low public expenditure. By the end of the 1930s, important elements of the family support system were introduced (among which family allowance, nurseries and childcare for factory workers, as well as subsidized housing loans for large families). Another important feature of the pre-WWII welfare system was the advantage of state employees over industrial workers and especially over agricultural workers (for whom the insurance system was poorly organized and participation was low). The pre-WWII system (similar to other countries in the region) can be regarded as a Bismarckian welfare system: less developed in terms of coverage, expenditure and legislation, but still broadly following the same general directions as Western European welfare regimes. A unique feature of the Hungarian welfare institutions, however, was their level of centralization, in line with the autocratic-conservative political system (Tomka, 2006; Szikra &Tomka, 2009).

After the communist takeover, a new socio-economic system was established, although many elements of the pre-war and wartime system were continued due to path dependence. In the first decade of the new communist country, the emphasis was on the elimination of the old ruling class, which was achieved by drastic measures (the persecutions of churches, show trials, nationalization, deportations, employment, university quotas, etc.). The forced industrialization policy, which ignored the living standards of the population, led to social conflict, resulting in massive demonstrations in Berlin and Warsaw, as well as the Hungarian Revolution of 1956. In the countries of Central and Eastern Europe under the influence of the Soviet Union, a more conciliatory totalitarian regime emerged from the late 1950s (Khrushchev's "socialism with a human face"), which, at least in its rhetoric, saw the raising of living standards as an important objective.

The new system that emerged could not be called a welfare society in the Marshallian sense because of the limited civil and political rights, although growing prosperity helped guarantee the stability of power. Welfare policies were based on an economic model in which, on the one hand, prices and incomes were determined by the state, and on the other, it guaranteed and also enforced full employment (i.e., punished those who did not work) (Szikra & Tomka, 2009, pp. 22–23). An important element of the state welfare system was the provision of a universal, centralized social security system, where forms of social support were closely linked to the payment of prior individual contributions (similar to the conservative welfare system, where higher previous income results in higher social support). In the 1960s, various forms of family support (such as long-term, maximum three years paid maternity leave and family allowances) were introduced. The social welfare system covering health, education, housing, old-age pensions and so forth, was provided by the state partly through companies and partly through local authorities ("councils"). However, this should not be interpreted as a withdrawal of the state from the provision of social services, as both companies and municipalities were under the direct control of the centralized one-party state.

State control of the economy can never be complete (Kornai, 1992). The "gaps" created by the lack of control are filled by private actors, whose behaviour and interactions result in new types of structures, different from the logic of the command economy. The informal or secondary economy plays an important role in the welfare provision of socialist systems.

According to the rhetoric of socialist politicians, social issues such as poverty, unemployment, and inequalities were (presumed) non-existent. Formal income inequalities were low, but there were large inequalities in access to essentially free and heavily subsidized services. The nomenclature, mostly consisting of more highly educated managerial workers and their families, benefited from better access to health care, education, holiday options, etc. (Deacon, 1992; Ferge, 2008). The quasi-market conditions of the informal economy also increased inequalities, although whether the inequalities of the informal economy compensated for or amplified the inequalities of the first economy is debated in the literature (Szelényi & Manchin, 1987).

Despite the corrective role of the family and the informal economy, significant social groups were excluded from welfare benefits under the state welfare system. The most obvious example was the Roma. The official view was that poverty could not exist in a socialist economy, it being merely the product of previous regimes, but special efforts were being made to eradicate it. Programs to eradicate the disadvantages experienced by the Roma were launched in the 1960s. Although their situation improved, the system could not eradicate poverty and policy had to address it through specific programs.

As a result of the regime change in 1990, a market economy and multi-party democracy were established. According to the prevailing neoliberal worldview, the task was to dismantle the command economy and build an efficient market economy. The transformation would entail an insurmountable social cost in that the two pillars of the socialist system – price control and full employment – were abandoned. By the 1980s, both principles had been violated: not only hidden but also visible inflation was appearing, and control over employment was far from complete, with unemployed and homeless people appearing, and large numbers of people were taking on jobs in the informal sector. Economic transition strongly exacerbated these processes in the 1990s and created unprecedented social problems for the newly elected governments. Between 1990 and 2010, a new welfare system was built against a backdrop of ideological debates, with incremental changes with often unpredictable consequences; measures were either closer to the neoliberal position or more solidly socially democratic.

An unemployment benefit was introduced, and the accompanying measures (early retirement for instance) were relatively generous, but the system was forced to move towards workfare by 2009 because welfare-dependent disadvantaged groups and people in the informal economy were stuck in deep poverty. The reform of the pension system was textbook neoliberal, but in its concrete implementation, the new private pension funds were able to impose their high cost of management, which made it unpopular even with liberal experts. Neoliberals and social democrats also clashed over the reform of the health care system, but a populist referendum in 2007 ended the experiment. However, the traditional system was gradually transformed, and a private healthcare market gradually emerged beside the state-managed public system without any major reforms, through a spontaneous quasi-privatization process.

The provision of benefits was delegated from the central state to the local municipalities. Decentralization was an important step in the politics of neoliberal society. Accountable municipalities with real autonomy could have become important players in the market economy. Yet local authorities have not been able to fulfil this role satisfactorily, partly because of funding anomalies and partly because of their political exposure. In the social sphere, the role of local authorities has been controversial, as they typically represented the interests of the local middle class, which led to serious conflicts with groups living in extreme poverty (many those of Roma ethnicity), which were reflected in the distribution of benefits, the fight against segregation in schools, and the controversies surrounding the organization of public works (Szalai, 2013).

A significant part of welfare spending is devoted to family policy, encouraging childbearing, and contributing to the costs of raising children. In principle, the support system can follow one of two logics: it either provides support in proportion to the number of children, irrespective of the mother's or family's income, or it provides support in the form of tax credits.

The period between 1990 and 2010 does not fit neatly into Esping Andersen's (1990) typology, and it is no coincidence that the literature has classified the nature of the emerging welfare regime in very different ways. In practice, different professional and political groups interacted – in a way, they fought a battle of professional ideas, which of course had political consequences for wealth and finance.

Post-2010, a populist authoritarian ("illiberal") regime has been built, which does not appear to have a clear vision on the nature of a desirable welfare regime, but rather mobilizes welfare provision in service of legitimizing and maintaining political power. The transformation of the welfare system in all areas is subordinated to this objective. The modification of the elements of the welfare system cannot be understood as a mere dismantling of the liberal welfare system (Vidra, 2018). The political rhetoric emphasizes family support as a catchphrase, including the empowerment of the "working middle class". In practice, it introduces the punitive intimidation of the poorest, often degrading conditions of public work, income-dependent subsidy schemes for family and housing benefits (where one can access higher subsidy amounts with higher income and savings), and the criminalization of homelessness. The public works program is effectively run by politically committed mayors essentially as a form of unemployment benefit, at a higher rate than the official unemployment benefit, but it leaves clients dependent on the local political leaders (mayors). Education is being centralized, which interestingly has not satisfied critics of decentralized education, as it has not eliminated segregated education, but instead reduces the role of desegregation programs by squeezing local autonomy. Private pension funds – previously part of a three-pillar system – were abolished in 2010, and integrated their accumulated funds into the central government budget to counterbalance the losses stemming from the financial crisis. This could have been seen as an attack on the neoliberal welfare system, but in practice, the main motivation of the state was the acquisition of budgetary resources.

In education and social care, religious organizations close to the government are emerging and taking over institutions and functions, squeezing out local authorities and NGOs with greater independence. Benefits and financial processes are becoming opaque.

Szalai (2013) does not seek to categorize the Hungarian welfare system in the same way as Esping-Andersen but argues that a dual system is emerging, which she calls a bifurcated welfare system, in which those in more influential positions in society successfully exploit or defend their welfare privileges against those who are deprived of them. The two groups have very different and unequal rights and entitlements in the distribution of wealth. However, for the post-2010 period, this theory in its present form is invalid. The reason is that Hungarian society cannot be described by a dual-class structure, as the middle class plays a dominant role in the welfare system (and in politics). So-called "unorthodox" social policies in welfare (and, as we will see, in housing policy) promise strong support for the working middle class, but this promise often proves empty. The key issue for the welfare system is the emergence of an insecure middle class (precariat), which is the social basis of the autocratic political establishment. It is this class that craves the security that socialism's low prices and full employment guarantees seemed to offer, which is essentially an illusion. It can only be temporarily guaranteed in market economies due to the scarcity of resources. The Orbán regime's populist policy is based on an insecure middle class and will remain so until its economic unsustainability is proven.

The interrelation between housing and welfare

Esping-Andersen's (1990) welfare systems theory gave a big boost to comparative housing research. Researchers clearly saw that housing was at least as important an area of the welfare system as pensions, education, or healthcare, and sought to extend welfare regime theory to include housing. However, research over the last 20–30 years has failed to show beyond doubt a link between welfare regime typology (conservative, social democratic, neoliberal) and housing (Hoekstra, 2003, 2005). The basic reason for this is that housing is a specific sector (a "wobbly pillar") within (and outside) the welfare system. At the same time, the close link between the housing and welfare systems has called into question whether the welfare functions of the housing system can be examined in isolation from the welfare system (e.g., general income support is part of both the housing and the welfare systems). The next section illustrates this link with two examples: the heating gas price subsidy, which was in force between 2007 and 2010, and the history of the "social policy" housing construction subsidy, linked to the number of (planned) children.

Manoeuvring between the general and targeted subsidy programs

One of the most important changes the transition induced in the post-socialist housing systems was a dramatic increase in the housing costs of households. In Hungary, housing-related expenditures increased from an average of 9 per cent of household budgets in 1989 to 21 per cent in 1993 and then stabilized at this level, while the income gap (the ratio of average income in the highest and lowest deciles) increased from 4.7 to 7.3 by the end of the 1990s. As a result, housing costs became a serious problem for the bottom quintile of the population, and the long-term housing security of even the lower middle class became uncertain. After the privatization of housing and the settlement of "old loans" (Hegedüs & Várhegyi, 2000), 90 per cent of homeowners were free of mortgage debt. Housing maintenance costs, especially energy costs, have become a major factor in housing affordability. In Hungary, the price of imported natural gas almost doubled between 2005 and 2008, so that the general price subsidy, which was then guaranteed to residential consumers, could no longer be financed by the central budget. Therefore, from the beginning of 2007, the previous subsidy was replaced by a means-tested subsidy scheme, which also meant that the price of gas and other household energy became significantly more expensive for the general population. Compared to 2003 prices, the price of household energy increased by 96 per cent, while gas prices rose by two and a half times. The targeted support scheme was also necessary because the increase in energy prices hit poorer households harder: while energy expenditure accounted for an average of 7 per cent of household income, its share for the lowest income group was between 10 and 14 per cent. The scheme introduced in 2007 set the eligibility threshold at a relatively high income level, and the impact of the price increase on a significant proportion of households was only phased in gradually, with more than half of consumers receiving income-related subsidies. The gas subsidy in 2007 was five times the amount of the means-tested housing maintenance subsidy, which had been in place since 1993. The gas subsidy (as opposed to the housing maintenance subsidy) was proportional to consumption, but it was indirectly well-targeted, with the bottom 20 per cent of households receiving 60 per cent of the subsidy. However, the gas subsidy was only intended to be temporary, and was phased out gradually: in 2008, HUF 74 billion (EUR 19 million) was allocated to the subsidy, in 2009 HUF 67.7 billion, while in the 2010 budget, only HUF 20 billion. In 2011, the government abolished the subsidy and instead extended the scope of the central housing maintenance subsidy, without increasing the amount of the subsidy.

Since 2010, the Fidesz (right-wing) government has replaced "traditional" austerity programs with measures based on "unorthodox" economic policies. This political paradigm shift had an impact on Hungary's emerging social policy, which focused primarily on strong support for the middle class. In 2015, the government abolished the national housing subsidy scheme, which distributed HUF 30 billion (per year) to 400,000 applicants (and typically targeted the lowest income quintile). It also transferred the responsibility for implementing housing subsidy schemes to local governments. These policies were in line with the general trend to devolve responsibility for social issues related to affordability (i.e. poverty) to the local level, leaving it to service providers, municipalities, and housing associations on the one hand, and consumers on the other, to resolve their affordability conflicts without central public assistance. Secondly, in 2015 the government launched a political campaign called the "war on utility costs": utility tariffs (for energy, water, waste management and other services) were cut and frozen across the board, in parallel with the partial renationalization and centralization of utility companies. As a result of this program, the household sector saved HUF 242 billion on utility costs in 2015 (MEKH, 2016), but the targeting of this "support" was regressive and did not offer a solution for the most deprived segment of the population. Our calculations, based on the data of the 2015 representative housing survey of the Hungarian Central Statistical Office (HCSO), show that the top two income quintiles benefited from this subsidy at a rate 34 per cent higher than the bottom two quintiles.

The rationale for this program was exactly the opposite of the gas price subsidy introduced in 2007 and was similar to the feature of the socialist welfare system whereby housing expenditures (rent, utility cost, and mortgage payments) for households were decommodified. However, maintaining such a program requires a significant budgetary outlay, and often leads to a rapid deterioration in the quality of services. In 2022, the budget deficit has increased so much that the government suspended this scheme, and introduced market tariffs above a centrally defined normative consumption level.

The "social policy" construction subsidy and its legacy

The close link between the welfare system and the housing system is demonstrated by the family policy elements of housing subsidies. The first program to link homeownership subsidies to family policy was the "social policy subsidy", introduced in 1971. It has been revised and renamed several times and had to be temporarily suspended following the 2008 financial and housing market crisis, but overall, it has remained relevant for most of the past 50 years. Its most recent version, the Family Housing Assistance (FHA) scheme, was relaunched in 2015 as a capital subsidy for home ownership, with the amount of the subsidy depending on the number of children (born or planned). Although the justification for the subsidy was to compensate for the additional burden on families with children, it has always been accompanied by changes to the subsidy package. In 1971, the direct subsidy for construction companies was abolished and replaced by the demand side "social housing subsidy"; in 1989, the increase in the previously 3 per cent fixed interest rate was to be compensated; in 1994, the introduction of VAT was to be compensated by an increase in the subsidy; in 2003/2004, another increase in the subsidy was meant to compensate for the reduction in the interest subsidy (Figure 12.2).

The most recent iteration of the subsidy was introduced six years after the global financial crisis. The Family Home Allowance (FHA) program was launched in 2015 for an indefinite period,

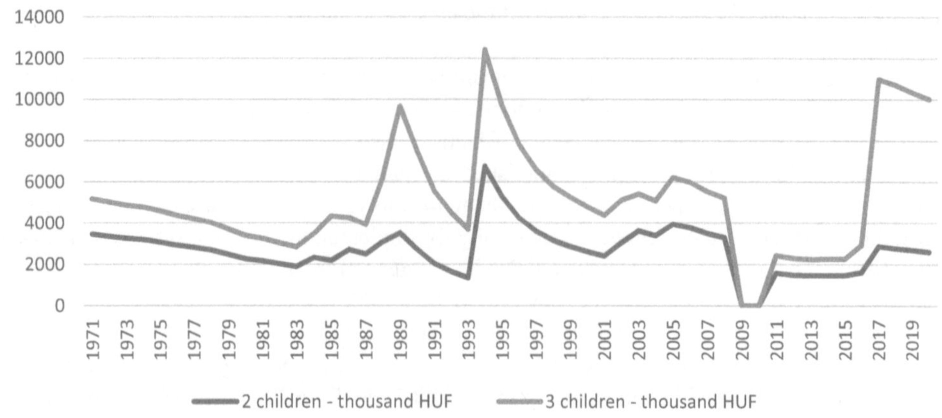

Figure 12.2 Amount of FHA in HUF (base year: 2018) in Hungary for a family with two children and a family with three children.

Source: MRI compilation based on housing policy laws.

and the VAT on construction was cut for a limited period of three years. Boosting housing construction was an obvious goal, and pro-ownership housing policy became even more predominant than previously. These measures also helped the recovery of mortgage lending.

One popular theory of the relationship between housing and welfare is the asset-based welfare system, which assumes that there is a trade-off between the share of owner-occupied housing and the share of welfare. In this sense, individual accumulation of housing wealth can create a source of welfare spending and relieve the burden on the state. This theory often goes hand in hand with financialization theory, which, in a simplified way, emphasizes the role of financial relations (mainly credit) in socio-economic relations (a new integration mechanism based on asymmetric power relations). The dynamics of mortgage lending have slowed down in the region and are clearly inferior to a well-functioning housing finance system. Moreover, the phenomena of formal and informal lending (e.g., usury) to the lowest income groups cannot be taken as evidence for financialization theory. Similarly, the broader spread of private property in itself cannot be taken as evidence of a wealth-based welfare system.

Regional outlook

The housing systems of CEE countries under the influence of the Soviet Union followed a similar model. In socialist political ideology, raising living standards (including access to and quality of housing) was an important legitimizing factor, but housing policy was influenced more by the practical needs of the planned economy than by ideological justification. The actual housing policy followed a common model, which is identified in the literature as the socialist or Eastern European housing model (Hegedüs, 1992; Hegedüs & Tosics, 1996). From an economic viewpoint, the housing system was part of the economic model in which shortage was a central feature (Kornai, 1992). In this system, bureaucratic coordination integrated the different sectors of the economy instead of market mechanisms. The allocation of resources (investment, credit, etc.) in the housing sector was controlled by the state, independently of supply and demand factors. The vast majority of services were provided in kind or below market prices and allocated on the basis of "merit". Artificially low, heavily subsidized housing costs created a permanent shortage, resulting in a dual housing market. In addition to the state-controlled housing sector, there was an informal economy consisting of owner-occupied housing, private transactions in the rental housing sector, a private real estate market, a submarket for renting, and a small, fully private rental sector (Alexeev, 1988, 1990; Hegedüs & Tosics, 1996).

However, many variations of the model have emerged as different countries have responded differently to the problems within the socialist economy. This led to "cracks" (Hegedüs, 1992) in the housing model: the behaviour of housing market actors to overcome the problems of the system no longer corresponded to centrally planned state policy, and the institutional and legal framework could not prevent this process. Countries reacted to the emergence of these "cracks" in different ways: either by introducing strict control mechanisms (e.g. Bulgaria, Russia, East Germany) or by tolerating quasi-market processes (Yugoslavia, Hungary). Thus, Yugoslav guest workers of the 1960s and 1970s brought home their income from Western Europe and invested it in private housing. Similarly, in Hungary, income from the informal economy, mainly from private agricultural production, was used to improve housing conditions. In theory, private renting did not fit into the socialist housing model, but it existed as a "crack" in the system in all countries due to housing shortages and the extreme pressure on urban housing markets (Hegedüs & Puzanov, 2017).

A review of the East European Housing Model shows that social housing policy was part of the broader welfare system and should be analyzed in this broader context. However, comparative studies have had difficulty in conceptualizing the Soviet "welfare system". Classifications such as the "industrial output-performance model" (Wilensky, 1975) or the "authoritarian welfare state" (Aidukaite, 2009) do not bring us any closer to understanding the welfare systems of socialist countries. Within the public sector, welfare benefits were not income-dependent because income differentials were small and services were either free or deeply subsidized and thus affordable for essentially all social groups. According to the official ideology (reflected in their constitutions, laws and policies) these societies were egalitarian, but detailed analysis shows that in reality individuals' life chances were strongly influenced by informal distribution mechanisms depending on their economic and political position. These may have been privileged positions within the one party-state (in terms of housing allocation, recreational opportunities, access to education and health care) or quasi-market mechanisms of distribution (Szelényi, 1983; Szelényi & Manchin, 1987; Alexeev, 1988).

The housing systems of CEE countries changed fundamentally after the regime changes in 1998/1990. The main periods of housing policy in the former socialist countries are very similar. The first period in the 1990s was dominated by housing privatization, the dismantling of socialist-era institutions (including the settlement of "old loans" and the completion of stalled construction projects), and the establishment of new state and market institutions. The second main period began around 2000, with increasing housing construction and the expansion of mortgage lending, to which countries provided varying degrees of state support. The main priority of housing policy was to support the owner-occupied housing sector (through a VAT tax credit, loan subsidies, or personal income tax allowance), but there have also been attempts to support specific public housing programs (social housing, youth housing, etc.). These programs are often short-term, based on policy incentives, and then stopped due to budget constraints. The 2008 crisis triggered a downturn, affecting individual countries to a very varied extent. The severity of the crisis was very much influenced by the size and quality of the housing loan portfolio (share of foreign currency loans and quality of lender underwriting). A new boom is expected in almost every CEE country within 3–7 years following the crisis, which will most likely be ended by the economic crisis following the war in Ukraine.

Unresolved social housing problems are an important issue in all post-socialist economies, with several shared elements

- The public rental housing stock was transferred to municipalities, and the privatization of housing has been significant. Privatization at a heavily discounted price has radically reduced the social housing stock. The remaining rents in municipal ownership do not cover the running costs, and municipalities have to cover the losses from their own revenues. This, combined with a lack of resources for renovation, has led to a decline in the public housing sector.
- Housing costs have risen rapidly in all countries due to the liberalization of utility and energy prices. Housing (and utility) allowances were introduced to reduce the burden. However, subsidies (due to income measurement problems and scarce budgetary resources) have not been able to offset the increase in the burden on households, and rising income inequalities in particular exacerbated it.
- Income inequalities have put housing for low-income (urban) families in a critical situation. The social rented housing sector has become limited, providing a solution for very few families. Low-income groups could find a solution in the urban private rented sector, in the low end of private market housing, or settlements further away from urban centres. Access to housing for young families depends strongly on intergenerational transfers.
- Much of the private rental market is informal. Liberal legal regulation has created serious risks in the system, for both landlords and tenants. The supply side is dominated by small-scale, casual landlords, although it is increasingly becoming a coveted investment form among higher-income families.
- Condominiums and multi-storey buildings (including pre-fabricated housing estates from the socialist era) in all countries are facing serious challenges, partly due to the quality of the housing stock (especially in terms of energy efficiency) and because of the low financial capacity of individual owners (Table 12.3).

Table 12.3 Comparative data, selected CEE countries

	Average annual dwellings built between 2010 and 2021 as a % age of housing stock of 2021	*Total outstanding residential loans to GDP ratio*		*Average transaction price of new dwellings (EUR/ sq.m.)*	*GDP 2021/2009*	*GDP PPP per capita*	*House prices in 2021 (2015 = 100)*	*Interest rates,*[18] *% (2021)*
		2008	*2021*					
Estonia	0.7%	37.3	29	n.a.	212	41,890	133	2.15
Poland	1.1%	12.7	20.5	1,581	183	37,320	130	3.06
Slovakia	0.9%	12.9	32.9	1,941	153	35,550	138	0.98
Hungary	0.3%	20.6	9.5	1,657	168	36,850	169	4.43
Romania	0.6%	3.6	7.6	1,332	204	35,066	110	4.12
Slovenia	0.4%	9	13.7	2,748	139	43,210	141	2.11

Source: Hypostat 2022, GDP per capita. https://en.wikipedia.org/wiki/List_of_sovereign_states_in_Europe_by_GDP_(PPP)_per_capita

Despite the commonalities, the differences between post-socialist housing systems have widened in the aftermath of the 2008 economic crisis. Housing finance must be built on a strong middle class that grows with stable economic growth and relies on a highly skilled, specialized workforce rather than an assembly line of cheap labour. Housing subsidies, if not targeted, will not lead to an efficient housing system, as the data in Hungary show. Another challenge facing the housing regimes in the new Member States is how to provide housing for the poor. Here too, differences are apparent (municipal housing sector, developing non-profit sector, rental agencies, etc.), but scale-up is still to come.

Concluding discussion: Competing interpretation of the post-socialist housing regimes

Housing policy literature in the past 30 years has been dominated by two competing paradigms. The first is the World Bank's "market-enabling" theory, which stems from a critique of state failures in developing countries, but was, in fact, shaped by the failure of socialist planned economies in the early 1990s (World Bank, 1993; Renaud, 1995). The essence of this approach is to make room for efficient market conditions; to get rid of unnecessary regulations, and to remove social and political interventions from the market. In the housing system, this has simply meant the dominance of private property and the introduction of market housing finance as well as specifically targeted housing and social assistance schemes for vulnerable groups. The other dominant paradigm was created as a reaction to market failures. The "Housing for All" paradigm argued that markets should be regulated, housing subsidies should be widely applied, through which private market rent volatility should be eliminated and predictable procedures should be used in the housing finance system; and was based on a positive mix of public interventions. Housing policies based on Kemeny's (1995) theory point in this direction, and Housing of Europe is at the forefront of this debate as it advocates a mixed-tenure, affordable housing policy alongside very social housing, the support of Limited Equity models, etc. These proposals essentially advocate a "Housing for All" policy.

An important difference between the two paradigms is the different understanding of the structure of society. The market enabling theory assumes that economic development creates strong middle classes whose stable labour market position does not require state intervention in housing. Under efficient market conditions, strong middle classes will also have secure housing. Of course, there will be those who fall behind, but they will be taken care of by society through targeted support programs. In the "Housing for All" paradigm, the middle class is not strong, instead, a large, precarious middle class emerges, whose members are exposed to the volatility of market conditions and whose position can therefore change rapidly for the better or the worse. This labour market volatility must be offset by housing policy, which, like education or health care, must not allow large sections of the population to suffer a deterioration in their position as a result of market volatility.

The failure of housing policy in the CEE EU Member States lies in the fact that they basically followed the first approach: where possible, they quickly introduced market conditions, which were ineffective due to the lack of stable social institutions (rule of law, corruption, etc.). In addition, and more importantly, the situation of the middle classes became very unstable; elements of the welfare system that had worked in Western Europe were not effective in CEE countries. The housing expectations of middle-class households in the new Member States are not significantly different from those of households in countries two to three times richer than those in Central and Eastern European countries, and if the economic development model does not consolidate the financial position of the majority of the middle class, a Western European-style welfare system will be too costly in relation to the performance of the economy. Furthermore, well-designed, targeted subsidy system runs into difficulties in measuring needs due to the size of the informal economy and the opacity of intergenerational transfers. A successful strategy for the new Member States will depend on their ability to regulate their markets in a way that minimizes the costs of intervention and introduces structural changes in the economy that improve the stability of the middle class.

Appendix: Source of Data for Table 1

I1 https://en.wikipedia.org/wiki/List_of_sovereign_states_in_Europe_by_GDP_(PPP)_per_capita#endnote_a

I2 ESTAT Population on 1 January [TPS00001], Population by sex, age group and country of citizenship [CENS_01NSCTZ__custom_4403926]

I3 ESTAT Foreign-born population [TPS00178]

I4 ESTAT Population on 1 January [TPS00001], Population by sex, age group and country of citizenship [CENS_01NSCTZ__custom_4403926]

I5 ESTAT Population on 1 January [TPS00001], Population by sex, age group and country of citizenship [CENS_01NSCTZ__custom_4403926]

I6 Average Transaction Price of a New Dwelling in Selected Countries (EUR/sqm), 2020 in Property Index Overview of European Residential Markets 10th edition, July Deliotte2021 https://www2.deloitte.com/content/dam/Deloitte/at/Documents/real-estate/at-property-index-2021.pdf;

I7 Average Monthly Rent (EUR/sqm) in capital cities, simple avarage in Property Index Overview of European Residential Markets 10th edition, July Deliotte2021 https://www2.deloitte.com/content/dam/Deloitte/at/Documents/real-estate/at-property-index-2021.pdf;

I8 Hypostat 2022

I9 ESTAT Share of population having moved to other dwelling within the last five-year period by tenure status and degree of urbanization [ILC_HCMP05$DEFAULTVIEW]

I10 Hypostat 2022 and ESTAT Population on 1 January [TPS00001], Population by sex, age group and country of citizenship [CENS_01NSCTZ__custom_4403926]

I11 ESTAT Distribution of population by tenure status, type of household and income group - EU-SILC survey [ILC_LVHO02__custom_4404467]

I12 ESTAT Distribution of population by tenure status, type of household and income group - EU-SILC survey [ILC_LVHO02__custom_4404467]

I13 Hypostat 2022 https://www.helgilibrary.com/indicators/housing-stock-number-of-dwellings/malta/; https://www2.deloitte.com/content/dam/Deloitte/at/Documents/presse/at-property-index-2022-final.pdf;https://www.economy.com/indicators/dwelling-stocks

I14 ESTAT Average number of rooms per person by tenure status and dwelling type from 2003 onwards - EU-SILC survey [ILC_LVHO03__custom_4404668]

I15 ESTAT Average size of dwelling by income quintile and tenure status [ILC_HCMH01$DEFAULTVIEW]

I16 ESTAT Overcrowding rate by age, sex and poverty status - total population - EU-SILC survey [ILC_LVHO05A$DV_569]

I17 ESTAT Severe housing deprivation rate by age, sex and poverty status - EU-SILC survey [ILC_MDHO06A$DV_657]

I18 ESTAT Share of housing costs in disposable household income, by type of household and income group - EU-SILC survey [ILC_MDED01$DEFAULTVIEW]

I19 ESTAT Arrears (mortgage or rent, utility bills or hire purchase) from 2003 onwards - EU-SILC survey [ILC_MDES05$DEFAULTVIEW]

I20 Medina, Leandro, and Mr Friedrich Schneider. Shadow economies around the world: what did we learn over the last 20 years?. International Monetary Fund, 2018. IMF Working Papers 2018,

Notes

1 This group comprises of the former communist countries which joined the EU since 2004, excluding Cyprus and Malta, referred to in the text as EU-11.
2 The indicators for each country group are weighted by population or housing stock.
3 See data sources in the Appendix.
4 Hypostat 2022 https://hypo.org/app/uploads/sites/2/2022/09/HYPOSTAT-2022-FOR-DISTRIBUTION.pdf
5 National Institute for Social Insurance (Országos Társadalombiztosítási Intézet – OTI)
6 Insurance Institute for Private Employees (Magánalkalmazottak Biztosító Intézete – MABI)
7 Then, in the 1960s, the process of tenancy consolidation began.
8 Imre Nagy, a centrist communist politician, replaced the Stalinist leaders in government after the death of Stalin.
9 J. Kádár was the first secretary of the communist party, and the de facto political leader, from 1956 to 1988.
10 Wage garnishment is a legal procedure in which a person's earnings are required by court order to be withheld by an employer for the payment of un-paid debt.
11 The counties had highest level of power within the one party system, right after the central comittee of the communist party.
12 The first administration led by PM Viktor Orbán, in 1998–2002.
13 The 'social policy grant' for housing construction was introduced in 1971. Its conditions were changed several times, but the main aim of the subsidy has remained the same: to help families with children to obtain home ownership by contributing to their down payment.
14 In decree LXXV (2011) key elements of the rescue programs were present, such as a moratorium on the number of evictions, an exchange rate cap, and conversion to HUF. However, repeated amendments of the law indicate uncertainty and a lack of long-term planning.
15 NET was established in 2015, when the average price of used dwellings was ca. HUF 11–12 million according to Hungarian Central Statistical Office data, see (in Hungarian): https://www.ksh.hu/stadat_files/lak/hu/lak0014.html, last accessed 20 November 2022.
16 1,000 HUF is Equivalent to 2,5 EUR (February 2023)
17 The maximum size of a unit eligible for the VAT allowance is 150 sq.m. for an apartment, or 300 sq.m. for a detached house.
18 Source: Hypostat 2022 Table 4. Representative Interest Rates on New Residential Loans, annual average based on monthly figures, percent.

References

Aidukaite, J. (2009). The Transformation of Welfare Systems in the Baltic States: Estonia, Latvia and Lithuania. In A. Cerami & P. Vanhuysse (Eds.), *Post-Communist Welfare Pathways* (96–111). London: Palgrave Macmillan. https://doi.org.10.1057/97802302458086

Alexeev, M. (1988). Market vs. Rationing: The Case of Soviet Housing. *The Review of Economics and Statistics, 70*(3), 414–420. https://doi.org/10.2307/1926779

Alexeev, M. (1990). Distribution of Housing Subsidies in the USSR, with Some Soviet-Hungarian Comparison. *Comparative Economic Studies, 32*(3), 138–157.

Csizmady, A., Hegedüs, J., & Vonnák, D. (2019). A Housing Regime Unchanged: The Rise and Fall of Foreign-currency Loans in Hungary. *Corvinus Journal of Sociology and Social Policy, 10*(2), 3–33. https://doi.org/10.14267/CJSSP.2019.2.1

Deacon, B. (1992). East European Welfare: Past, Present and Future in Comparative Context. In Bob Deacon (Ed.), *The New Eastern Europe: Social Policy Past, Present and Future* (pp. 1–31). Sage.

Esping-Andersen, G. (1990). *The Three Worlds of Welfare Capitalism*. Polity.

Ferge, Z. (2008). Is There a Specific East-Central European Welfare Culture? In W. van Oorscot, M. Opielka, & B. Pfau-Effinger (Eds.), *Culture and Welfare: Values of Social Policy from a Comparative Perspective* (pp.141–161). Edward Elgar.

Hegedüs, J. (1992). Self Help Housing in Hungary. In K. Matey (Ed.), *Beyond Self-Help Housing* (pp. 217–231). Profil Verlag.

Hegedüs, J. (2010). Towards a New Housing System in Transitional Countries: The Case of Hungary. In P. Arestis, P. Mooslechner, & Wagner (Eds.), *Housing Market Challenges in Europe and the United States* (pp. 178–202). Palgrave Macmillan.

Hegedüs, J. (2018). Lakásrezsimek Rendszerváltás Előtt és Után a Posztszocialista Államokban [Housing Regimes Before and after the Transition in Post-Socialist States]. In A. Bozóki & K. Füzér (Eds.), *Lépték és Irónia: Szociológiai kalandozások [Scale and Irony: Adventures in sociology]* (pp. 73–116). Hungarian Academy of Sciences, Centre for Social Science.

Hegedüs, J., Horváth, V., & Somogyi, E. (2017). *Affordable Housing in Central and Eastern Europe: Identifying and Overcoming Constraints in New Member States*. Metropolitan Research Institute. https://ec.europa.eu/futurium/en/system/files/ged/2018.10.22_affordable_housing_in_central_ and_eastern_europe.pdf

Hegedüs, J., Mark, K., & Tosics, I. (1996). Uncharted Territory: Hungarian Housing in Transition. In R. Struyk (Ed.), *Economic Restructuring in the Former Soviet Block: The Case of Housing* (pp. 71–137). The Urban Institute Press.

Hegedüs, J., & Puzanov, A. (2017). The Private Rental Sector Under Socialism. In J. Hegedüs, M. Lux, & V. Horváth (Eds.), *Private Rental Housing in Transition Countries – an Alternative to Owner Occupation?* (pp. 41–69). Palgrave.

Hegedüs, J., & Somogyi, E. (2005). Evaluation of the Hungarian Mortgage Program 2000–2004. In J. Hegedüs & R. J. Struyk (Eds.), *Housing Finance: New and Old Models in Central Europe, Russia and Kazakhstan* (pp. 177–208). OSI/LGI.

Hegedüs, J., & Tosics, I. (1983). Housing Classes and Housing Policy: Some Changes in the Budapest Housing Market. *International Journal of Urban and Regional Research, 7*(4), 467–494. https://doi.org/10.1111/j.1468-2427.1983.tb00403.x

Hegedüs, J., & Tosics, I. (1996). Hungary. In P. Balchin (Ed.), *Housing Policy in Europe* (pp. 244–271). Routledge.

Hegedüs, J., & Várhegyi, É. (2000). The Crisis in Housing Financing in Hungary in the 1990s. *Urban Studies, 37*(9), 1610–1641. https://www.doi.org/10.1080/00420980050085469

Hoekstra, J. (2003). Housing and the Welfare State in the Netherlands: An Application of Esping-Andersen's Typology. *Housing, Theory and Society, 20*(2), 58–71. https://doi.org/10.1080/14036090310000634

Hoekstra, J. (2005). Is there a Connection between Welfare State Regime and Dwelling Type? An Exploratory Statistical Analysis. *Housing Studies, 20*(3), 475–495. https://doi.org/10.1080/02673030500062509

Kemeny, J. (1995). *From Public Housing to the Social Market: Rental Policy Strategies in Comparative Perspective*. Routledge.

Kornai, J. (1992). *The Socialist System: The Political Economy of Communism*. Oxford University Press.

Mitra, P. (2001). *Transition—the First Ten Years: Analysis and Lessons for Eastern Europe and the Former Soviet Union*. The World Bank

Ruisz, D. (1941). Lakáspolitikánk Időszerű Kérdései [Current Issues in our Housing Policy]. *Az Ország útja, 1941*(9), 308–312.

Szalai, J. (2013). Hungary's Bifurcated Welfare State: Splitting Social Rights and the Social Exclusion of Roma. *Adam Smith Foundation Working Papers* 2013:5. University of Glasgow.

Szelényi, I. (1983). *Urban Inequalities under State Socialism*. Oxford University Press.

Szelenyi, I., & Manchin, R. (1987) Social Policy Under State Socialism: Market, Redistribution, and Social Inequalities in East European Socialist Societies. In M. Rein, G. Esping-Anderson, & L. Rainwater (Eds.), *Stagnation and Renewal in Social Policy* (pp. 102–139). M.E. Sharpe.

Szikra, D., & Tomka, B. (2009). Social Policy in East Central Europe: Major Trends in the Twentieth Century. In A. Cerami & P. Vanhuysse (Eds.), *Post-communist Welfare Pathways* (pp. 17–34). Palgrave Macmillan.

Tomka, B. (2006). East Central Europe and the European Social Policy Model: A Long-Term View. *East European Quarterly, 40*(2), 135.

Vidra, Z. (2018). Hungary's Punitive Turn: The Shift From Welfare to Workfare. *Communist and Post–Communist Studies, 51*(1), 73–80. https://doi.org/10.1016/j.postcomstud.2018.01.008

Wilensky, H. (1975). *The Welfare State and Equality: Structural and Ideological Roots of Public Expenditures*. University of California Press.

World Bank. (1991). *Hungary - Housing Policy Reform*. World Bank Group. http://documents.worldbank.org/curated/en/164301468254705513/Hungary-Housing-policy-reform

13

HOUSING AND WELFARE IN CHINA

Julie Tian Miao

The current state of housing in China

China operates dual urban-rural land and population management systems underpinned by its household registration system. Urban land is owned by the state and its use rights can be sold, transferred, or leased. Rural land is collectively owned by farmers, and its usage is confined to three major types: residential plots for farmers as a form of social welfare, public facilities, and township or village enterprises. Rural land cannot be used for urban development nor transacted on the open market until it is acquired by the state. In this case, rural land is effectively converted into urban land, and rural households become urban households (Lin & Ho, 2005). This chapter focuses on the features and evolution of urban housing and welfare in China.

Housing tenure structure in urban China used to be predominantly rental (82 per cent before 1980) and was provided by municipal governments and/or work units and let at a token rent (Kirkby, 1985). Distribution was geared towards senior public officials and employees in key industrial sectors deemed crucial for economic growth. In the following four decades, China transformed its housing system from one which was largely controlled by the state to one that was mediated by the market. Private developers now provide over 80 per cent of new housing in urban areas. Individual households have become the primary purchasers, and urban homeownership surged to 87 per cent by 2019. More than 20 per cent of Chinese households own multiple homes, higher than in many developed nations (Huang, Yi, & Clark, 2020). This achievement is in sharp contrast to the recent decline of homeownership rate in the U.S., much of Western Europe and other developed countries.

Along with the booming real estate market came, however, severe unaffordability challenges. The house price-to-average income ratio reached 9.2 at the end of 2015. In major cities such as Shenzhen, Beijing, and Shanghai, the price-to-income ratio reached 27.7, 20.8, and 18.1, respectively, if affordable housing subsidized or regulated by the government is excluded (Cao, 2016). Increasingly, people take to social media to express their indignation at deteriorating affordability. Expressions such as 'panic' and 'housing slave' are used to capture the pressures faced by the Chinese middle class, let alone those lower-income families in climbing up the housing ladder.

Consequently, solving the urgent housing needs of 7.5 million low-income urban households and the 12 million families living in shantytowns became a government priority at the 17th

DOI: 10.1201/9781003212690-15

Congress of the Chinese Communist Party (CCP) in 2007 (Wang & Murie, 2011). Prompted by the Great Financial Crisis, there has been a noticeable return of government attention to housing issues since 2011. Given the variegated housing subsidies used in China, this chapter uses "public housing" (*gongfang*), to denote housing directly provided by the government to lower-income families. It includes low-rent housing (*lianzu fang*) and public-rental housing (*gongzu fang*). The official collective term "welfare-based housing" (*baozhangxing zhufang*) refers to all types of housing aimed at enhancing affordability as defined by the Chinese government. This includes subsidized rental housing; subsidized for-sale housing such as Cheap and Comfortable housing (*jingjishiyong zhufang*), which is provided by developers using land leased by the government at concessionary rates; below-market-rate housing such as Talent Housing (*rencai zhufang*); and resettlement housing (*anzhixing zhufang*). Welfare-based housing is therefore the broadest category intervened by the governments in its production and provision process. The Housing Security Plan proposed in the 12th five-year plan, for example, promised 36 million units of welfare-based housing and shantytown upgrading housing between 2011 and 2015, covering 20 per cent of urban households (Housing Construction Bureau, 2016). The Central Economic Committee of China further formulated the "houses are for living, not for speculation" (HFL) principle at the end of 2016, which was strongly supported and reiterated by Primer Xi on several occasions. This principle, which is to build a housing system composed of private and subsidized owner-occupier and rental accommodation for consumption but not speculative gain, arguably prescribes the most influential real estate macro-regulation in China to the property market in the new century.

Until the 1980s, housing production was largely financed through public budgets as a type of social welfare for urban employees (Wang & Murie, 1996). Privatization and commercialization quickened in the 1990s resulting in the transfer of the bulk of housing production to the private sector. There are six channels through which real estate developers in China assemble capital, including self-finance, pre-payment, domestic loans, foreign loans, personal mortgage loans, and others. Among these, self-finance and pre-payment from purchasers have always been the most important finance sources. In 2020, for example, these two channels contributed to 32.8 per cent and 34.5 per cent of total assembled development finance, respectively (SSB, 2021). Finance for welfare-based housing, however, is facing a serious shortage. At the moment, there are three major means through which local authorities assemble funds, including (1) the financial channels (account for around 60 per cent). These include sales of welfare-based housing; loans from banks by construction agencies; donations; and others; (2) locally assembled revenue (account for approx. 30 per cent), including the investment return of housing provision fund; up to 10 per cent of the net revenue from land lease; and local budget allocation; and (3) support from upper-level government (around 10 per cent), primarily from central and provincial budget redistribution. Land finance is one of the most important mechanisms through which local governments could increase their revenues to cover various public spending, including those for housing (Wu, 2020).

The rest of this chapter will further illustrate the reform trajectory of China's housing system, the characteristics of its welfare regime, and how China compares with its East Asian neighbours. The conclusion summarizes the main findings and points to several key areas for further exploration.

The historical development of China's housing regime

Good accounts of China's housing system transition already exist, and there is a consensus among scholars regarding the close relationship between China's housing regime and its wider economic and social development rationales. Moreover, China's incremental and pragmatic transformation makes it difficult to pigeonhole this country into the conventional regime categories. Miao and Maclennan (2017) divided China's long journey towards housing regime (re)construction into four broad stages: (1) the "experimental period" between 1978 and 1988, when its socialist housing system was transformed nearer to a social democratic regime where market transaction was closely monitored and sometimes subsidized to ensure social stabilization; (2) the "two-track" period in the next decade, when China's housing regime exhibited a hybrid of non-market (sustained public work unit housing), social democratic (disposal of state housing) and corporate (emerging private housing) regimes; (3) the "marketization" period from 1998 to 2010, when China moved away from its socialist legacy and embraced market principles comprehensively; and (4) the "rebounding" period after 2010, when there was a noticeable shift from promoting private homeownership to a renewed interest in public rental housing (PRH) provision.

China gradually established a socialist housing system after 1949, in which the private house-building industry was effectively eliminated and the private rental sector was systematically brought under state control (Wang, 1995). Mang and Murie (1996) identified several characteristics of the Chinese housing regime during this period. First, housing building, distribution, maintenance, and management were decentralized to individual public institutions (work units). To provide the necessary housing for their growing workforce, each public institution included housing investment as an integral part of capital investment for a particular project. Second, all public-sector housing was freely distributed among their employees. The only condition for housing access was 'need', which, however, was different from the needs assessment in welfare capitalist systems but based on government-determined standards related to work status or merit. Finally, very low rents were charged, as housing was regarded as part of the socialist welfare system. Between 1949 and 1990, rent in most Chinese cities accounted for only 2–3 per cent of total household income (Wang & Murie, 1996).

An "experimental period" started in 1976 when a series of economic reforms were launched, allowing market forces and private enterprises to play a bigger role. Housing provision was not an exception. Various new policies were introduced to commercialize and reform the public-sector-dominated housing system. In the early 1980s, the State Council took further steps to protect private property rights and raise housing prices and rents. This paved the way for an expansion of private housing construction and house purchase (Wang, 1992). At the same time, local experiments were carried out to commercialize the existing urban public housing sector (Kirkby, 1985). The involvement of work units in housing construction and allocation, however, was retained and even reinforced during this stage.

The need to quicken and scale-up reform was established in 1988 with the issue of the "Implementation Plan for a Gradual Housing System Reform in Cities and Towns" (State Council, 1988). This marked the start of the "two-track" period in China, in which a social housing supply mechanism was to meet the needs of middle and low-income households and a commercial housing supply mechanism was for high-income families on the market. Privatization of public housing was quickened, and measures to improve the housing investment, management, distribution, provision, and finance were proposed. In this setting, work units continued to play an important role by buying or commissioning "commercial" apartments before selling them to their employees at hugely discounted prices (Wu, 1996). At this same time, a functioning financial system to underpin the commercialization of public housing did not exist. Chinese banks were more willing to provide construction funds rather than housing mortgages (Han, 1999). Consequently, the bulk of urban dwellers purchasing homes had to resort to their savings or borrow from relatives.

It was the 1998 "Notification on Further Deepening the Reform of the Urban Housing System" that laid down a comprehensive strategy and marked a watershed in the "marketization" period. From 1998 to 2010, China witnessed the maturation of housing commodification with

Table 13.1 Shares of urban housing tenures in China (% of households)

	China					
	1950–1970s	*1984*	*1994*	*2000*	*2010*	*2020*
Homeownership: new commercial housing	0		~33	9.2	26.0	34.4
Homeownership: second-hand commercial housing				0.0	5.0	12.1
Homeownership: public housing		~2		29.4	17.3	7.8
Homeownership: ECH or Two-cap housing				6.5	5.1	4.1
Homeownership: self-built	~10	9.4		26.8	16.4	10.1
Private rented	0		~1–2	6.9	23.1	21.8
Public/social rented	~90	88.2	~66%	16.3	2.7	3.7
Others				4.8	4.5	5.2

Sources: Author's calculation based on the Urban Household Survey database and China Statistics Yearbook of the National Bureau of Statistics, various years.

differentiated consumption options among commercial, economic and comfortable housing (ECH) and cheap rental housing (CRH) (Ye, Wu, & Wu, 2006). The latter two were the main forms of welfare-based housing in China, but their role had been gradually marginalized. Investment in ECH among all housing investments, for example, had dropped from over 10 per cent in 1998 to 4 per cent in 2007 (Man, 2011). Some 31 per cent of households achieved homeownership by purchasing commercial housing in 2010, an increase of 366 per cent compared to 2000 (Table 13.1). Moreover, banks became proactive in financing the lucrative housing industry. In 1997, outstanding residential mortgage loans to GDP was 0.2 per cent, which increased to 10.9 per cent by 2007 (Shen & Yan, 2009). Oversupply of money along with loose regulation had resulted in the first overheating of the Chinese real estate market.

Since 2010, the Chinese housing market has undergone noticeable fluctuations fuelled by speculations and greater financial penetrations (Wu, 2015). On the surface, China's current housing system is seemingly commodified and has become akin to the liberal interventionist system. However, as argued by Chiu (2008), while China's system of housing provision by and large operates on commercial terms, the two major factors of production – land and financial institutions – remain state-owned. In fact, through government control of commercial lending, planning, and the speed and scale of land supply, housing consumption in China is influenced by the central and local governments to a significant degree. Miao (2022) therefore, argued that housing financialization in China is unfolded through at least three stages, each intertwined with unique state-market relations. The "houses are for living" principle, introduced earlier on, signals the CCP's recognition that over-exploiting the financial value of housing might conflict with its other necessary functions (Wu, Chen, Pan, Gallent, & Zhang, 2020). The "rebounding" phase, therefore, represents a rebalancing of the state and market's role in the housing sector.

The welfare regime and its development

Today China operates an extensive public welfare system. By the end of 2020, 1.36 billion population (around 95 per cent of the total) were covered by basic medical insurance; 1.03 billion had basic pensions, and 0.23 billion and 0.27 billion population were covered by unemployment and injury insurance (Jin, 2021). This achievement was gradually reached through three major stages.

In stage 1 (1950s–1970s), a basic urban welfare support system was established, which was strongly geared towards employees in state-owned enterprises or work units. However, this embryonic system was sabotaged from the late 1960s until the late 1970s by the Cultural Revolution. Stage 2 (1980s–1990s) featured a painful trial-and-error process of outsourcing social responsibility from the state (work units) to individuals. However, the old support system was dismantled early (at the end of stage 1) while the new system emerged much later and tentatively. This interim therefore was a period of "policy neglect, social chaos and misery" (Ringen & Ngok, 2013, p. 7). Substantial progress has been made in stage 3 (since the new century) and China is approaching near-universal access to basic social protection, although the system is still conservative and fragmented. Housing used to be a key component of the urban welfare system, but reforms in these two domains have progressed at different paces with differential reliance on market forces.

Looking through the whole reform trajectory, it is clear that the overriding objective was to replace the state (work unit) welfare with a social welfare system that is shared between the state, employers and individuals. There is consistency in what is being built: relatively low-level social insurance supplemented by state basic social assistance and welfare services. Funding, management, and implementation of social welfare have been decentralized to such an extent that its efficiency, equity, probity, and transparency are often under threat (Ringen & Ngok, 2013). There has been a National Social Security Fund in operation since 2000, which is now managed by the Ministry of Finance. This fund operates partially as an international sovereign investment fund and absorbs other capital that the central government can use to meet its obligations for social security (SSF, 2022). But, overall, financial tools are insignificant in China's welfare provision.

In more detail, a state welfare program was initiated in China in the early 1950s as was the case in many other newly independent agrarian nations. Transplanted directly from the Soviet Union, the initial purpose of China's welfare system was to prepare for the country's extensive industrialization which would lead to the formation of an urban working class. This contrasted with the welfare systems in Europe and North America, which were formed out of the clashes between labour, capital, and the state. Based upon the presumption of universal lifetime employment, urban state sector workers received comprehensive, work-based social welfare coverage, including generous retirement pensions, free health care, housing, food subsidies, and support for the sick and disabled (Mark Selden & You, 1997). This ambitious urban social welfare system was made possible by the rigorous divide between urban and rural populations through the enforcement of the household registration system (*hukou*). Separate rural welfare programs were financed and administered by local communities with widely varying resource endowments and provided benefits that were a fraction of those enjoyed by the urban population (Selden, 1993).

The centrepiece underpinning this nationwide welfare system was the Labour Insurance Regulations (LIR) issued in 1950. It covered employees in state-operated, joint state-private, cooperative and private factories and mines with 100 or more workers and staff (LIR, 1961, Art.2, p11). Participating enterprises contributed 3 per cent of their monthly payroll to a labour insurance fund. A total of 70 per cent of the monthly contribution was deposited in the account managed by the enterprise trade union to pay for pensions, disability allowances, and subventions to dependents. The remaining 30 per cent was deposited with the Chinese Federation of Trade Unions as a general labour insurance fund for redistribution among enterprises of varying financial strength and liabilities (Art. 9). The Cultural Revolution abolished the Federation, and the system disintegrated (Dixon, 1981). A 1969 Ministry of Finance directive ordered enterprises to pay "labour insurance expenses" directly to eligible workers (Chow, 1988, p. 24).

Stage 2 reform started with Deng taking power in the late 1970s. At that time, urban China faced a severe unemployment challenge. As a countermeasure, the state encouraged early retirement from mid-1978 by relaxing eligibility requirements, offering more generous benefits, and agreeing to hire one unemployed child for each retiring worker (Selden & You, 1997). The number of retirees therefore increased tenfold from 3.1 million in 1978 to 30.9 million in 1995. The cost of total insurance and welfare rose from 7.8 billion yuan in 1978 to 236 billion yuan in 1995, increasing from 13.7 per cent to 29.2 per cent of the total wage bill in state enterprises (SSB, 1996, p. 733). The consequential financial burden in the face of spiralling pension and medical costs had been the key driver to reform state enterprises as well as the welfare programs. A myriad of local approaches to welfare finance and distribution emerged. In June 1991, the State Council issued the Decision on Reform of Retirement Insurance for Enterprise Employees (ZLB, 1991), which proposed three important measures: first, a three-tier retirement insurance system was to be established to reduce the overreliance on state enterprises. This system was composed of a basic retirement program managed by the state, supplementary retirement programs funded by enterprises, and individual savings retirement programs chosen by each employee. Second, a unified financing program was to be centralized from the city and county to the provincial level and third, extending the program coverage to enterprises with different ownerships, so as to provide pension coverage for all urban employees. This three-tier system was further affirmed by State Council Circular No. 6, issued in 1995 (State Council, 1995), which also offered provinces the option to gradually increase contributions from individuals while reducing enterprise contributions. Yet instead of promoting a unified national system, this Circular allocated the power of determining the contribution levels to the prefectural authorities and left open the choice of provident fund or social pooling and the minimum pension level. Many places, therefore, chose the provident fund approach which eliminated pension fund obligations for enterprises and placed the responsibility for saving entirely on workers. Since many system problems and fraud were reported, this Circular was terminated by the State Council in 1996, which also stipulated that a unified pension system at the provincial level would be established (China Commercial Times, 1996).

Two decades of trial and errors have produced the basic structure of China's current welfare system. It is made up of a social insurance system (which is the main body of social welfare), social assistance (a nationwide safe net), and welfare services (locally provided supplementary support) (Zheng, 2008). Stage 3 reform features a continuous expansion and improvement of this structure.

Social insurance is the backbone of China's welfare and covers five main categories: pensions, medical, work injury, unemployment, and maternity (Table 13.2). Basic social insurance is state-run and owned, but different places might set different standards. Citizens can purchase additional private insurance, but they cannot opt out of the state system. The general structure follows the three-tier model whereby employers, employees, the self-employed and the state contribute to the social funds, while entitlements are gained by contributions. The taken-up of private insurance is limited in China. Among the total pension contributions in 2019, for example, 6.29 trillion (65.5 per cent) yuan were state pensions, 1.8 trillion (18.7 per cent) yuan were pensions from enterprises and occupational annuities, and 1.52 trillion (15.8 per cent) yuan were private pensions. The state, therefore, has initiated various incentives to raise the share of private pensions.

Social assistance is a centuries-old tradition in China. The main provision is the Minimum Subsistence Guarantee. This gives households access to a locally determined minimum subsistence level of cash support, conditional on family income. The relief is locally managed and funded, with some central or provincial government subsidy. Urban residents with resident permits have, in principle, been covered since 1999 and rural residents since 2008. Migrant workers are excluded from this scheme but can be partially covered, depending on local practices. Some other discretionary assistance may be available, such as emergency relief, assistance for homeless people, education, medical and housing aid, again depending on local practices. The level of minimum subsistence is low and with large local variations both between and within urban areas. The newly adjusted minimal subsistence in Beijing since July 2021 for example, is 1,245 yuan per capita per month, but the

Table 13.2 Features of social insurance in China

Category	*Features*
Pension	Four basic types designated for employees, other urban residents, government and CCP employees, and rural residents, respectively. Standards and contribution rates differ at different places. Employees (1) have access to a basic pension insurance, paid by both employers and employees; (2) may have access to a supplementary enterprise pension insurance, either run by the enterprise or run jointly by pools of enterprises, for their own employees; and (3) are encouraged to purchase additional personal (commercial) pension. In Beijing for example, employers are required to contribute 20% of employees' gross salary. Individual employees are required to contribute 8% of the city's base-level pension standard (calculated based on the average gross salary in the city) Other urban residents, such as migrant workers, and workers in part-time or irregular employment, can also access the basic pension, contributed by themselves and employers at lower rates (14% by employer and 6% by individuals, for example). But there is normally no supplementary insurance. Self-employed have to cover their pension by themselves. Government and CCP employee pensions are fully covered by government budgets and arranged as defined final salary benefit schemes. These employees have privileged pension entitlements compared to enterprise employees. Pensions for rural residents were gradually rolled out after 2003, but lag in both coverage and generosity to those of urban residents. They are based on a similar structure to the basic enterprise pension, including personal payments into personal accounts, and a collective and government subsidy in the place of the employer contribution.
Medical	Similar types to that of pension. The basic insurance is set up to cover a part, but not all, of the participants' medical expenses. The portion covered is settled directly between the social insurance agency and the provider. The rural cooperative medical insurance scheme started in 2003 and by the end of 2011, 97% of the rural population was covered. It is a voluntary scheme. Contributions are from central and local governments, and rural collectives, and premiums are paid by participating persons. Reimbursements from the scheme are very low compared to the basic medical insurance for urban enterprise employees.
Work injury	Obligatory for employing enterprises and funded fully by employer contributions, set at an adequate level to cover running expenses and vary across regions, sectors and enterprises. It does not cover the consequences of permanent loss of working capacity or of self-inflicted injuries.
Unemployment	Obligatory for employing enterprises and contributed by employers and employees. It provides subsistence – living and, conditionally, medical expenses – for unemployed workers for a duration of up to 24 months. Unemployment insurance for migrant workers, who have labour contracts with urban institutions, has been introduced in many cities with variegated contribution rates and procedures.
Maternity	Funded fully by employers at a regionally differentiated rate of no more than 1% of the employer's salary costs. The birth allowance entitlement is set at a local average salary level and for a minimum of 90 days.

Source: Compiled from Ringen and Ngok (2013).

average monthly spending per capita is 3,898 yuan (Zhang, 2021). In Zhengzhou, the minimal subsistence is 740 yuan, 59 per cent of that in Beijing (Zhengzhou Civil Affairs Bureau, 2021). Public assistance can be supplemented by authorized quasi-governmental and quasi-voluntary charities mainly in the form of assistance in kind, but again it depends on local practices.

Welfare services are, traditionally, a local responsibility and have been provided to the most disadvantaged groups, such as the poorest elderly, orphans and disabled people. The provision continues to vary between local government areas. Social services for the elderly, for example, are targeted at needy or extremely needy groups and have been integrated into the urban and rural social assistance systems. Services are mainly distributed through institutions (e.g., senior citizens' homes), which provide accommodation, in-house services, emergency aid, daytime care, health and recreational services.

All things considered, China today operates an extensive system of public welfare. This is primarily a social insurance system, with the state playing a decisive role, whilst the family, kinship and networks fill important and substantial gaps (Sander, Schmitt, & Kuhnle, 2012). The social insurance system design favours the urban population with minimal and disjointed supplementary provisions from social assistance and welfare services. Although the basic structure is unified nationwide, the coverage, contribution rates, and base levels vary substantially on the local level, inferring that there is not one welfare model in China but many (Ringen & Ngok, 2013). The great divide is between the urban and rural populations. Moreover, the large number of migrant and irregular workers, over 292.5 million in 2021, have notably inferior social rights compared to regular urban residents.

The interrelation between housing and welfare

Corresponding to the development of its housing and welfare systems, China's welfare-based housing policy has evolved through several phases, with the key components summarized in Table 13.3. Phase 1, from the 1950s to the 1980s, housing was a central component of social support. Phase 2, noticeably from the late 1980s to the early 2000s, marked the privatization and commercialization of public housing, a reduction in welfare-based housing supply, and a downplaying of its policy significance. During Phase 3, especially since 2010, welfare-based housing has re-emerged as a major policy concern.

The current welfare-based housing system in China is much more comprehensive, involving four main instruments: ECH, CRH, control-priced commercial housing (CPCH), and PRH. CRH can be regarded as the descendant of the old socialist PRH and part of the nationwide system of social assistance. It is heavily subsidized by the national government while the tenants only pay a symbolic rent. CPCH, or "90/70" housing – construction space under $90m^2$ and sale price limited to no more than 70 per cent of comparable commercial housing – is a new type of affordable, market housing introduced in 2010. PRH is also a new policy category introduced in 2010, which extends beneficiaries to medium to lower medium-income households, regardless of whether they are local residents or migrants (Zhou & Ronald, 2017).

ECH is one of the earliest forms of subsided commercial housing. It differs from ordinary commercial housing in that the land is administratively allocated, and the price is regulated by the government. Profit made by developers on top of construction costs is restricted to a maximum of 3 per cent (State Council, 1998). The construction of ECH peaked when it was introduced in 1998, and by 1999, the share of ECH within housing investment was 16.6 per cent. However, since then, investment in ECH has been steadily decreasing. Investment in 2005 for example, registered a notable decrease of 14 per cent. The design of the ECH policy is partially to blame here. The responsibility

Table 13.3 Key welfare-based housing in China

Cornerstones	*Type*	*Target*	*Method*	*Source of supply*
1994 The decision on deepening the urban housing reform	ECH	Medium- and low-income families	Sale only	Newly-built
1998 A notification from the state council on further deepening	CRH	Lowest-income families	Rent only	Newly built, purchased, refurbished, and donation
the reform of the urban housing system and accelerating housing construction	ECH	Medium- and low-income families	Sale only	Newly built
2007 State council on solving the housing difficulties of low-income families	CRH	Expanded from lowest-income families to low-income families	Rent only	Newly built, purchased, refurbished and donation
	ECH	Low-income families	Sale only	Newly built
2010 Notice on promoting the stable and healthy development	ECH	Low-income families	Sale only	Newly built
of the real estate market	CRH	Low-income families	Rent only	Newly built, purchased, refurbished, and donation
	CPCH	Lower medium-income families	Sale only	Newly built
	PRH	Lower medium-income families	Rent only	Newly built, purchased, refurbished, and long-term market renting

Source: Author.

of providing ECH, for example, is devolved to local governments. Yet because land charged for ECH is minimal, local governments rarely gain any revenue from ECH projects. At the same time, fiscal reforms in 1994 left local governments with an obligation to meet nearly 80 per cent of total government expenditure, but with direct receipt of only 47 per cent of total government revenues (Man, 2011). Such fiscal imbalances have left many local governments short of construction funds, and reliant on land leasing fees for revenue to finance infrastructure investment and economic development (Qiao, 2007). As a result, local governments prefer offering state-owned land to the highest bidder in auctions involving market developers, rather than using the land for ECH projects. When pushed harder by the central government, local governments tend to find inconvenient locations for ECH development (Wang et al., 2011). Infrastructure and ancillary facilities provided on ECH developments are often of a poor standard, or even non-existent (Zhou & Logan, 1996).

The financing of ECH derives partially from the Housing Provident Fund (HPF), which was introduced first in Shanghai in 1991 and rolled out nationally in 1994. Most enterprises (except some small private firms) and their employees (regardless of hukou status) contribute to this fund. Five per cent is the minimal contribution rate set by the state. Cities with stronger economic performance are allowed to increase this rate. Individuals can only access their HPF account to purchase, construct, renovate, or repair their houses or to make housing loans or interest repayments. Renters can access the fund to pay rents that exceed the normal rent-to-income ratio. Although individuals can apply for a mortgage from HPF, the permitted proportion is limited and needs to be topped up with a commercial mortgage (Deng, Yan, & Chen, 2021). Funds assembled by HPF can

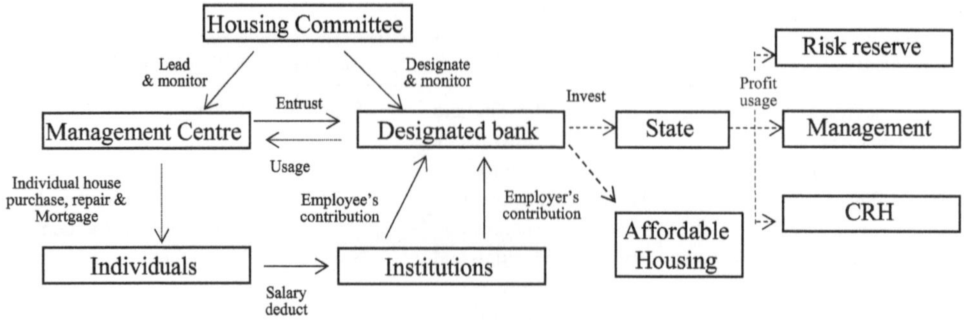

Figure 13.1 Management system and financial flow of HPF in China.
Source: Author.

be invested in state bonds. Profits gained through this investment must be used either as reserves to cover HPF management costs, or to support welfare-based housing construction (Figure 13.1).

However, the deposits of HPF are often unstable and inadequate. Other sources of financing, such as the 10 per cent of funds from the net profit of land transfer fees earmarked for low-rent housing construction, have often failed to be collected. Moreover, the eligibility criteria for ECH are either too broad or their enforcement is lax. As a result, ECH often benefits high and middle-income families, rather than lower-income households. A high vacancy rate for ECH projects is also an "open secret", as many high-income households managed to buy these properties for investment purposes (Deng, Shen, & Wang, 2011).

The low supply of subsidized housing and the continuous house price rises since 2000 accentuated housing problems, which grew into a prominent social issue (Wang & Murie, 2011). The 2007 State Council regulation on 'Solving the Housing Difficulties of Low-income Families' announced a new round of policy stimulation for welfare-based housing construction. CCP promised that "all Chinese citizens should have a place to live" and a Housing Act was drafted (Hou, 2009) with the intention of establishing a universal 'basic housing right'. Local governments were required to assist low-income households to acquire and/or live in decent homes, regardless of their financial capability. This policy, at least superficially, revives some commitment to the principle of housing decommodification by repositioning housing as a basic human right. To support this goal, the latest Ten-Year Housing Plan required local authorities to produce their annual public housing plans in tandem with land use and urban development. The public budget would be the main source of finance. The target was to use 1 per cent of the total GDP every year for welfare-based housing construction, shared by central, provincial and county/district level governments (Hou, 2009). Moreover, a noticeable policy change since 2011 has been the emphasis on PRH instead of ECH. The 12th five-year plan (2011–2015) announced that PRH would become the backbone of China's affordable housing system. To support this policy, the central government planned to construct 10 million units of affordable housing in five years. Among these units, around half were earmarked for renting, either through PRH or CRH. The 13th and 14th five-year plans reaffirmed the "housing is for living" principle and continued to support the rental market.

Regional outlook

Systematic differences have been observed between the East Asian model and the prototype Esping-Andersen welfare regime. Variously labelled as Confucian (Slote & Vos, 1998), productivist

(Holliday, 2000), and developmental (Midgley, 1986), East Asian welfare regimes feature a strong state with a state-led economy. Economic growth is the supreme goal, and social policy is used as an instrument to achieve this goal. Social security tends to be occupation-based, introduced early in the development process to enhance regime legitimacy. The state oversees the development of welfare legislation and regulations, whereas provision tends to be outsourced to businesses in the form of occupational welfare, and voluntary agencies in the form of social services. As a result, there is a low level of state social spending and a continued reliance on families for welfare and security. China could be broadly classified as a developmental state, at least until 2010, when economic growth and efficiency are emphasized as the priority whereas social security is provided to safeguard the growth machine. The separation of housing welfare from general welfare provision and the promotion of the housing market as a key pillar of its economic growth, for example, exemplify its developmental orientation.

Yet, as studies progress, there are growing recognitions of the interregional divergence as well as temporal variations within this region, making it difficult to apply a single label to the East Asian welfare regimes. Lee (2003) for example, compared the housing culture across Hong Kong, Singapore, South Korea and Taiwan. While acknowledging that in all four economies, housing is regarded as both a commodity and social welfare good and the state tends to monopolize its production and exchange, two aspects of housing policy could be noted, including the state-driven model (e.g., Singapore) and the state-facilitated model (e.g., Taiwan). In Doling's (1999) attempt to classify the housing policy regimes of the newly industrialized countries as exemplified by the 'little tigers', he focused on the balance between the state and the market along the whole housing provision chain. Three general housing policy regimes of liberal, communist and corporatist were suggested, and the 'litter tigers' featured a communist regime in development, a corporatist regime in production, and a liberal regime in consumption. Again, the underlying logic of this unique regime is the State ownership of land and the dominance of economic growth as political goals. Focusing on a policy regime typology of government intervention in housing across five Asian economies, Chiu (2008) classified Singapore as a liberal socialist system, Hong Kong as a liberal interventionist system, South Korea as a regulated market system, Taiwan as a free market system, and China as a marketized socialist system.

Ringen et al. (2011) argued that developmental welfare is less of a permanent model but more of a transitory state that is associated with the early take-off period of economic development. As an economy's development progresses, its welfare state could emerge fully rather than subordinate to economic policy. In South Korea, for example, the developmental state was rolled back and social policy has taken a big step forward following the democratic consolidation and economic crisis in 1997. In China, Mok and Qian (2019) argued that a new phase of Chinese social policy development started in 2002–2003 because of the leadership succession. Since then, the "pro-growth" slogan has been gradually replaced by a "pro-poor growth" narrative, and several new social policy programmes were launched, which took China's social policy regime beyond the occupation-based social insurance model. Yet these authors noticed a continuation of China's "paternalistic welfare pragmatism" instead of a new welfare regime with a fundamentally different philosophy and governance policies. Partially echoing this view, Chiu (2013) argued that the Asian financial crises and the global economic crisis had reduced the differences in housing policies among the major Asian economies. Recognizing the speculation risk, there was a loss of impetus for home ownership and a new emphasis on rental subsidies. Nonetheless, the economic revival since mid-2009 has caused the re-introduction of home ownership subsidies making use of the strong economic conditions. This general shift can also be found in China, as illustrated above.

Concluding discussion

The nature of housing, and its relationship with the state in general, and with the social welfare regime in particular, is continuously debated and evolving in China. While there is a consensus that housing provision cannot be completely left to the market, debates continue regarding the relative responsibility between the state and the market, and across different levels of government. Housing reform in the past 50 years has cultivated the institutional architecture, the culture, and the financial instruments to support the housing market which contributes to wealth accumulation both for individuals and for the nation as a whole. But it also leads to skyrocketing prices of owner-occupied housing and a shortage of affordable rented housing across China, especially in major urban areas. Socio-economic inequity and segregation are increasingly linked to housing wealth accumulation and distribution, and current serious challenges for further development and social coherence. Inadequate engagement of communities and NGOs in welfare provision featured prominently in China during the COVID-19 pandemic (Mok, Ku, & Yuda, 2021), all of which call for urgent action and further reform.

There has been a noticeable return of government commitment towards housing since 2010. As exemplified by the slogan, "housing is for living, not for investing", there seems to be a potential return of a government commitment to providing a basic housing right to its citizens, echoing Article 25 of the UN Declaration of Human Rights, under which everyone has a right to adequate housing. But as this chapter demonstrates, the latest policy concern on housing is not meant for de-commodification, but for a stronger public intervention in the market operation and a wider coverage of housing needs. In China, it seems that the social cost of rising unaffordability is implicitly accepted – at least for now – so long as such costs are sufficiently compensated by the benefits of market efficiency and economic prosperity. The expansion of welfare-based housing in the past decade represents a re-balancing of the market and the state as China's housing system matures on its commodification process. The strategic objective of the government was to constrain over-speculation, increase housing market efficiency, and enhance affordability efficacy. An auxiliary purpose is to reduce social disparity. The mechanism to achieve this target is clearly stated as "government takes the lead; the market operates the system". Different measures have been adopted to realize the strategy, among which the most noticeable ones were to distinguish the use of housing finance, taxation, and land supply between housing for living in and housing for investment; and to encourage housing rental. But these measures risk further aggravating the gaps between renters and homeowners, especially at a time when house prices rise more quickly than interest rates.

Amid this rebalancing of the state and market's role in housing development is the missing role of civil society in China. The Chinese state has always been cautious of delegating authority to non-governmental organizations, especially for such crucial sectors as housing. Another key feature of China's housing and welfare regime is the scale and scope of decentralization. So far, the state is largely confined to making strategies and legislations, but the provision of welfare in general, and welfare-based housing in particular, has largely fallen on the shoulders of local governments. This calls for a theoretical and empirical re-think of the approach to investigate housing systems or regimes, where categories tend to be developed at the national level while ignoring local variations. Local diversity in societal contexts could be a determinant of the intermediate situation of the housing system between state and market, but it might well be an outcome of the growing national housing regime itself. Maybe it is time we move beyond national housing and welfare regimes towards studying regional and local housing systems within a broad national context, especially for vast countries like China.

References

Cao, N. (2016). *The Ranking of Housing Price to Income Ratio Among 35 Large and Medium-Sized Chinese Cities in 2015*. http://bj.bendibao.com/news/2016420/222966_2.shtm

Chiu, R. L. H. (2008). Government Intervention in Housing: Convergence and Divergence of the Asian Dragons. *Urban Policy and Research, 26*(3), 249–269. https://doi.org/10.1080/08111140802301781

Chiu, R. L. H. (2013). A Comparison of Housing Welfare Policies Among Major Asian Countries in the Modern Era. *LHI Journal, 4*(1), 23–31. https://doi.org/10.5804/LHIJ.2013.4.1.023

Chow, N. (1988). *The Administration and Financing of Social Security in China*. Center of Asian Studies, University of Hong Kong.

Deng, L., Shen, Q., & Wang, L. (2011). The Emerging Housing Policy Framework in China. *Journal of Planning Literature, 26*(2), 168–183. https://doi.org/10.1177/0885412210390220

Deng, L., Yan, X., & Chen, J. (2021). Housing Affordability, Subsidized Lending and Cross-City Variation in the Performance of China's Housing Provident Fund Program. *Housing Studies, 36*(4), 455–478. https://doi.org/10.1080/02673037.2019.1585521

Dixon, J. (1981). *The Chinese Welfare System, 1949–1979*. Praeger.

Doling, J. (1999). Housing Policies and the Little Tigers: How Do They Compare with Other Industrialised Countries? *Housing Studies, 14*(2), 229–250. https://doi.org/10.1080/02673039982939

Han, Q. (1999). The Dilemma and Solution in China's Housing Financial System. *China's Real Estate,* (1), 58–61.

Holliday, I. (2000). Productivist Welfare Capitalism: Social Policy in East Asia. *Political Studies, 48*(4), 706–723. https://doi.org/10.1111/1467-9248.00279

Hou, X. M. (2009). *Preliminary Research Findings on Establishing the Basic Housing Security System*. Paper presented at the International Symposium on Basic Housing Security System Beijing.

Housing Construction Bureau. (2016). *The 12th Housing Security Plan*. http://www.shaanxijs.gov.cn/chengxianggh/

Huang, Y., Yi, D., & A. V. Clark, W. (2020). Multiple Home Ownership in Chinese Cities: An Institutional and Cultural Perspective. *Cities, 97*(20), 1–13. https://doi.org/10.1016/j.cities.2019.102518

Jin, W. (2021). *The Transformation of Social Insurance from Broad Coverage to Full Coverage*. http://www.mohrss.gov.cn/SYrlzyhshbzb/dongtaixinwen/buneiyaowen/rsxw/202107/t20210701_417519.html.

Kirkby, R. (1985). *Urbanization in China: Town and Country in a Developing Economy 1949–2000 A.D.* Columbia University Press.

Lee, J. (2003). Is There An East Asian Housing Culture? Contrasting Housing Systems of Hong Kong, Singapore, Taiwan and South Korea. *of Comparative Asian Development, 2*(1), 3–19. doi:10.1080/15339114.2003.9678369

Lin, G. C. S., & Ho, S. P. S. (2005). The State, Land System, and Land Development Processes in Contemporary China. *Annals of the Association of American Geographers, 95*(2), 411–436. doi:https://doi.org/10.1111/j.1467-8306.2005.00467.x

LIR. (1961). *Labor Insurance Regulations of the People's Republic of China*. Foreign Languages Press

Man, J. Y. (2011). Affordable Housing in China. *Lincoln Institute of Land Policy*, 16–20 January. https://center4affordablehousing.org/wp-content/uploads/2021/12/Cina-Affordable-housing-in-China.pdf.

Miao, J. T. (2022). The Temporal Variation of Housing Financialisation: A Case Study of China. *Transactions in Planning and Urban Research, 1*(1–2), 50–65. https://doi.org/10.1177/27541223221113948

Miao, J. T., & Maclennan, D. (2017). Exploring the 'Middle Ground' Between State and Market: The Example of China. *Housing Studies, 32*(1), 73–94. https://doi.org/10.1080/02673037.2016.1181723

Midgley, J. (1986). Industrialization and Welfare: The Case of the Four Little Tigers. *Social Policy and Administration, 20*(3), 225–238. https://doi.org/10.1111/j.1467-9515.1986.tb00253.x

Mok, K. H., Ku, Y.-W., & Yuda, T. K. (2021). Managing the COVID-19 Pandemic Crisis and Changing Welfare Regimes. *Journal of Asian Public Policy, 14*(1), 1–12. doi:https://doi.org/10.1080/17516234.2020.1861722

Mok, K. H., & Qian, J. (2019). A New Welfare Regime in the Making? Paternalistic Welfare Pragmatism in China. *Journal of European Social Policy, 29*(1), 100–114. https://doi.org/10.1177/0958928718767603

Qiao, L. (2007) Chinese Property Rights Law: Old Wine in a New Bottle? *Lawasia Journal,* 163–186. https://ssrn.com/abstract=1687614

Ringen, S., Kwon, H.-J., Yi, I., Kim, T., & Lee, J. (2011). *The Korean State and Social Policy*. Oxford University Press.

Ringen, S., & Ngok, K. (2013). *What Kind of Welfare State Is Emerging in China?* Geneva: UNRISD Working Paper, No. 2013-2.
Sander, A., Schmitt, C., & Kuhnle, S. (2012). Towards a Chinese Welfare State? Tagging the Concept of Social Security in China. *The Perspective of the World Review, 4*(2), 9–35.
Selden, M. (1993). *The Political Economy of Chinese Development*. M. E. Sharpe.
Selden, M., & You, L. (1997). The Reform of Social Welfare in China. *World Development, 25*(10), 1657–1668.
Shen, B., & Yan, L. (2009). Development of Consumer Credit in China. BIS Papers No 46: https://www.bis.org/publ/bppdf/bispap46g.pdf
Slote, W. H., & Vos, G. A. D. (Eds.) (1998). *Confucianism and the Family*. State University of New York Press.
SSB. (1996). *China Statistical Yearbook 1996*. Beijing: National Bureau of Statistics.
SSB. (2021). *China Statistical Yearbok 2021*. Beijing: National Bureau of Statistics.
SSF. (2022). *Institution Introduction*. Retrived from National Council for Social Security Fund of China: http://www.ssf.gov.cn/portal/index.htm.
State Council. (1988). *Implementation Plan for a Gradual Housing System Reform in Cities and Towns*. CCTV News. http://news.cctv.com/science/20081007/106663.shtml
State Council. (1995). Circular of the State Council on Deepening the Reform of the Pension Insurance System for Enterprise Workers and Staff. *Gazette of the State Council,* 147–150.
Wang, Y. (1992). Private-Sector Housing in Urban China Since 1949: The Case of Xian. *Urban Studies, 7*(2), 119–137. https://doi.org/10.1080/02673039208720729
Wang, Y. (1995). Public Sector Housing in Urban China Since 1949: The Case of Xian. *Housing Studies, 10*(1), 57–82. https://doi.org/10.1080/02673039508720809
Wang, Y., & Murie, A. (1996). The Process of Commercialisation of Urban Housing in China. *Urban Studies, 33*(6), 971–989. https://www.jstor.org/stable/43100335
Wang, Y., & Murie, A (2011). The New Affordable and Social Housing Provision System in China: Implications for Comparative Housing Studies. *International Journal of Housing Policy, 11*(3), 237–254. https://doi.org/10.1080/14616718.2011.599130
Wu, F. (1996). Changes in the Structure of Public Housing Provision in Urban China. *Urban Studies, 33*(9), 1601–1627. https://www.jstor.org/stable/43083396
Wu, F. (2015). Commodification and Housing Market Cycles in Chinese Cities. *International Journal of Housing Policy, 15*(1), 6–26. https://doi.org/10.1080/14616718.2014.925255
Wu, F. (2020). Land Financialisation and the Financing of Urban Development in China. *Land Use Policy, 112*(January), 1–10. https://doi.org/10.1016/j.landusepol.2019.104412. https://doi.org/10.1016/j.landusepol.2019.104412
Wu, F., Chen, J., Pan, F., Gallent, N., & Zhang, F. (2020). Assetization: The Chinese Path to Housing Financialization. *Annals of the American Association of Geographers, 110*(5), 1483–1499. https://doi.org/10.1080/24694452.2020.1715195.
Ye, J., Wu, D., & Wu, J. (2006). A Study on the Chinese Housing Policy During Social Transition: Practice and Development. *Housing Finance International, 20*(3), 50–58. http://www.housingfinance.org/uploads/Publicationsmanager/0603_Chi.pdf
Zhang, W. (2021). *Monthly Minimal Subsistence per Capital in Beijing is Increased from 1170 Yuan to 1245 Yuan*. https://baijiahao.baidu.com/s?id=1705408959035669852&wfr=spider&for=pc.
Zheng, G. C. (2008). *China's Social Security in Thirty Years (in Chinese)*. Beijing: People's Press.
Zhengzhou Civil Affairs Bureau. (2021). *Notice on Adjusting and Raising the Minimum Living Security Standard and the Basic Living Standard*. http://mzj.zhengzhou.gov.cn/notice/6146639.jhtml (Accessed: 26 June 2022).
Zhou, M., & Logan, J. (1996). Market Transition and the Commodification of Housing in Urban China. *International Journal of Urban and Regional Research, 20*, 400–421.
Zhou, J., & Ronald, R. (2017). Housing and Welfare Regimes: Examining the Changing Role of Public Housing in China. *Housing, Theory and Society, 34*(3), 253–276. https://doi.org/10.1080/14036096.2016.1223165
ZLB. (1991, 2nd July). State Council Decision on Reform of Retirement Insurance for Enterprise Employees. *Zhongguo Laodong Bao (Chinese Labor Newspaper)*.

14

HOUSING AND WELFARE IN THE WIDER LATIN AMERICAN CONTEXT

The Chilean experience

Valentina Abufhele and Nicolás Angelcos

The current state of housing in the region

Chile's current housing crisis must be understood in the context of 40 years of market-oriented housing and urban policies. Particularly in the past decade, these long-term processes have coincided with increased migration in Latin America, a political crisis in Chilean institutions and economic constraints arising from the COVID-19 pandemic. This has placed an additional burden on the existing housing stock, worsening the shortage. Both long and short-term processes, however, are not merely a matter of a housing shortage or the failure of housing policy to address growing demand. Rather, they reflect the exhaustion of a neoliberal-oriented model of housing production, provision and distribution. In this chapter, we examine the current crisis and the limitations of the Chilean housing model, looking at the rationales of government underpinning both the welfare state regime and the housing system. We attempt to show that welfare regimes and housing policies do not respond to uniform political and economic orientations but to rationales of government between which there is tension. Conflicting rationales of government provide entry points for analysing the entanglement of the housing system, political actors' disputes about the status of housing as a dimension of welfare, the roles of the state, the market and civil society in its provision and the quality and coverage of the benefits delivered.

The symptoms of the current crisis in Chile arise partly from the neoliberal restructuring of the economy during the military dictatorship (1973–1989). Since the mid-1970s, the urban and housing sectors have been reformed according to neoliberal principles, moving from a state-led provision of housing to a market-led policy. Under this model, a "social housing"[1] programme was implemented, based on demand-side subsidies delivered selectively to eligible applicants for the acquisition of new housing units built by private providers (Kusnetzoff, 1987; Richards, 1995; Rojas, 2001; Gilbert, 2002). This model of housing production and the social housing programme were maintained after the transition to democracy and their market-oriented foundations have remained unaltered (Richards, 1995; Rodríguez and Sugranyes, 2005; Özler, 2012; Posner, 2012).

By 1990, the housing shortage had peaked at approximately 800,000 units, equivalent to over 20 per cent of the Chilean population (Ducci, 1997; MINVU, 2004, 2006; Caldera, 2012). The coalition of democratic parties that took office in 1990 – the Concertación – expanded the capacity of the market-oriented housing policy and incorporated a "social democratic" rationale in a bid to

 DOI: 10.1201/9781003212690-16

counteract the exclusionary effects of neoliberal policies and provide social housing for the poorest sectors of the population. The measures included higher public expenditure, increased and accelerated private construction and expanded efforts to target low-income groups more efficiently.

The first post-dictatorship decade (1990–2000) saw improvements in housing outcomes. For the first time in Chilean history, new housing was provided at a faster rate than the formation of new households, making Chile the first Latin American country to reduce its shortage (Ducci, 1997). This success was coupled with positive assessments of the policy's economic efficiency compared to previous state-led models (Gilbert, 2004a). Targeting also proved effective in distributing limited state resources to low-income groups (Posner, 2012). In addition, housing informality had been reduced by the mid-2000s. These accomplishments served to legitimize the structural market-oriented principles and direction of housing policy.

However, before the end of the 1990s, the system of housing construction had proved to have a "dark side" (Ducci, 1997) in terms of the adverse urban effects and housing quality deficiencies of a model which (until then) had been considered successful (MINVU, 2004). The policy's market-oriented rationale and construction companies' drive to maximize profits resulted in low-quality homes. In addition, large-scale public housing projects were built on low-cost, peripheral land, with no urban infrastructure and little access to social services, thereby increasing segregation by concentrating poor residents on the urban margins (Ducci, 1997; Rodríguez and Sugranyes, 2005). In light of this, both urban scholars and residents began to criticize the housing policy in the late 1990s.

Despite a general demobilization of the homeless in the 1990s, an organized movement of poor urban residents – the *movimiento de pobladores* – re-emerged in the mid-2000s. It questioned the peripheral location of housing projects, the resulting segregation, the deficient quality of housing and the state-regulated channels of participation in the design and construction of housing projects (Angelcos, 2012; Cortés, 2013, 2014; Angelcos and Pérez, 2017; Pérez, 2017a, 2017b, 2018).

Despite the criticisms, the housing policy's structural and predominant neoliberal orientation has remained unchanged. Since the 2000s, improvements have been made, seeking to increase quality standards and reduce urban segregation, but the adverse effects of the system of housing production have persisted (Özler, 2012; Dohnke et al., 2014; Rasse et al., 2021).

In 2000, the authorities finally acknowledged an increase in the housing deficit,[2] which had reached 438,113 units (MINVU, 2020). In 2015, scholars and private institutions, such as think tanks and advocacy NGOs, had already warned of a rising housing deficit which peaked in 2017 at 497,560 units, with rates as high as they had been in 1998 (Fundación Vivienda, 2018). Rates of housing informality, which are not included in the calculation of the housing deficit, have also increased steadily since the mid-2000s: most recently, by 20.3 per cent between 2018 and 2020 to 969 settlements (Techo, 2021; MINVU, 2019). Again, current rates are as high as in the late 1990s. Rates of home-sharing[3] (*allegamiento*) increased to 300,158 families in 2017, up by 90 per cent from the 2015 survey (Fundación Vivienda, 2018). Although 1,315,099 new housing units were approved and their construction began between 2009 and 2018 (MINVU, 2019), the current housing deficit, informality rates and the pressing demands of the *pobladores* movement have finally put the escalating housing crisis on the agenda.

The social uprising that took place in Chile in October 2019, the economic crisis generated by the COVID-19 pandemic and increasing rates of migration from other Latin American countries have stressed the capacity of the system of housing production, revealing the exhaustion of the market-led housing model. Moreover, land prices have increased considerably, and the value of subsidies is insufficient to build low-income housing units in central or pericentral municipal districts (Dohnke et al., 2014). Rents have also risen, leading some renters to seek lower-cost housing solutions in the informal market.

The Chilean case is particularly relevant within the Latin American context. In addition to the perceived success of Chilean policy in the early 1990s, international development agencies saw a match between Chile's model and the "enabling approach" associated with the Washington Consensus, prompting efforts to transfer it to other countries (World Bank, 1993; Richards, 1995; Held, 2000; Gilbert, 2002, 2004a; Posner, 2012). Aided by the World Bank, the Inter-American Development Bank and USAID, key features of Chile's housing policy were implemented in several other Latin American countries (Gilbert, 2002). Thus, a closer examination of the benefits, limitations and risks of the Chilean experience may inform processes underway in the Latin American region at large.

A brief overview of the historical development of the housing regime

Massive rural-urban migration in the early 20th century unleashed a pressing demand for housing. Between 1907 and 1960, 960,298 people migrated to the capital, Santiago. The rate of urbanization accelerated and, between the 1940s and the 1970s, Santiago's population almost tripled from 1,075,000 to 2,779,499 inhabitants (Murphy, 2015). Other Latin American countries, such as Argentina and Brazil, also experienced similar urban growth, leading to the first "housing crisis" in each country.

Population growth exceeded Santiago's capacity to accommodate new residents, resulting in the emergence of numerous informal settlements on the banks of the Zanjón de la Aguada water course and the Mapocho River. These settlements became known as poblaciones callampas or "mushroom shantytowns" because of the resemblance of their seemingly overnight emergence and subsequent growth to that of fungi in humid conditions. In 1952, the first national housing census estimated that 30 per cent of the population lacked adequate housing, 60.1 per cent of the urban poor lived in tenements and 13.8 per cent in informal settlements (De Ramón, 2007). Conditions were dramatic. The pobladores built shacks of light materials, without minimal sanitary infrastructure, such as running water, electricity or sewers. The media denounced these precarious and inhumane living conditions and a consensus emerged on the need to solve the housing crisis (Giannotti, 2014).

In the 1940s and 1950s, the state's incapacity to resolve the housing problem led to the first land occupation in Santiago, including the emblematic 1957 *Toma de la Victoria* (La Victoria Land Grab) (Cortés, 2014). In contrast to other informal settlements, land occupations were organized or led by (usually left-wing) political parties and involved the one-off, sometimes violent, seizure of usually state-owned land (De Ramón, 1990). In the 1950s and 1960s, *pobladores* organized land takeovers in order to demand land ownership, marking the start of a lasting tradition of low-income groups' demands for property titles (Giannotti, 2014).

This situation led the government to formulate a National Housing Plan in 1953, supported by the Chilean Chamber of Construction[4] (*Cámara Chilena de la Construcción*). Under it, the state created a Housing Corporation to promote the construction of social housing. However, the Plan's outcomes were constrained by the country's weak economic situation. The approximately 40,000 housing units built with state support between 1953 and 1958 were fewer than originally planned (Hidalgo, 2019). From the 1950s through to the present, land and homeownership have been the predominant forms of housing provision in Chile (Giannotti, 2014; Murphy, 2015).

The intensification of the housing shortage in the 1960s and early 1970s, together with political polarization, triggered various government measures. The Christian Democrat government of Eduardo Frei Montalva (1964–1970) saw the development of the Ministry of Housing and

Urbanization (MINVU) and a new housing programme – *Operación Sitio* – consisting of the provision of a serviced plot with only a bathroom and kitchen. Unlike previous state approaches, this policy gave central importance to popular organizations, integrating the *pobladores* into the solution of their housing problem and expecting them to complete the house through self-construction (De Ramón, 1990).

The socialist government of Salvador Allende (1970–1973) proclaimed that housing was a social right and its provision a state responsibility (CORVI, 1972). His government promised to build 100,000 housing units per year, but without significantly altering the existing institutional structure and capacity. It faced considerable obstacles of which the most important was a boycott by the Chilean Chamber of Construction, whose affiliated companies refused to build a single house during the government's first seven months. However, in 1971, the construction of 73,000 housing units began – 40,000 of them with public funds – representing the highest rates of housing construction in Chilean history (Castells, 1973).

On the part of *pobladores,* land seizures intensified as a strategy for securing access to housing. Seizures peaked in 1971 when 350 land occupations were recorded in Santiago (Murphy, 2015). Between 1967 and 1972, 14 per cent of the city's population participated in these mobilizations (Schneider, 1995). This positioned housing as a central socio-political issue and generated the conditions for the recognition of the *pobladores* as legitimate political actors by political parties and the government (Garcés, 2002).

The military coup in 1973 led to the repression of the *pobladores* movements. Their political organizations were partially dismantled and land seizures declined. Efforts to "eradicate" informal settlements became the predominant housing policy through to the mid-1980s (Bruey, 2012) and their residents were relocated to social housing projects. These efforts targeted "extreme poverty" groups and territories through means-tested assessments. This top-down approach served to curtail the participation of *pobladores* (Murphy, 2015).

In line with its neoliberal structural reforms of the economy in general, the military regime reformed the urban and housing system, laying the institutional foundations for the market-led housing policies that have endured since then. The National Urban Development Policy, introduced in 1979, declared that urban land was not a scarce resource and could be a source of profit. This reform and the creation of a social housing subsidy programme in 1978, permitted the commodification of urban land and housing through deregulation of the land market, privatization of social housing construction and the financialization of housing provision.

The new social housing programme was based on demand-side subsidies for the acquisition of housing by eligible applicants. It was financed through a three-tier system: a capital subsidy provided by the government, the beneficiary's own savings and a mortgage taken out by the beneficiary (Kusnetzoff, 1987; Richards, 1995; Rojas, 2001; Dohnke et al., 2014). In other words, housing was no longer understood as a right to be guaranteed by the state, as in preceding governments, but as a commodity to be acquired through the market. Housing was built by private providers and the state's role was confined to subsidizing the demand for housing, but only for those deemed to require assistance. Eligibility depended on a means-tested assessment of applicants' socioeconomic status and their compliance with savings and loan requirements. In other words, state assistance was conditional on citizens demonstrating they were poor and committed to helping themselves by contributing to the cost of their homes (Gilbert, 2002). As in previous housing programmes, the families selected became the owners of the new housing unit. However, the regime's failure to address housing demands, together with an incipient political opening, meant that, in 1983, *pobladores* again began to carry out land seizures (Morales and Rojas, 1986).

In 1989, at the end of the dictatorship, the poverty rate was running at 38.1 per cent (ECLAC, 1990) and the housing deficit at over 20 per cent. This had, in turn, resulted in high rates of home-sharing (MINVU, 2004). With the restoration of democracy, the Concertación faced the threat of a return of land seizures (Hipsher, 1996; MINVU, 2004). Amidst a fragile political transition, the imperative for the authorities was to maintain political stability (Oppenheim, 1993). This shaped how the new government appropriated the neoliberal-oriented housing policy and redefined its goals. The Concertación advocated social-democratic principles and strengthened social policies to address the military government's "social debt" to the most disadvantaged sectors of the population (Oppenheim, 1993; Han, 2012). However, the Concertación decided not to reform the housing policy, leaving its market-oriented foundations unaltered (Richards, 1995; Rodríguez and Sugranyes, 2005; Özler, 2012; Posner, 2012). Convinced of the potential of the inherited system, it strengthened its capacity, increasing housing production and diversifying housing programmes to target the urban poor more effectively.

Social housing construction increased[5] (MINVU, 2009) and, as mentioned above, the housing shortage was reduced considerably under the first post-dictatorship administrations. Between 1992 and 2002, it decreased from 771,935 to 543,542 units, representing a 29.6 per cent reduction at an average annual rate of 3.5 per cent (MINVU, 2006). Both the government and scholars considered this a successful result (Ducci, 1997; Rojas, 2001; MINVU, 2004, 2009; Rodríguez and Sugranyes, 2005). In the following intercensal period (2002–2017), the housing shortage dropped further to 393,613 units in 2017 (MINVU, 2017) according to official estimates, however, non-governmental institutions began to question.[6]

Despite the reduction in the housing deficit, the trend in informal settlements was less clear. They dropped from 712 settlements in 1996 to 490 in 2007 (MINVU-INVI, 1997; MINVU, 2007), but subsequent surveys showed a steady increase from 657 in 2011 to 969 in 2020 (MINVU, 2013, 2019; Techo, 2021). In 2013, an MINVU report argued that, despite government efforts, informal settlements "were not absorbed naturally by housing programmes as expected" (MINVU, 2013: 15). On the contrary, studies showed that the housing policy had contributed to the reproduction of informality. The creation of informal settlements as "targeted territories" and priority access to housing subsidies for their inhabitants encouraged property-less people to seek homeownership through informal urbanization (Abufhele, 2018).

Scholars have criticized the housing policy on several grounds. Housing production in the 1990s and 2000s created homogeneously poor and segregated neighbourhoods, contributing to social exclusion (Ducci, 1997; Sabatini, 2000; Sabatini et al., 2001; Márquez, 2004; Rodríguez and Sugranyes, 2005; Hidalgo, 2007).[7] Other studies have shown that housing policies reduced citizen participation, enhanced competition among applicants, eroded social cohesion and discouraged collective action (Özler, 2012; Posner, 2012). Overall, scholars have found that policies were poorly equipped to address the needs of the urban poor (Haggard and Kaufman, 2009) and that homeownership did not necessarily improve life chances or quality of life, let alone integration into society (Márquez, 2004; Rodríguez and Sugranyes, 2005). The model of targeted subsidies has reinforced the stratification of middle and low-income households according to relative need through access to differentiated subsidy programmes and housing markets that determine the location and typology of the housing to which each group is entitled (Table 14.1).

The welfare regime and its development

According to the traditional Esping-Andersen (1990) typology of welfare states, Chile moved from a liberal to a corporatist model (1930–1960s) and then gradually towards a social democratic

Table 14.1 Evolution of housing tenures in Chile, 1952–2002

	Chile					
	1952	*1960*	*1970*	*1980*	*1992*	*2002*
Housing tenure (% stock)*						
Homeownership	32.7	43.4	54.2	65.0	69.9	(-)
Tenants	67.3	56.6	45.8	35.0	30.1	(-)
*Housing tenure (% households**)*						
Homeownership (owned outright)					52.9	52.8
Homeownership (mortgaged)					19.6	15.4
Private rented					17.6	17.9
Granted					4.7	6.7
Other					5.1	7.0

(*) Tenure over permanent housing (houses, apartments, tenements), excluding irregular housing.
(**) Based on Census data.

model (1960s–1973). After 1973, a liberal welfare state was installed and persisted even after the restoration of democracy (1990-present) (Arellano, 1985; Richards, 1995; Posner, 2012). While this application of Esping-Andersen's typology to the Chilean case is accurate in general terms, we show that each of these periods was marked by conflicting rationales of government in two areas: the state's role in economic and social development, and the dynamics leading to the political incorporation or exclusion of different class-based groups in the coverage of the welfare regime.

This section shows that, from the 1960s through to 1973, the development and consolidation of a corporatist model as the hegemonic rationale was called into question and strained by a social-democratic/socialist rationale that sought to advance social rights for previously excluded sectors of the population. Similarly, as other scholars have shown, the liberal welfare regime that replaced the previous model in 1973 required constant adjustment, despite its clear-cut neoliberal underpinning, because the orthodox neoliberal blueprint did not work as originally planned (Richards, 1995). Conflicting principles of government are especially evident between 1990 and the current decade. While the welfare regime has remained "liberal" according to the Esping-Andersen typology, this period, under neoliberal principles of government, has also been infused with social-democratic rationales.

Starting in the 1930s, the state assumed a more active role in economic and social development. Different Latin American countries implemented an "import substitution industrialization" model (ISI[8]) (Gonçalves et al., 2016). Its traits included a developmental discourse promoting the state's participation in industrialization and a corporatist logic of political incorporation. In Chile, the state was active in the development and protection of local industry and in promoting the construction of infrastructure. Thus, developmental plans included housing construction (Arellano, 1985; Posner, 2012; Murphy, 2015).

Between the 1930s and 1950s, public expenditure on social protection increased and welfare benefits expanded significantly. Leftist political parties and labour unions also played an active role in the political incorporation of the working classes. In line with the corporatist model, this expansion benefitted mostly public employees, professionals and unionized workers in strategic sectors of the economy, such as mining, and large companies. Thus, agricultural workers, and low-income urban sectors – dependent mainly on informal employment – were largely excluded from this phase of development.

Both the state's role in economic development and the political incorporation of actors advanced further from the mid-1960s. Under a Christian Democrat approach, President Eduardo Frei Montalva (1964–1970) reformed several strategic areas. His administration's agrarian reform sought to modernize agricultural production through the redistribution of land and the unionization of agricultural labourers. It also reformed the mining sector, acquiring assets from international mining companies and initiating the nationalization of copper. For the urban poor, the government implemented a Popular Promotion policy, aiming to integrate poorer sectors through grassroots organizations. In 1965, it created the Ministry of Housing and Urbanization (MINVU) to frame a new housing policy in line with the broader goal of economic and social development. In its underlying principle, this new policy viewed housing as a necessity to which every family had a right (Hidalgo, 2019).

President Salvador Allende's socialist government (1970–1973) further strengthened the state's role in economic and social development, deepening the polarization of political forces. By 1970, state participation in the economy was the highest in Latin America (Arellano, 1985). Key reforms included the nationalization of mineral resources, giving the state property rights over mining deposits. In the case of political incorporation, the government dramatically increased public spending on social welfare (by over 70 per cent) and, particularly, housing (Posner, 2012). Government spending on social programmes reached some 20 per cent of GNP, again above the average for Latin America (Arellano, 1985). In addition, Allende proclaimed that housing was "a fundamental right of all Chilean families, independent of their income level" (Kusnetzoff, 1987, p. 162) and the state assumed the provision of housing as its obligation (MINVU, 2004).

In sum, two trends emerged during this phase: an expansion of the state's role in key areas of the economy and the start of the political incorporation and mobilization of previously excluded groups, such as the agricultural labourers and the urban poor. The dictatorship reversed both these trends.

Together with its radical neoliberal economic policies – including structural adjustment policies, deregulation of the economy and its opening to foreign trade and the privatization of state companies – the dictatorship transformed the state's role in promoting development. It reduced the state's previous entrepreneurial role as the key agent in fostering industrialization and production, allowing these economic processes to be the result of decisions by private agents in free and open markets (Ffrench-Davis, 2003).

In addition, the 1980 Constitution, written by the dictatorship, established a new contract between the state and private capital under which the state would play a subsidiary role in economic development. In line with neoliberal tenets, the principle of subsidiarity reduced the state's economic role to a minimum: basically, that of guaranteeing the security of individuals and the fulfilment of contracts between private parties and of promoting and maintaining the operation of private markets to ensure their efficiency and individuals' freedom of choice. This principle gave the state the power to create markets where none existed but excluded it from participation in them, except when they were not functioning. In line with these principles, the education, health, social security, urban and housing sectors were privatized and state participation in them was reduced.

The progress achieved in previous periods on the political incorporation of citizens was reversed. Political organizations, such as left-wing parties and grassroots organizations, including the *pobladores* movement, were dismantled and severely repressed. Social rights that had previously been understood as part of the state's responsibility towards citizens, such as education, health, housing and social security, were now regarded as market goods. The creation of segmented markets for different classes contributed to processes of socioeconomic stratification.

On taking office in 1990, the Concertación coalition's priority was to maintain the country's fragile political stability. Politically, it focused on managing the tensions between the government and the armed forces and on disarticulating leftist groups. It avoided engagement in confrontational politics and felt dependent on the cooperation of the right, business interests and the military. While popular mobilizations had been a crucial component of the political culture in previous periods, grassroots activists were excluded from the political process (Schneider, 1995) and decision-making (Oppenheim, 1993; Paley, 2001).

The first Concertación government headed by Patricio Aylwin (1990–1994) avoided structural transformations of the neoliberal model. However, in an attempt to distance itself from the dictatorship, it sought to combine free market and redistributive social policies – a strategy it referred to as "growth with equality" (Ffrench-Davis, 2003; Arellano, 2005). This strategy illustrates the combination of rationales that shaped the strategy adopted by Concertación governments: a neoliberal orientation under which the state's role was to create markets to make room for private profit and foster macroeconomic growth, accompanied by a social-democratic rationale to counteract the detrimental effects of neoliberal policies on the most disadvantaged sectors of the population (Garretón, 2012). This latter rationale included an increase in redistributive social spending and the expansion of paternalist assistance to the poor. This combination of rationales created hybrid practices that informed social policies in all areas, including housing (Fuster-Farfán, 2019).

The interrelation between housing and welfare

Based on the previous sections, we identify three forms of relationship between the housing system and the welfare state. They reflect not only the political rationales underpinning the different welfare regimes but also how these rationales inform the design and operation of housing policy. We describe how conflicting political rationales are built into housing policy and look at the tensions and conflicts to which housing policy has been subject over the past 30 years. We argue that a better understanding of the relationship between housing and welfare regimes should include how housing policies are informed and reflect conflicting principles of governmental action.

During the first phase (1930–1973), characterized by the gradual development of a corporatist welfare regime, tensions were introduced by a social-democratic approach and a socialist drive for greater state involvement in economic and social development and more decisive political incorporation of marginalized groups. In this context, housing policy was a key area for promoting both economic *and* social development. By the mid-1960s, housing was understood as a social right to which every family should have access regardless of socioeconomic status. Although the state had an interest in the construction of housing for low-income groups, in practice, most housing units built during this period were for the middle classes (Arellano, 1985; Posner, 2012; Murphy, 2015).

The 1973 coup brought a dramatic halt to state-led development and the gradual incorporation of different groups into the incipient welfare state. This led to a second phase (1973–1989), characterized by a neoliberal regime of social provision, in which the development model was transformed in line with a market-oriented blueprint. This reduced the state's participation in the market to a subsidiary role and organized the provision of basic goods according to the principles of supply and demand. However, the assumption that there would be a balance between supply and demand was not met and the market for low-income housing failed to function, necessitating a policy modification: "The government also had to assume a more active role than that originally envisaged when it became evident that developers and banks were not interested in building and financing houses for households earning less than US$200 per month" (Rojas, 2001, p. 466). Thus, contrary to the initial plan, the state took on the task of contracting out the construction of low-income housing to private companies and allocating it to eligible beneficiaries. Compared to the previous model, targeted housing subsidies were effective in reaching low-income groups. While some scholars suggest that the military government was interested in tackling poverty (Richards, 1995), we contend that targeting was mostly driven by a need to provide incentives for the functioning of low-income markets. In this model, housing was regarded as a market commodity that would promote economic growth.

Finally, during the third phase (1990–2022), a hybrid welfare regime emerged. It has remained neoliberal in its basic tenets, particularly as regards the state's subsidiary role. Housing production remains private and continues to be regarded as a key activity for the economy. Housing is also still regarded as a commodity, not a right, and social groups are channelled into socioeconomically segmented markets. Moreover, state housing subsidies are distributed selectively, not universally (Bengtsson, 2001).

However, to counteract the negative effects of neoliberalism, social-democratic principles were incorporated into the model. Public expenditure on housing increased and the system was reinforced in a bid to provide housing more quickly and effectively. The changes included a diversification of subsidized housing programmes as a way of targeting different social groups. As previous analysis of liberal regimes has shown, the implementation of differentiated programmes contributed to the segmentation of access to benefits between and even within classes (Posner, 2012).

Initially, the housing production system was thought to articulate both neoliberal and social democratic principles effectively, given the reduction achieved in the housing deficit and positive evaluations of targeting mechanisms. In accordance with the neoliberal blueprint, the system was effective in creating a low-income housing market. As housing construction increased during the 1990s and 2000s, this proved to be a profitable business for private companies (Rodríguez and Sugranyes, 2005; Imilan, 2016). As a result, alternative forms of housing provision were not implemented and the government preferred turnkey projects to programmes of self-built homes and renting. In addition, the housing deficit indicator proved instrumental for the proper functioning of the state-subsidized housing market, with the estimation of housing "demand" helping to give the private housing market financed by the state stability and predictability (Abufhele, 2018).

At the same time, the housing production system reflected the government's social democratic orientation. According to longstanding tradition, housing policy and residents themselves have fuelled "the dream of homeownership". The housing subsidy functioned as a form of asset-based welfare, that is, a permanent form of financial assistance. Because it funds a durable good, the subsidy counts as an asset and serves as a rent that reduces monthly expenditure on housing, freeing up income for other necessities (MINVU, 2004). In addition, the system of targeted subsidies proved effective in meeting the housing needs of low-income groups. Despite criticisms, the housing construction system has, therefore, remained in place. However, despite its social-democratic orientation in attending to the needs of low-income groups, it remained weak in counteracting the model's segregating and stratifying effects.

Alongside these overarching rationales of government, the relationship between housing and welfare regimes also reflects government institutional practices that are played out through social policies. This has been particularly relevant during the past 30 years of democratic administrations in Chile (1990–2020). As mentioned above, the authorities' priority during the first post-dictatorship decade (1990–2000) was to overcome poverty. The housing system's paramount importance in achieving this goal justified asset-based housing provision and the reinforcement of subsidies targeting the poor. This selective form of incorporation implied that poor groups were compelled to request public assistance and were subject to strict means-testing mechanisms to obtain benefits.

The model's stratification effects were reinforced in subsequent decades, mostly through the diversification of housing programmes for the urban poor. The late 1990s brought the emergence of conflict over the repayment of house mortgages. A substantial number of social housing beneficiaries were not making repayments and the movement of housing debtors, ANDHA Chile, demanded the cancellation of these debts (Guzmán, 2014). The authorities responded by creating a new programme ("Debt-free Social Housing") in 2002, targeting exclusively extremely poor sectors and abolishing the mortgage requirement. Here, too, contradictory rationales of government were in play. While this programme was regarded as a welfare improvement for the extremely poor, it continued to segment and stratify access to housing.

In the second post-dictatorship decade (2000–2010), the orientation shifted from a "poverty approach" to a vulnerability and rights-based approach (Rojas, 2019). As the poverty rate dropped, the focus shifted to a social protectionist approach to guard against families falling back into poverty. This approach was implemented differently in each subsystem. While the education and health systems began to move towards "universal" rights of access, the housing system remained selective.

The model's segregating effects have proved difficult to counteract. According to the neoliberal orthodoxy, the Chilean model used an "enabling approach" to facilitate the expansion of the private sector in housing provision (Gilbert, 2002). The model's assumptions were met on the macro level in that it reduced public spending and increased its effectiveness. At the same time, the model served to secure a housing "demand" that provided incentives for construction companies to cater for low-income groups. The model of targeted subsidies assisted only those citizens who could not access the private housing market by their own means (Richards, 1995). That is, through targeted subsidies, the state helped to create a private housing market for poorer sectors. On the micro level, however, the model's assumptions were not met. By subsidizing demand, the programme was supposed to enable low-income families "to participate with economic freedom in the marketplace" (Richards, 1995, p. 519). However, low-income households could not choose their house or its location. When declared eligible, they were assigned a housing unit based on the availability of housing projects built by construction companies, mostly on peripheral low-cost land.

Several improvements have sought to increase the quality of housing and integrate public housing projects into the urban fabric. Importantly, they have included the implementation of a conditional subsidy *(Subsidio diferenciado a la localización)* added to the regular subsidy based on the location of housing projects, thereby creating incentives for developments with a better location within the city. Yet, as shown in previous sections, the desegregation effect of this subsidy has been weak (Dohnke et al., 2014; Rasse et al., 2021), since the costs of building on better-located land proved incompatible with the profitability for construction companies required for the model to function.

Conflicting rationales of government are expressed in normative and political orientations regarding the state's role in promoting economic and social development, the "width" of access to welfare by different class-based groups (Bengtsson, 2001) and the institutional design of housing policy. Some scholars posit that the Chilean housing system represents an original and innovative synthesis of liberalization policies within Chile's state-dominated tradition (Smolka and Sabatini, 2000; Rojas, 2001). By contrast, others suggest that it is less a synthesis than the implementation of a state-led neoliberal model (Rodríguez and Sugranyes, 2005; Imilan, 2016). Despite efforts to incorporate a social-democratic logic into the market-based system, the state has reinforced the housing policy's neoliberal pillars. Although the state continues to advance towards better-quality social housing and has striven to expand access and improve the policy's efficiency, it continues to treat housing as a commodity governed by market forces and to provide stratified and selective access to housing subsidies for middle and low-income groups.

Regional outlook

The Chilean model of housing production is relevant beyond the country's borders. In the 1990s, Washington-based developmental agencies began to promote its housing policy in the region and beyond as "best practice" (Gilbert, 2004a). It was even featured as a "showpiece of social policy" in Time Magazine (Richards, 1995, p. 516). Consequently, several aspects of the Chilean experience were adopted by other Latin American countries, such as Colombia, Costa Rica, Ecuador, Panama, Paraguay, Mexico and Brazil and even by South Africa (Held, 2000; Gilbert, 2004a, 2004b).

The spread and adoption of market-led housing policies were explained by several factors. Despite criticism, Chile's housing policy was regarded as an improvement on previous state-led models (Gilbert, 2002). There was consensus about the financial and technical failings of earlier state institutions and interventions, which included limited production, low-quality housing units, high administrative costs, ineffective targeting and opaque political criteria for the distribution of housing (Arellano, 1985; Kusnetzoff, 1990; Rojas, 2001; Gilbert, 2002, 2004b). In Colombia, Paraguay and South Africa, there was a particular need to control the spread of land invasions and informal urbanization, symptoms of the state's inefficiency in meeting growing demand (Gilbert, 2004a, 2004b; Amarilla and Ruiz-Tagle, 2018). By contrast, Chile had initially been successful in increasing the pace of housing construction, reducing the housing deficit and controlling informality. Even more importantly, the model of targeted subsidies had proved effective in meeting the main rationale of neoliberal orthodoxy: to limit public expenditure while controlling the housing demands of the urban poor through more effective and transparent targeting (Held, 2000; Gilbert, 2002, 2004b). According to Gilbert (2004b), the country's economic "miracle" immediately after the transition to democracy and the apparent success of privatizations positioned Chile as an important "source of lessons for the rest of Spanish America" (p. 202).

In addition, the Chilean model fitted perfectly the developmental Washington approach, promoted by the World Bank, the IMF, the Inter-American Development Bank and USAID. The post-Washington Consensus approach acknowledged that the neoliberal orthodoxy of minimal states and private markets was limited to producing broader economic growth and international agencies subsequently incorporated "poverty alleviation" into their objectives and advice. Specifically, the Chilean experience was in line with the "enabling approach" to housing policy and was held up as an exemplary case. The infamous World Bank (1993) report "Housing: Enabling Markets to Work" outlined the main features of the enabling strategy that developing countries should follow to stimulate demand, facilitate supply and organize the overall housing sector to ensure that housing markets would work more efficiently while promoting macroeconomic growth (Pugh, 1994a). In this model, the state's role was to provide the institutional and regulatory conditions to enable the privatization of housing production, aiding and directing demand for housing towards the market. Key to this model were the state's regulation and direction of mortgage financing toward the urban poor and the provision of limited, transparent and well-targeted subsidies (World Bank, 1993; Pugh, 1994a; Gilbert, 2002).

Following a perceived crisis of the state-led model of housing provision, several Latin American countries have, therefore, embraced some features of Chile's "enabling approach" (Gilbert, 2004a, 2004b; Amarilla and Ruiz-Tagle, 2018), although scholars have shown that the spread and adoption of foreign models was not a linear and unidirectional process. The adaptation of policies from other countries to reflect local features (Gilbert, 2004b) and the influence of Washington depended upon country-level negotiations between particular aid agencies and governments (Pugh, 1994b; Gilbert, 2002).

As early as 1987, Costa Rica created a National System for Housing Finance (*Sistema Financiero Nacional para la Vivienda*), acknowledging the role of mortgages in financing social housing. It also introduced the model of direct housing subsidies to stimulate demand (Held, 2000). In the early 1990s, Colombia adopted the housing policy promoted by the World Bank, redirecting the state from contracting public housing construction to facilitating supply through the private sector and introducing targeted subsidies for the urban poor (Held, 2000; Gilbert, 2004b). Like Chile, these two countries continue to embrace the model, although with substantive improvements. Paraguay took a different course. In 1990, it used an Inter-American Development Bank loan to implement a system of targeted subsidies through the National Housing Council *(Consejo Nacional de Vivienda)*. However, the abolition of the Housing Council by the end of the 1990s weakened the system's implementation. A second attempt was made in 2006, this time with formal cooperation from the Chilean Ministry of Housing and Urbanization (MINVU). In 2010, Paraguay finally reformed its housing institutions, oriented housing policy according to market principles and readopted the system of targeted subsidies (Amarilla and Ruiz-Tagle, 2018).

Evaluations of these countries' housing policies point to similar achievements and deficiencies. They succeeded in expanding the supply of social housing and reducing housing shortages. In Costa Rica, the housing deficit dropped from 21 per cent in 1985 to 3.8 per cent in 1989 and 1.9 per cent in 1997 (Held, 2000). A similar trend was seen in Paraguay in the 2010s. The adoption of the model increased housing production, helped to regulate informal settlements and reduced the housing shortage (Amarilla and Ruiz-Tagle, 2018). Housing deficit reduction in Colombia was initially not as impressive as in the other two cases, due mostly to low levels of housing expenditure (Gilbert, 2004a). Nonetheless, the housing shortage fell by 247,000 units between 1993 and 1997 (Held, 2000). In the long run, all three examples show the model is effective in stimulating housing production. In addition, all three countries have been successful in directing housing subsidies more effectively to the target populations.

Yet, as in Chile, the model has shown severe shortcomings. Most importantly, the model contains a core contradiction: it seeks to facilitate housing construction while reducing public expenditure. This has led to higher levels of housing construction, but much of it is of low quality and on low-cost land, thereby increasing the qualitative housing deficit (Held, 2000) as well as segregation and social stratification. As Gilbert (2004a) found, "housing subsidies may be superior to providing housing through public agencies, but they are totally powerless to overcome the fundamental contradictions underpinning their adoption" (p. 35).

Concluding discussion

This historical review illustrates the interrelation between housing systems and the different welfare regime configurations in Chile. Scholars have previously shown that housing systems respond and should be examined in relation to high-level forces as well as institutional details (Stephens, 2016). In line with this, an examination of the Chilean case shows that to gain a better understanding of the housing system and its position within the welfare state, closer attention should be paid to the rationales of government that inform the welfare state regime, the institutional design of the housing system and in particular the execution of housing policies.

We identify three distinct welfare regimes between the 1920s and the 2020s. We contend that, while Esping-Andersen's (1990) ideal types are useful for analysing the general features and contours of welfare regimes, a more nuanced analysis reveals that welfare regimes are not monolithic epochal configurations, but are influenced by political actors, with contested definitions of welfare, how it should be provided and to whom. Thus, we show that each welfare regime is strained by conflicting rationales of government that are expressed in two ways: how the regime defines the roles of the state and the market in promoting economic and social development and its approach to the political incorporation of different population segments into welfare benefits, that is, the width of the social strata the regime covers.

Between the 1930s and 1970s, Chile's welfare regime was predominantly corporatist, tending towards a social-democratic welfare regime, tensioned by political parties and civil society, which were urging greater state participation in economic and social development and the gradual expansion of welfare to excluded groups. By the end of this period, housing was being promoted as a social right, which it was the state's responsibility to guarantee, regardless of citizens' socioeconomic condition. However, the provision of housing still followed a corporatist logic and did not acquire a universal form. The welfare state's expansion was interrupted by the military dictatorship's neoliberal restructuring reforms. The following period (1973–1989), characterized as a liberal welfare state, adhered to hegemonic neoliberal principles, yet implemented market-based social policies, requiring constant adjustment and greater state participation. Under this regime, housing was regarded as a good to be obtained through the market, with state subsidies only for targeted groups. Thus, housing was provided selectively according to socioeconomic status. Finally, in the third period analysed (1990–2020), we identify a hybrid welfare regime. It remained neoliberal in its basic tenets but infused with a social-democratic rationale that sought to counteract the model's adverse effects. However, social-democratic efforts were deployed at the level of institutional practices, without altering the housing system's structural foundations. As a result, the model continues to reproduce and reinforce housing policy's segregating and stratifying outcomes.

Based on this framework, we show that the current housing crisis in Chile is a result not only of current political and economic factors but also of the long-term processes described above. The current crisis reflects the exhaustion of a neoliberal-oriented model of housing production. The modifications and improvements introduced over the past 30 years, which have legitimated the system's continuation, have failed to prevent the reproduction of housing inequalities, let alone keep up with growing demand. Finally, given the adoption of the Chilean model in other Latin American countries and beyond, we contend that its shortcomings in addressing the current crisis and solving persistent housing inequalities should be borne in mind by countries moving in the same direction.

Acknowledgements

This work was supported by the Fondo Nacional de Desarrollo Científico y Tecnológico through grants ANID/FONDECYT/3200926 and ANID/FONDECYT/11190211.

Notes

1 "Social housing" refers to the affordable and subsidized housing sponsored or financed by governments that emerged in the 1980s, after neoliberal restructuring, to solve the exclusion of low-income groups from private house markets (Dohnke, 2014; Ozer and Jacoby, 2022).

2 "Housing deficit" is the main indicator used to assess the effectiveness and challenges of housing policy (MINVU, 2018). The quantitative housing deficit indicates the number of housing units needed to replace run down housing and provide housing to households living in overcrowded conditions.

3 Home-sharing is the cohabitation of two or more households in the same house or site, generally in overcrowded conditions (MINVU, 2004)

4 The CChC, created in 1951, is a business association that represents the main private construction companies.
5 Social housing construction increased from an annual average of 39,134 units between 1980 and 1989 to 76,602 between 1990 and 1999 and 97,370 between 2000 and 2009 (MINVU, 2009).
6 Based on alternative data, the National Socioeconomic Characterization Survey - CASEN (1990–2017), scholars confirm this long-term trend toward reducing the housing shortage, but only up to 2015 (391,546 units in 2015). The 2017 survey estimated a rebound to 497,615 units (Irarrázabal, 2019; Medel, 2020). The Cámara Chilena de la Construcción (CChC) estimated the 2017 deficit at 739,603 units, corresponding to 2.2 million Chileans (CChC, 2017).
7 Between 1980 and 2000, only 15 per cent of social housing was located in central or pericentral areas of the Santiago Metropolitan Region, while 77 per cent was located on the periphery where residential segregation was greatest. Between 2001 and 2017, this trend remained largely unchanged: 16 per cent of social housing was built in central and pericentral areas, while 51 per cent was built on the periphery. In addition, 30 per cent was located in periurban sectors which, in the previous period, accounted for less than 6 per cent (Rasse et al., 2021). By contrast, subsidized middle-class housing tended to be located in pericentral municipal districts (Dohnke et al., 2014).
8 Import substitution industrialization was a development model, implemented in developing countries, based on state protection of nascent local industries which aimed to reduce dependence of developed economies. It was introduced in Chile in response to the local effects of the 1930s crisis, which were caused by dependence on overseas economies (Arellano, 1985).

References

Abufhele, V. (2018) *The State's Poverty Trap: The Reproduction of Informal Settlements in Santiago, Chile 1990–2017* [PhD Dissertation. New School for Social Research].

Amarilla, X., & Ruiz-Tagle, J. (2018). Desde la Informalidad Hacia la Masividad del Mercado. *Ciudades, 29*(120), 53–60. https://doi.org/10.13140/RG.2.2.34692.78728

Angelcos, N. (2012). Lucha por la Vivienda y Politización de las Trayectorias Individuales. *Polis (Santiago), 11*(31), 17–38. https://doi.org/10.4067/S0718-65682012000100002

Angelcos, N., & Pérez, M. (2017). De la "Desaparición" a la Reemergencia: Continuidades y Rupturas del Movimiento de Pobladores en Chile. *Latin American Research Review, 52*(1), 94–109. http://doi.org/10.25222/larr.39

Arellano, J. (1985). Social Policies in Chile: An Historical Review. *Journal of Latin American Studies, 17*(2), 397–418. https://doi.org/10.1017/S0022216X0000794X

Arellano, J. (2005). Políticas Sociales Para el Crecimiento con Equidad: Chile 1990-2002. *Serie Estudios Socio/Económicos*, No. 26. https://www.cieplan.org/wp-content/uploads/2019/12/serie-est-socioec-26.pdf

Bengtsson, B. (2001). Housing as a Social Right: Implications for Welfare State Theory. *Scandinavian Political Studies, 24*(4), 255–275. https://doi.org/10.1111/1467-9477.00056

Bruey, A. J. (2012). Limitless Land and the Redefinition of Rights: Popular Mobilisation and the Limits of Neoliberalism in Chile, 1973–1985. *Journal of Latin American Studies, 44*(03), 523–552. https://doi.org/10.1017/S0022216X12000399

Caldera, A. (2012). Building Blocks for a Better Functioning Housing Market in Chile. OECD *Economics Department Working Papers*, No. 943. https://doi.org/10.1787/5k9fj3hgsnvh-en

Castells, M. (1973). Movimientos de Pobladores y Lucha de Clases en Chile. *EURE, 3*(7), 9–35. https://www.eure.cl/index.php/eure/article/view/834/694

CEPAL. (1990). *Una Estimación de la Magnitud de la Pobreza en Chile, 1987*. http://hdl.handle.net/11362/29716

Cortés, A. (2013). A Struggle Larger Than a House: Pobladores and Favelados in Latin American Social Theory. *Latin American Perspectives, 40*(2), 168–184. https://doi.org/10.1177/0094582X12467763

Cortés, A. (2014). El Movimiento de Pobladores Chilenos y la Población La Victoria: Ejemplaridad, Movimientos Sociales y el Derecho a la Ciudad. *EURE, 40*(119), 239–260. http://doi.org/10.4067/S0250-71612014000100011

CORVI. (1972). *60 mil Viviendas. CORVI en el Centro de la Construcción*. Corporación de la Vivienda.

De Ramón, A. (1990). La Población Informal. Poblamiento de la Periferia de Santiago de Chile. 1920–1970. *EURE, 16*(50), 5–17. http://www.eure.cl/index.php/eure/article/view/1049

De Ramón, A. (2007). *Santiago de Chile (1541-1991): Historia de una Sociedad Urbana*. Catalonia.

Dohnke, J. et al. (2014). Achieving a Socio-Spatial Mix? Prospects and Limitations of Social Housing Policy in Santiago de Chile. *Housing Studies*, *30*(6), 839–857. https://doi.org/10.1080/02673037.2014.982516

Ducci, M. (1997). Chile: El Lado Oscuro de una Política de Vivienda Exitosa. *EURE*, *23*(69), 99–115. https://www.eure.cl/index.php/eure/article/view/1164

Esping-Andersen, G. (1990). *The Three Worlds of Welfare Capitalism*. Princeton University Press.

Ffrench-Davis, R. (2003). Chile, Entre el Neoliberalismo y el Crecimiento con Equidad. Nueva Sociedad. *Democracia y Política en América Latina,* (183), 70–90. https://static.nuso.org/media/articles/downloads/3100_1.pdf

Fundación Vivienda. (2018). *Déficit Habitacional Cuantitativo*. Censo 2017. Publicación N°5. http://ceschile.org/wp-content/uploads/2020/11/Informe-4-D%C3%A9ficit-Habitacional-y-Censo.pdf

Fuster-Farfán, X. (2019). Las Políticas de Vivienda Social en Chile en un Contexto de Neoliberalismo Híbrido. *EURE*, *45*(135), 5–26. https://www.eure.cl/index.php/eure/article/view/2906

Garcés, M. (2002). *Tomando su Sitio: El Movimiento de Pobladores de Santiago, 1957-1970*. LOM Ediciones.

Garretón, M. A. (2012). *Neoliberalismo Corregido y Progresismo Limitado: Los gobiernos de la Concertación en Chile, 1990-2010*. Editorial ARCIS-CLACSO-PROSPAL.

Giannotti, E. (2014). Una Ciudad de Propietarios. El Caso de la Población La Victoria. *Revista AUS*, (15), 40–45. https://doi.org/10.4206/aus.2014.n15-08

Gilbert, A. (2002). Power, Ideology and the Washington Consensus: The Development and Spread of Chilean Housing Policy. *Housing Studies*, *17*(2), 305–324. https://doi.org/10.1080/02673030220123243

Gilbert, A. (2004a). Helping the Poor Through Housing Subsidies: Lessons from Chile, Colombia and South Africa. *Hábitat International, 28*(1), 13–40. https://doi.org/10.1016/S0197-3975(02)00070-X

Gilbert, A. (2004b). Learning from Others: The Spread of Capital Housing Subsidies. *International Planning Studies*, 9, 197–216. https://doi.org/10.1080/1356347042000311776

Gonçalves, R. et al. (2016). Vulnerabilidad Estructural Externa en América Latina. In E. Arceo & E. Basualdo (Eds.), *Los Condicionantes de la Crisis en América Latina. Inserción Internacional y Modalidades de Acumulación* (pp. 119–138). CLACSO. http://biblioteca.clacso.edu.ar/clacso/gt/20160219035440/07gonca.pdf

Guzmán, S. (2014). *To Pay or to Protest: Consent and Resistance to Social Housing Debt in Chile* (Doctoral Dissertation. The New School for Social Research, New York).

Haggard, S., & Kaufman, R. R. (2009). Poverty, Inequality, and Democracy (II): How Regions Differ. *Journal of Democracy, 20*(4), 64–78. https://doi.org/10.1353/jod.0.0122.

Han, C. (2012). *Life in Debt: Times of Care and Violence in Neoliberal Chile* (1st ed.). University of California Press.

Held, G. (2000). Políticas de Viviendas de Interés Social Orientadas al Mercado: Experiencias Recientes con Subsidios a la Demanda en Chile, Costa Rica y Colombia. *CEPAL Serie Financiamiento del Desarrollo*, (96). http://hdl.handle.net/11362/5304

Hidalgo, R. (2007). ¿Se Acabó el Suelo en la Gran Ciudad? Las Nuevas Periferias Metropolitanas de la Vivienda Social en Santiago de Chile. *EURE, 33*(98), 55–75. https://doi.org/10.4067/S0250-71612007000100004

Hidalgo, R. (2019). *La Vivienda Social en Chile y la Construcción del Espacio Urbano en el Santiago del Siglo XX*. RIL editores.

Hipsher, P. (1996). Democratization and the Decline of Urban Social Movements in Chile and Spain. *Comparative Politics*, *28*(3), 273–297. https://doi.org/10.2307/422208

Imilan, W. (2016). *Políticas y Luchas por la Vivienda en Chile: El Camino Neoliberal*. Working Paper Series. Contested Cities, (V). http://contested-cities.net/working-papers/2016/politicas-y-luchas-por-la-vivienda-en-chile-el-camino-neoliberal/

Irarrázabal, G. (2019). *Déficit habitacional en Chile: Desafíos de política pública* (Serie Informe Social 179). Libertad y Desarrollo. https://lyd.org/wp-content/uploads/2019/09/serie-informe-social-179-agosto.pdf

Kusnetzoff, F. (1987). Urban and Housing Policies under Chile's Military Dictatorship 1973-1985. *Latin American Perspectives*, *14*(2), 157–186. https://doi.org/10.1177/0094582X8701400203

Kusnetzoff, F. (1990). The State and Housing in Chile: Regime Types and Policy Choices. In G. Shidlo (Ed.), *Housing Policy in Developing Countries* (pp. 48–66). Routledge.

Márquez, F. (2004). Márgenes y Ceremonial: Los Pobladores y las Políticas de Vivienda Social en Chile. *Política Revista de Ciencia Política*, No. 43, 185–203. https://revistapolitica.uchile.cl/index.php/RP/article/view/55769/58684

Ministerio de Vivienda y Urbanismo, Instituto de la Vivienda INVI. (1997). *Catastro de Campamentos y Asentamientos Irregulares*. Facultad de Arquitectura y Urbanismo, Universidad de Chile. http://repositorio.uchile.cl/handle/2250/140433

Ministerio de Vivienda y Urbanismo. (2004a). *Chile: Un Siglo de Políticas de Vivienda y Barrio*. Pehuén Editores.

Ministerio de Vivienda y Urbanismo. (2004b). El Déficit Habitacional en Chile. Medición de Requerimientos de Vivienda y su Distribución Espacial. *Revista INVI, 20*(53), 231–233. https://doi.org/10.5354/0718-8358.2005.62274

Ministerio de Vivienda y Urbanismo. (2006). *Atlas de la Evolución del Déficit Habitacional en Chile 1992–2002*. Pehuén Editores.

Ministerio de Vivienda y Urbanismo. (2007). *Catastro Nacional de Campamentos 2007*. Centro de Investigación Social Un Techo para Chile. http://datos.techo.org/dataset/5fee0aa3-7a2c-48a7-ad3c-b2f-cd2974ff0/resource/5a102dab-9604-44c5-b1da-1f588240b0c7/download/cnc_2007.pdf

Ministerio de Vivienda y Urbanismo. (2009). *Déficit Urbano Habitacional. Una Mirada Integral a la Calidad de vida y el Hábitat Residencial en Chile*. Comisión de Estudios Habitacionales y Urbanos. CEHU. https://www.mppn.org/wp-content/uploads/2018/02/Deficit-Urbano-Habitacional01-1.pdf

Ministerio de Vivienda y Urbanismo. (2013). Mapa Social de Campamentos. Secretaría Ejecutiva de Campamentos. *Secretaría Ejecutiva de Campamentos*, No. 339. https://www.desarrollosocialyfamilia.gob.cl/btca/txtcompleto/mapasocial-campamentos.pdf

Ministerio de Vivienda y Urbanismo. (2018). *Resultados Déficit Habitacional Cuantitativo*. Censo 2017 Comisión de Estudios Habitacionales y Urbanos. http://ceschile.org/wp-content/uploads/2020/11/Informe-4-D%C3%A9ficit-Habitacional-y-Censo.pdf

Ministerio de Vivienda y Urbanismo. (2019). *Catastro Nacional de Campamentos*. https://storymaps.arcgis.com/stories/dfe1fe1afd334ec790f879e736a5af5e

Ministerio de Vivienda y Urbanismo. (2021). *Déficit Habitacional Casen en Pandemia 2020: Resultados y Notas Metodológicas*. Centro de Estudios de Ciudad y Territorio. https://catalogo.minvu.cl/cgi-bin/koha/opac-retrieve-file.pl?id=3628046fcb74059efb3db8a011d80369

Morales, E. & Rojas, S. (1986). Relocalización Socio-Espacial de la Pobreza. Política Estatal y Presión Popular. 1979–1985. FLACSO-Santiago. *Documento de Trabajo,* (280). https://flacsochile.org/biblioteca/pub/memoria/1986/000856.pdf

Murphy, E. (2015). *For a Proper Home: Housing Rights in the Margins of Urban Chile, 1960-2010*. University of Pittsburgh Press.

Oppenheim, L. (1993). *Politics in Chile: Democracy, Authoritarianism, and the Search for Development*. Westview Press.

Ozer, S., & Jacoby, S. (2022). The design of subsidized housing: Towards an interdisciplinary and cross-national research agenda. *Housing Studies*. https://doi.org/10.1080/02673037.2022.2045005

Özler, Ş. (2012). The Concertación and Homelessness in Chile: Market-Based Housing Policies and Limited Popular Participation. *Latin American Perspectives, 39*(4), 53–70. https://doi.org/10.1177/0094582X10397917

Paley, J. (2001). *Marketing Democracy: Power and Social Movements in Post-Dictatorship Chile*. University of California Press.

Pérez, M. (2017a). A New Poblador Is Being Born: Housing Struggles in a Gentrified Area of Santiago. *Latin American Perspectives, 44*(3), 28–45. https://doi.org/10.1177/0094582X16668318

Pérez, M. (2017b). Reframing Housing Struggles. Right to the City and Urban Citizenship in Santiago, Chile. *City, 21*(5), 530–549. https://doi.org/10.1080/13604813.2017.1374783

Pérez, M. (2018). Toward a Life with Dignity: Housing Struggles and New Political Horizons in Urban Chile. *American Ethnologist, 45*(4), 508–520. https://doi.org/10.1111/amet.12705

Posner, P. (2012). Targeted Assistance and Social Capital: Housing Policy in Chile's Neoliberal Democracy. *International Journal of Urban and Regional Research, 36*(1), 49–70. https://doi.org/10.1111/j.1468-2427.2011.01059.x

Pugh, C. (1994a). The Idea of Enablement in Housing Sector Development: The Political Economy of Housing for Developing Countries. *Cities, 11*(6), 357–371. https://doi.org/10.1016/0264-2751(94)90068-X

Pugh, C. (1994b). Housing Policy Development in Developing Countries: The World Bank and Internationalization, 1972–93. *Cities, 11*(3), 159–180. https://doi.org/10.1016/0264-2751(94)90057-4

Rasse, A. et al. (2021). Desde la Segregación a la Exclusión Residencial ¿Dónde Están los Nuevos Hogares Pobres (2000- 2017) de la Ciudad de Santiago, Chile? *Revista de Urbanismo*, (44), 39–59. https://doi.org/10.5354/0717-5051.2021.55948

Richards, B. (1995). Poverty and Housing in Chile: The Development of a Neo-Liberal Welfare State. *Habitat International*, *19*(4), 515–527. https://doi.org/10.1016/0197-3975(95)00043-F

Rodríguez, A., & Sugranyes, A. (Eds.) (2005). *Los con Techo: Un Desafío Para la Política de Vivienda Social*. Ediciones SUR.

Rojas, C. (2019). *Ayudar a los Pobres. Etnografía del Estado Social y las Prácticas de Asistencia*. Ediciones UAH.

Rojas, E. (2001). The Long Road to Housing Sector Reform: Lessons from the Chilean Housing Experience. *Housing Studies*, *16*(4), 461–483. https://doi.org/10.1080/02673030120066554

Sabatini, F. (2000). Reforma de los Mercados de suelo en Santiago, Chile: Efectos Sobre los Precios de la Tierra y la Segregación Residencial. *EURE*, *26*(77), 49–80. https://doi.org/10.4067/S0250-71612000007700003

Sabatini, F. et al. (2001). Segregación Residencial en las Principales Ciudades Chilenas: Tendencias de las Tres Últimas Décadas y Posibles Cursos de Acción. *EURE*, *27*(82), 21–42. https://doi.org/10.4067/S0250-71612001008200002

Schneider, C. (1995). *Shantytown Protest in Pinochet's Chile*. Temple University Press.

Smolka, M., & Sabatini, F. (2000). The Land Market Deregulation Debate in Chile. *Landlines*, *12*(1), 1–3. https://www.lincolninst.edu/sites/default/files/pubfiles/312_linc_landlines%201.00.pdf

Stephens, M. (2016). Using Esping-Andersen and Kemeny's Welfare and Housing Regimes in Comparative Housing Research. *Critical Housing Analysis*, *3*(1), 19–29. https://doi.org/10.13060/23362839.2016.3.1.250

Techo. (2021). *Catastro Nacional de Campamentos 2020-2021. Techo, Fundación Vivienda, CES*. https://ceschile.org/wp-content/uploads/2020/11/Catastro%20Campamentos%202020-2021%20TECHO-FV.pdf

World Bank. (1993). Housing: Enabling Markets to Work, World Bank Policy Paper. *World Bank Group*. https://doi.org/10.1596/0-8213-2434-9

15

HOUSING AND WELFARE IN SOUTH AFRICA

Elelwani Mmbadi, Lochner Marais and Job Gbadegesin

The current state of housing in South Africa

The history of housing in South Africa is closely related to apartheid and colonial planning, which prevented black people from accessing urban housing and deliberately kept them in rural areas. This policy began to change in the mid-1980s, during the late apartheid era, but its legacy remains a problem.

Despite attempts during late apartheid to redress the historical imbalances, in 1994 the housing shortage was estimated at 1.5 million units (Department of Housing, 1994). Consequently, the post-apartheid government embarked on a housing delivery programme using a capital subsidy. This subsidy originated in World Bank thinking and was first applied in the mid-1980s by the Urban Foundation, after which the Independent Development Trust (IDT) used it between 1992 and 1994. The National Housing Forum (a non-governmental organization) developed South African housing policy during the transition period (Rust & Rubenstein, 1996). The size of the subsidy was debated: some argued that the subsidy should support the purchase of a large house (a single house on a site), while others argued that it should be for the stand-alone. The final decision was a compromise, which Tomlinson (1998) described as a victory for breadth over depth. The subsidy provided a small product to as many people as possible, giving poor households a starter home, homeownership and basic services. Using the subsidy to include a house was a deviation from the World Bank approach.

Between 1994 and 2018, the government subsidy programme delivered about 3.2 million houses (Department of Human Settlements, 2019). However, there has been a rapid decline since the annual peak delivery of 250,000 houses in 1999. The rate of housing production declined to about 80,000 in 2018. In the past four years, the number has dropped even further, with no clear direction from the government on the continuation of the housing subsidy programme. The decline has been caused by government budget cuts that have reduced the funds available for housing subsidy over the past decade, and secondly because, despite the initial victory for breadth over depth, the subsidy amount has increased more than inflation since the government instituted it in 1995. Over the past decade, depth rather than breadth has dominated the housing approach (Tomlinson, 2015). Yet, in contrast, there is evidence of some acceptance of providing "site and services", i.e. a plot of land with basic infrastructure, but without a house.

DOI: 10.1201/9781003212690-17

In addition to the drive towards homeownership, in 1994, the post-apartheid government also introduced a social housing programme. Independent social housing institutions (third-sector institutions as opposed to the government) provide rental housing. The third sector approach largely resembles the approach adopted in the Netherlands, where the government supported independent social housing institutions (Scheba & Turok, 2021). Many Netherlands consultants have provided technical support to the South African government. However, progress has been slow. Only about 35,000 houses have been constructed in more than 25 years (Turok, Scheba, & Visagie, 2022), and the social housing programme remains smaller than its homeownership counterpart, where the government has facilitated ownership of between 3 and 4 million houses.

South Africa is about 65 per cent urbanized, with approximately 35 million people living in urban areas (Stats SA, 2016). Although government housing programmes have made inroads into the housing shortage, about 13 per cent of urban households still live in informal housing with limited housing rights (Stats SA, 2016). Although this percentage is much lower than the figure in 1996, the housing backlog remains large (Stats SA, 1996). Figures on land tenure in South Africa are not accurate and should be treated with caution. Nonetheless, Stats SA (2016) found that 61 per cent of urban households have title to land (this percentage may be lower, as many people say they own their house when legally this may not be the case), 25 per cent rent and a further 14 per cent live rent-free. Rural housing tenure varies. Commercial farmers have land ownership, but generally, few farmworkers enjoy property rights (Marais & Lenka, 2021). In traditional areas, households have housing rights but do not own the land. In most cases, the state is the landowner, but in the KwaZulu-Natal province, the Ingonyama Trust is the largest landowner – owning nearly 30 per cent of land in the province. The Trust was funded in 1994 (through an Act of the KwaZulu Bantustan) and made the king of the Zulu nation the trustee of the land.

The production of new housing is financed in four main ways. First, mortgage finance is available to middle- and high-income households based on affordability criteria (Marais & Cloete, 2017). About 20 per cent of houses in South African urban areas are owned with a mortgage. The government established the National Housing Finance Corporation in 1996 to make housing affordable to middle-class households. Second, micro-lending is a common way of financing housing. The Rural Housing Loan Fund provides micro-finance to several intermediaries which fund small-scale housing extensions (RHLF, 2012). Third, the government funds housing at the bottom end of the market. This is done through a capital subsidy that ensures that most low-income households receive a small starter home of about 40m^2. The subsidy value per house in 2018 was R150,000[1] per unit. However, the future of the capital subsidy is uncertain and delivery has largely come to a standstill. As mentioned earlier, the government has also funded the social housing programme through social housing institutions. Fourth, many households fund their own houses. This applies to people in informal housing and middle- and high-income households.

The capital subsidy has created policy tension. There has been a long-standing trade-off between the breadth and depth of subsidy (Tomlinson, 1998). As indicated above, the shift over the last 15 years has been from breadth to depth, with larger subsidies awarded to fewer households. A second conflict has been between homeownership and rental housing. This conflict started at the National Housing Forum (1992–1994), where community groups advocated for state rental housing and the construction industry favoured state-led construction. The latter approach would see the state taking the risk and construction companies building the houses. Despite these minority voices, homeownership became the dominant housing approach. The homeownership programme linked to the capital subsidy provided many more houses than the social housing programme. A third point of contention was the argument that the capital subsidy programme is a neoliberal housing approach and as such is instrumental in keeping the racial structures of South African cities intact (Huchzermeyer, 2004, 2010). The main tension is caused by disagreement over whether a housing programme should promote the commodification of housing. This includes debates about the role of the private sector in housing delivery. The strict application of the capital subsidy has resulted in the location of housing projects on the urban fringes, where land costs are the lowest, perpetuating the racial structure of South African cities. In place of this inflexible arrangement, a larger subsidy could have been provided to well-located land (Huchzermeyer, 2004). However, the fixed amount of capital subsidy and a focus on housing size have reinforced historical urban patterns. This is an example of how the commodification of housing has perpetuated the racial structure of the apartheid city. (In this subchapter we use the race terminology that is used by Statistics South Africa (Stats SA): black, coloured, white and Indian/Asian).

Historical development of South Africa's housing regime

Colonial and apartheid housing regimes

South Africa's historical race-based housing regimes have been part of its politics for the past century (Mabin, 2020). Homeownership has been a dominant form of tenure for whites, but not always for blacks. The colonial and apartheid governments left South Africa with racially segregated suburbs and uneven access to housing based on race. (However, it should be noted that urban planning informally segregated suburbs on the basis of race long before this was formalized by the Group Areas Act of 1950). This uneven access to housing in urban areas resulted from racially based policies, long periods of migration control and the demolition of informal settlements (Harrison, 1992). For example, after 1948, the apartheid government rescinded the black homeownership that had been permitted in a small number of urban areas and resettled many of these households elsewhere (without homeownership). Forced removals like this were common. From the early 1950s to 1983, urban homeownership was available only to the white population.

By 1950, many black suburbs (known as 'townships') were in poor condition. Consequently, the state embarked on a large-scale rental housing programme between 1950 and 1970, delivering about 500,000 rental houses. This programme had two political aims: to clean up township areas and curb black urbanization. Once townships had been 'cleaned up', the apartheid government did not allow further black urbanization (Mabin, 1991). The emphasis on rental housing in the former black townships showed that the apartheid government viewed blacks as temporary visitors in urban areas. In some cases, it meant redirecting black urbanization to the periphery of cities to reserve the centre for whites (Mabin, 1991).

Resistance to apartheid made many former black townships ungovernable by the early 1980s (Chaskalson, Jochelson, & Seekings, 2007). For example, civic movements initiated rent strikes, in consequence of which local governments neglected housing maintenance. In July 1983, the apartheid government introduced the 'Big Sale', allowing residents to buy state rental houses. Very few households took this opportunity (Mabin & Parnell, 1983), mainly because they did not have the capital, so a discount benefit scheme was introduced in 1992. This scheme allowed the residents to take up ownership for free if the house value was R7,500 or less (most houses fell in this bracket). The post-apartheid government continued with this policy. Today, most houses have been transferred to individual ownership, although around 20,000–30,000 units are still in state hands.

Migration control policies under colonial and apartheid rule also influenced the provision of black housing by preventing black urbanization and the development of informal settlements. Furthermore, from 1968, local authorities were not allowed to provide more land for blacks.

Consequently, many people urbanized illegally and were often housed illegally in the backyards of existing houses. The reinstatement of homeownership for blacks in 1983 allowed for the first extension of township areas since 1968. These houses were for middle-income earners. No provision was made for poor households and the introduction of the policy of orderly urbanization meant that the government actively prevented the development of informal settlements (Harrison, 1992). Consequently, even more poor households resorted to living in backyards. Since the mid-1980s, the government has slowly introduced site and services programmes. Although these were common in former black areas (also known as 'toilets in the veld'), this was the first attempt to apply this housing approach in the main urban areas. With the release of Nelson Mandela in February 1990, a spirit of freedom arose across South Africa and thousands of householders who lived in backyards invaded open land (Marais & Ntema, 2013). The apartheid government was unable to prevent these informal settlements from developing. In 1992, the apartheid state introduced the IDT, which started with a capital subsidy programme financing 100,000 households with a stand, services and ownership (but no housing structure). The IDT programme was a typical site and services programme.

The IDT borrowed its idea of a capital subsidy from the Urban Foundation (a private-sector think tank) and the World Bank in the mid-1980s (World Bank, 1993). The focus on ownership was also the result of the failure of the rental housing programme in former black townships by the mid-1980s. Rent payments largely came to a halt. The Big Sale in 1983 was the first attempt to move away from a rental housing regime to a homeownership regime for blacks. This also meant that mortgage finance became available to black households in former black suburbs (Tomlinson, 2007), although high interest rates at the end of the 1980s resulted in the banks repossessing many houses. By the late 1990s, the banks held approximately 33,000 houses as properties in possession (Servcon Housing Solutions, 2002). The discount benefit scheme (1992) and the IDT (1992) were further steps toward encouraging ownership.

Under apartheid rule, South Africa had a dual housing approach. Blacks could not own their urban houses and the government provided state rental housing. Whites had a wide range of possibilities and homeownership was dominant. Thus, in addition to the many ideological and practical reasons for promoting black homeownership, providing homeownership to blacks was also a way of addressing the racial disparities that apartheid created.

Post-apartheid housing development

The post-apartheid government continued using the capital subsidy and the discount benefit scheme, and the Housing White Paper in 1994 confirmed the centrality of homeownership in housing policy. Landownership linked to the capital subsidy became a prominent feature of this policy.

Data on housing and tenure in South Africa are fragmented and sometimes contradictory. Stats SA provides information based on census data and community surveys. The Centre for Affordable Housing and Finance in Africa (CAHF, 2020) uses data from the deeds register. The 2016 Community Survey (Stats SA, 2016) reported 7.5 million urban households. It estimated that 61 per cent of urban households were homeowners, of whom 20 per cent had outstanding mortgages. The Centre for Affordable Housing in Africa reported 6.6 million urban residential properties valued at R5.5 trillion in South Africa in 2019. About 50 per cent of these properties were valued below R600,000. Of the residential properties, only about 2 million received a government subsidy. This was because, contrary to the policy guidelines, not all government-subsidized housing received full title.

South Africa has an active housing market, with 75 per cent of housing transactions being re-sales, and also an informal housing market. Homeownership for the black population was intended

Table 15.1 Changing forms of tenure in South Africa

Criteria	*Pre-1945*	*1960*	*1980*	*2000*	*2020*
Urban homeownership: policy changes	Available for blacks and whites	Mainly available for whites	Black homeownership restored in 1983	54% of urban households are homeowners	61% of urban households are homeowners
Urban homeownership (CAHF), 2020	n/d	n/d	n/d	12.6 million people; 4 million HH	14.4 million people; 4.5 million HH
Urban homeownership (Stats SA)	n/d	n/d	n/d	12.6 million people 4 million HH	22.1 million people 7.1 million HH
Cooperative ownership	None	None	None	3,200 people 1,000 HH	4,500 people 1,500 HH
Private rented	n/d	n/d	n/d	7.6 million people 2.1 million HH	5.1 million people 1.9 million HH
Government rented (estimates)	n/d	300,000	500,000	64,000 20,000 HH	32,000 10,000 HH
Live rent-free*	n/d	n/d	n/d	3.8 million 1.2 million HH	5.1 million 1.6 million HH
Social rented	0	0	0	38,400 12,000 HH	112,000 35,000 HH

Sources: Compiled from Stats SA, 2016; CAHF, 2020; Turok et al., 2022.
* Many households live with families or in backyards of existing stands and do not pay rent

to be a way to redress the historical unfairness. To take an example, in Mangaung, one of the eight metropolitan municipalities, individual landownership in former black suburbs increased from 10,000 in 1985 to 100,000 in 2016 (Marais, Awuzie, Cloete, Napier, & Greyling, 2020). Nevertheless, market imbalances remain (Marais et al., 2020). In Mangaung, 42 per cent of properties in formerly white suburbs were mortgaged in 2018 compared to 7 per cent in former black suburbs. The housing market in formerly white suburbs was also more active. The percentage of properties that changed ownership in five years in formerly white suburbs in Mangaung was above 20 per cent over five years. In former black suburbs, the percentage was only about 5 per cent.

The post-apartheid government instituted a social rental programme, but delivery has been poor (Turok et al., 2022). Only about 35,000 units have been constructed since 1994. If we compare this to the nearly 2 million households in rental housing, the lack of scale becomes even more evident. Table 15.1 shows that the number of people in private rentals has not grown. This lack of growth contrasts with the substantial growth in rent-free accommodation. It is believed that the COVID-19 lockdown increased this percentage even further.

The welfare system and its development

South Africa has the most extensive welfare system in Africa. Yet compared to many countries in the Global North its benefits remain small and underdeveloped. To assess the nature of the welfare system in South Africa, we use a framework developed by Esping-Andersen (1990). Included in this framework are the main providers of social welfare, the welfare systems, the levels of commodification and stratification, and the distribution of income. We map the South African welfare system to the three main welfare regimes identified by Esping-Andersen: liberal, corporatist and social-democratic.

South Africa's current population is estimated at 60.6 million with about 90 per cent black (including the coloured and Indian populations). In 2016 population growth was at about 1.3 per cent per annum (Stats SA, 2016). Urbanization increased from 55 per cent in 1996 to 65 per cent in 2016 (Stats SA, 1996, 2016). Despite active redress policies since 1994, racial disparities remain, and South Africa's Gini coefficient (0.65 in 2015) is among the world's highest (Stats SA, 2020). The racial disparities can be blamed on historical policies that prevented blacks from owning land and obtaining qualifications, excluding them from certain jobs and preventing urbanization. The racial disparities are also visible in labour participation and wage levels. Black unemployment was 38.6 per cent in the second quarter of 2022 (Statistica, 2022). The corresponding percentages were 10 per cent for whites, 25.9 per cent for coloureds and 17.1 per cent for Indians/Asians (Statistica, 2022). Stats SA (2020) indicates that the top 10 per cent of the population spent 8.6 times more than the bottom 40 per cent in 2006, dropping to 7.9 in 2015 and that the bottom 60 per cent depends mainly on social grants (Stats SA, 2020). There are also substantial labour market income differences, with educational attainment being the most prominent reason for these differences (Mabuza, 2020). Overall, blacks earn substantially less than whites and women less than men.

The post-apartheid social welfare system had to buffer the poor against these inequalities. Social welfare, health and education receive approximately half of the South African budget of R2.1 trillion in 2022. The Department of Human Settlements receives a further 5 per cent while 4 per cent is budgeted for the equitable share, which subsidizes poor households to help them pay for municipal services. The 5 per cent allocated to the Department of Human Settlements is similar to the percentage in other middle-income countries and largely provides government subsidies, for social housing and homeownership for poor households. Compared to the other social services, housing receives a very small percentage. Table 15.2 shows the grants that make up social welfare in South Africa.

The older person's grant is the grant with the longest history in South Africa. It was initiated to provide welfare for whites in 1928, but since 1944 it has also been available to blacks (Sagner, 2000). All the other grants have been instituted since the start of democracy in 1994. The future of the two grants introduced in response to the distress caused by the lockdown during the COVID-19 pandemic is unclear. There is some discussion about transferring these grants into a basic-income grant, but this is unlikely as the South African budget is under strain because of inflation and high levels of inappropriate spending and corruption by the ANC government between 2009 and 2017. All these grants are means-tested and contribute to further socio-economic stratification.

Education expenditure supports both basic education and higher education and training. Since 2016, education expenditure also entails a bursary scheme for university students, the National Student Financial Aid Scheme (NSFAS). Access depends on a means-test based on the household income of the student's parents or guardians. Although free basic health services are available in South Africa, the system is largely stratified between private and public systems. Access to the private system depends on income and the ability to afford health insurance. Only about 17 per cent of the South African population has access to health insurance. The equitable share funds municipalities, but indirectly it funds poor households that cannot afford to pay for their municipal services. The equitable share contributes 10 to 20 per cent of municipal income.

Analysing the South African welfare system against the framework of Esping-Andersen (1990) points to multiple characteristics. The state, rather than the market and private organizations, plays the dominant role in providing welfare. The level of stratification and the commodification of

Table 15.2 Overview of social welfare grants in South Africa (2022)

Name	*Aim of the grant*	*Entitlements*	*Monthly amount*	*Estimated number of beneficiaries (2020)*
Child support grant	Support to caregivers of children in need	A parent or primary caregiver of children aged 18 and under. SA citizens + permanent residents + refugees	R460	12.7 million children
Foster child grant	Income support to caregivers of children in foster care (court order is essential)	Foster parents of children under 18 or up to 21. SA citizens + permanent residents + refugees.	R1,050	365,000 children
Care dependency grant	Income support to caregivers providing permanent care to children with a severe mental or physical disability (must have a medical assessment)	Parent or caregiver or foster parent of children 1–18. SA citizens + permanent residents + refugees.	R1,980	155,000 children
Older person's grant	Income support to older men and women	60 years and older. SA citizens + permanent residents + refugees.	R 1,980 R 2,000 (over 75)	3.6 million people
War veteran's grant	Income support to older men and women who served.	60 years and older. SA citizens + permanent residents.	R1,910	Not available but included in the older person's Grant
Grant-in-Aid +	Income support to people (already receiving older person's, war veteran's or disability grants) who need full-time care from someone	60 years and older. SA citizens + permanent residents.	R460	Not available
Temporary social relief of distress – also became the COVID-19 Grant	Unemployed people who receive no other government assistance	SA citizens + permanent residents 18–59 years.	R350	10 million people
COVID-19 caregiver's grant	Caregivers receiving a child support grant	Parent or primary caregiver of children aged 18 and under. SA citizens + permanent residents + refugees.	R500	Not available

Sources: Compiled from: Kohler and Bhorat, 2020; UCT, 2022.
Note: COVID-19 grants could overlap with other grants.

welfare point to a liberal regime. There is substantial stratification in the South African welfare system. A means test is required to access any grant from the Department of Social Development (Table 15.2), a bursary from the NSFAS or support from the equitable share at a municipal level. Only those individuals or households with low incomes qualify for these grants. The health system is also stratified between the public and private sectors. Those who can afford health insurance can access high-quality healthcare from private providers. Those who cannot access medical insurance depend on a poorer level of service in public services. The focus on the poor shows that welfare is commodified and not inclusive. However, there are also elements of universalism in welfare spending. For example, in basic and higher education, substantial parts of the funding reach students and schoolchildren without a means test (although the bursaries only go to students from poor households). These social investments in higher education might perpetuate inequalities in South African society. The fact that South Africa is one of the most unequal countries in the world (a Gini coefficient of 0.65) plays a role in how welfare has been designed. Much of this inequality arises from the labour market and the tax and social security systems play only a small role in reducing it. There is evidence of support from extended families, but intergenerational transfers have been limited because of the exclusion of black people under apartheid. Consequently, intergenerational asset accumulation for these households has been difficult. Considering these inequalities and the stratification of welfare, South Africa is a good example of a liberal welfare regime.

The interrelation between housing and welfare

In this section, we analyse the interrelationship between housing and welfare. There are two opposing trends concerning the commodification of housing policy in South Africa. On the one hand, there has been a heavy emphasis on housing as a way to overcome poverty and on the other hand to create wealth. Private-sector finance and the emphasis on homeownership with households investing in their own housing are the most important elements of this policy. These aspects point to the importance of a liberal regime, though in South Africa the policy has often been labelled neoliberal (Huchzermeyer, 2004). At the same time, there are also elements in practice and policy that decommodify housing. For example, in 2005, only about 50 per cent of the houses provided under the capital subsidy scheme obtained title deeds despite the policy directive emphasizing homeownership (Department of Human Settlements, 2015). The reason for this deviation is complex. The shift from large-scale developers to small-scale developers has been the main reason. This process bypassed conveyancing attorneys. But it also resulted from the administrative and political process that did not consider registering titles important. The result was that these houses largely remained the property of the government. Despite the emphasis in policy on homeownership and the development of a second-hand market (promoting the selling and buying of houses and climbing the housing ladder to create wealth), this aspect has been neglected in practice. Furthermore, in the early 2000s, an earlier attempt to require households that benefit from the subsidy to contribute to the cost was dropped in favour of a full subsidy. The increase in the subsidy amount was also substantially more than the increase in inflation.

The South African housing market is stratified according to race and household income. Years of racial planning, separate suburbs and exclusion have created areas where the market works and where it does not. The housing market in Mangaung, which we referred to earlier, shows how unevenly the property market operates in former white and former black suburbs in many South

African cities. Although many black high-income households have been able to move out of former black suburbs, the stratification based on household income continues. Yet some former black suburbs have thriving housing markets, although a fair amount of this housing market is informal, with not all transactions being reported in the deeds register.

The above evidence shows that the market, government and families play prominent roles in housing. Although the government is a central provider of government subsidies and wants to see households climbing the housing ladder (selling and buying houses) (related to liberal regimes), there is also contradictory evidence. The large-scale housing subsidy programme providing about 3.2 million houses over the past 25 years shows how dominant the government has been in the housing process. Yet despite this large-scale state involvement in subsidies, the subsidies reflect both corporatist and liberal regime characteristics. For example, like many other welfare programmes, a means test is required, classifying it as liberal. In addition to focusing on the working-age poor household with dependents, it also focuses on older people and cannot be accessed by an individual.

Government involvement in the housing market goes beyond the financing of housing and several strict housing regulations are in place. For example, contractors must be registered with the National Home Builders Registration Council. This registration is an attempt to ensure high-quality workmanship. Despite this requirement, there are many examples of shoddy quality in practice. There are also regulations preventing homeowners who received a subsidy from selling their houses in five years.

The state does not intervene in any price setting for housing (this includes private and social housing, although local governments determine the rents for the small number of public rental houses). This is left to market forces. A fair number (up to 20 per cent in former black township areas) of housing transactions are informal, without full registration on the deeds register. The private sector and individuals are directly involved in the construction processes. Private-sector involvement in construction varies over time. Initially, large developers and contractors were involved. By the early 2000s, this changed, and the focus was on small-scale developers. Since 2008, there has been again a return to large-scale developers and mega-projects. Spatial planning is less stringent than it was under apartheid, but it is still applied in many areas. Local governments have found it difficult to enforce these regulations. Poor households and voters of the ruling party often transgressed these regulations. Furthermore, the eroding of local government capacity also meant that transgressions were seldom monitored.

Finally, informal housing still accounts for about 13 per cent of households. Informal housing is usually self-built and people provide their own services. This approach generally falls outside the formal regimes, but Griggs and Kemp (2012), in a study of European welfare regimes, refer to this as a rudimentary form of welfare provision. This accords with the "Southern European" housing regime (Allen, 2006), but more recent studies demonstrate that informality and self-build also play a significant role in former socialist countries, particularly in south-east Europe (Stephens et al., 2015).

Regional outlook

South Africa is an exception in housing policy and welfare provision in Africa (Bah, Faye, & Geh, 2018; Gbadegesin & Marais, 2020). Its extensive subsidization of housing is unprecedented in Africa. This is also true of the country's wider welfare system. Many African countries do not have a welfare system in place at all and those that do exist are rudimentary (Luiz, 2013).

Consequently, housing subsidies outside South Africa are uncommon. This is despite substantial subsidies in other African countries up to the 1980s – mainly for the construction of public housing (Stren, 1990).

The inability of African governments to respond to housing needs originates from the failure of public housing in the 1950s and 1960s and a shift towards informal settlement upgrading in the 1970s. The World Bank has largely financed informal settlement upgrading and, although cost recovery measures were central, they seldom materialized. Over the past 40 years, most housing in African cities has been financed by the private sector or households. Both housing finance and land systems have operated below capacity, and much more can be done to improve these systems. A liberal but highly stratified housing system has developed without any government involvement while the focus on self-build resembles familial elements from southern Europe. High-income households either access housing finance or use their funds to develop high-quality housing, while poor people occupy land without the owner's permission, build their own houses and install their own services.

Considering the precarious situation of public finance in many African cities and the role of structural adjustment programmes, it is highly unlikely that more stringent government regulations or active involvement will occur in the foreseeable future. South Africa's strained fiscal position has already meant a rapid decline in the number of subsidized houses. Although the government has not formally declared that it has discontinued the housing subsidy programme, this has largely happened in practice.

Conclusion

This chapter assessed South African housing policy against Esping-Andersen's (1990) state welfare theory and the place of housing in the country's welfare system. South Africa has an extensive welfare system unknown to the rest of Africa. The system targets the poor (there is no universal system and eligibility is based on a means test), leading to stratification. Housing remains a small part of the overall welfare spending in South Africa and housing subsidies have declined rapidly over the past ten years – partly because of an increase in the subsidy per house (see introduction). Furthermore, there is a close relationship in policy between government spending on housing and linking houses to the market. The idea is that the housing subsidy and ownership should enable people to trade in the market. Since 2004, one of the main ideas has been that housing should develop as a tradeable asset and that people should actively participate in the housing market. Government housing subsidies should provide a platform for poverty alleviation and wealth creation.

It is difficult to classify the South African housing policy as a particular state welfare type. Largely, it contains elements of liberal, corporatist and familial regimes. In South African literature, the policy is often labelled 'neoliberal', emphasizing the private sector's dominant role, the focus on housing as something that should create wealth, and the role of the capital subsidy (as opposed to more flexible subsidy instruments). The inability to change the uneven spatial characteristics of the South African urban system results from the inflexible capital subsidy. The other liberal element is the requirement for a means test for the housing subsidy. Thus, several elements of a liberal housing regime are present in the South African policy. One of the outcomes is that policy promotes the commodification of housing. Consequently, the historical social stratification created under apartheid and colonial rule continues amongst others because labour market outcomes remain unequal despite the substantially larger number of black people involved in the formal labour market.

The government has subsidized extensively and the subsidy has outstripped inflation over the past decade, pointing to the fact that the initial victory of breadth over depth has been reversed Yet these subsidies aim to create assets and wealth for poor households by providing ownership. Only a very small percentage of state housing funds goes to rental housing. There are also discrepancies between policy and practice. For example, despite the policy focus on homeownership, many people living in houses constructed by the housing subsidy have not acquired homeownership. Many of these houses remain the government's property because a deed registration has not occurred.

Finally, the precarious position of South Africa's public finances means that one of the first places where austerity measures are applied is housing subsidies. The former high level of subsidization is unlikely to resume. The first signs of cost-cutting measures are already visible. This might result in less funding for the starter home and a return to the site and services model that was despised by the post-apartheid government when it took office in 1994.

Applying state welfare theory to the South African housing system has several benefits. Firstly, it provides a framework against which the complexity of welfare and housing welfare can be assessed (Venter, Marais, Hoekstra, & Cloete, 2015). Too often, the reflections on the South African housing systems use simplistic political and economic frameworks without considering the complexities associated with the housing system. One needs to accept that the welfare and the housing regime in South Africa are far more complex than labelling it neoliberal or government-driven. Second, it provides a framework against which to consider the role of welfare in stratifying inequalities in South African society. In one of the more unequal countries in the world, the question is whether it is possible to find alternatives. Thirdly, as an exception in the African context, the South African welfare system remains the most prominent example. It is highly unlikely that similar welfare systems will soon be implemented in the rest of Africa.

Note

1 At the time of going to press (October 2023). 1 euro = approximately R20.

References

Allen, J. (2006). Welfare Regimes, Welfare Systems and Housing in Southern Europe. *European Journal of Housing Policy*, 6(3), 251–277. https://doi.org/10.1080/14616710600973102

Bah, E., Faye, I., & Geh, Z. (2018). *Housing Market Dynamics in Africa.* Palgrave Macmillan.

CAHF (Centre for Affordable Housing and Finance in Africa). (2020). *South African Housing Market Report, 2020.* CAHF.

Chaskalson, M., Jochelson, K., & Seekings, J. (2007). Rent Boycotts, the State, and the Transformation of the Urban Political Economy in South Africa. *Review of African Political Economy, 14*, 47–64.

Department of Housing. (1994). *White Paper on Housing: A New Housing Policy and Strategy for South Africa.* Government Gazette No. 16178. Government Printer.

Department of Human Settlements. (2015). *Report on the Evaluation of Housing Assets.* Department of Human Settlements.

Department of Human Settlements. (2019). *Housing Delivery Statistics.* Department of Human Settlements.

Esping-Andersen, G. (1990). *The Three Worlds of Welfare Capitalism.* Princeton University Press.

Gbadegesin, J., & Marais, L. (2020). The State of Housing Policy Research in Africa. *International Journal of Housing Policy, 20*(4), 474–490. https://doi.org/10.1080/19491247.2020.1780186

Griggs, J., & Kemp, P. (2012). Housing Allowances as Income Support: Comparing European Welfare Regimes. *International Journal of Housing Policy, 12*(4), 391–412. https://doi.org/10.1080/14616718.2012.711987

Harrison, P. (1992). The Policies and Politics of Informal Settlement in South Africa: A Historical Background. *Africa Insight, 22*(1), 14–22.

Huchzermeyer, M. (2004). *Unlawful Occupation: Informal Settlements and Urban Policy in South Africa and Brazil.* Africa World Press.

Huchzermeyer, M. (2010). A Legacy of Control? The Capital Subsidy for Housing and Informal Settlement Intervention in South Africa. *International Journal of Urban and Regional Research, 27*(3), 591–612. https://doi.org/10.1111/1468-2427.00468

Kohler, T., & Bhorat, H. (2020). *Social Assistance During South Africa's National Lockdown: Examining the COVID-19 Grant, Changes to the Child Support Grant, and Post-October Policy Options.* Working Paper 202009. Development Policy Research Unit (DPRU), University of Cape Town.

Luiz, J.M. (2013). A Review of Social Welfare in Sub-Saharan Africa: From the Colonial Legacy to the Millennium Development Goals. *Econ Pap, 32*(1), 110–121. https://doi.org/10.1111/1759-3441.12013

Mabin, A. (1991). The Dynamics of Urbanisation Since 1960. In M. Swilling, R. Humphries, & K. Shubane (Eds.), *Apartheid City in Transition* (pp. 33–47). Oxford University Press.

Mabin, A. (2020). A Century of South African Housing Acts 1920–2020. *Urban Forum, 31*, 453–472. https://doi.org/10.1007/s12132-020-09411-7

Mabin, A., & Parnell, S. (1983). Recommodification of Working-Class Home Ownership: New Directions for South African Cities. *South African Geographical Journal, 65*(2), 148–166. https://doi.org/10.1080/03736245.1983.10559681

Mabuza, N. (2020). *Salary Disparities in South Africa: An Analysis on Race and Gender in the Labour Market* [PhD dissertation, University of Cape Town].

Marais, L., Awuzie, B., Cloete, J., Napier, M., & Greyling, E. (2020). A Tale of Two Markets: Unequal Access to the Property Market in a South African City. *Tijdschrift for Economische en Sociale Geografie, 111*(1), 80–98. https://doi.org/10.1111/tesg.12367

Marais, L., & Cloete, J. (2017). Housing Policy and Private Sector Housing Finance: Policy Intent and Market Directions in South Africa. *Habitat International, 61*, 22–30. https://doi.org/10.1016/j.habitatint.2017.01.004

Marais, L., & Lenka, M. (2021). Urban Housing for Rural Peasants: Farmworker Housing in South Africa. *Development Southern Africa, 38*(3), 391–403. https://doi.org/10.1080/0376835X.2020.1796596

Marais, L., & Ntema, J. (2013). The Upgrading of an Informal Settlement in South Africa: Two decades onwards. *Habitat International, 39*, 85–95. https://doi.org/10.1016/j.habitatint.2012.11.001RHLF (Rural Housing Loan Fund). (2012). *Annual Report, 2011/12.* RHLF.

Rust, K., & Rubenstein, S. (1996). *A Mandate to Build: Developing Consensus Around a National Housing Policy in South Africa.* Ravan.

Sagner, A. (2000). Ageing and Social Policy in South Africa: Historical Perspectives with particular Reference to the Eastern Cape. *Journal of Southern African Studies, 26*(3), 523–553. https://doi.org/10.1080/713683587

Scheba, A., & Turok, I. (2021). The Role of Institutions in Social Housing Provision: Salutary Lessons from the South. *Housing Studies, 38*(6), 1132–1153. https://doi.org/10.1080/02673037.2021.1935765

Servcon Housing Solutions. (2002). *Servcon Housing Solutions Report to Stakeholders – 31 March 2002.* https://static.pmg.org.za/docs/2002/appendices/020429servstake.htm

Statistica. (2022). *Unemployment Rate in South Africa from Q1 2019 to Q1 2022, by Population Group.* https://www.statista.com/statistics/1129481/unemployment-rate-by-population-group-in-south-africa/

Stats SA. (1996). *Census Data, 1996.* Statistics South Africa.

Stats SA. (2016). *Community Survey.* Statistics South Africa.

Stats SA. (2020). *How Unequal Is South Africa*? Pretoria: Statistics South Africa. https://www.statssa.gov.za/?p=12930

Stephens, M., Lux, M., & Sunega, P. (2015). Post-Socialist Housing Systems in Europe: Housing Welfare Regimes by Default? *Housing Studies, 30*(8), 1210–1234. https://doi.org/10.1080/02673037.2015.1013090

Stren, R. (1990). Urban Housing in Africa: The Changing Role of Government Policy. In P. Amis & P. Lloyd (Eds.), *Housing Africa's Urban Poor* (pp. 35–54). Routledge.

Tomlinson, M. (1998). South Africa's New Housing Policy: An Assessment of the First Two Years, 1994–1996. *International Journal for Urban and Regional Research, 22*(1), 137–146. https://doi.org/10.1111/1468-2427.00128

Tomlinson, M. (2007). The Development of a Low-Income Housing Finance Sector in South Africa: Have We Finally Found a Way Forward? *Habitat International, 31*(1), 77–86. https://doi.org/10.1016/j.habitatint.2006.04.004

Tomlinson, M. (2015). South Africa's Housing Conundrum. *Liberty: The Policy Bulletin of the Institute of Race Relations, 20*(4), 1–7.

Turok, I., Scheba, A., & Visagie, J. (2022). Can Social Housing Help to Integrate Divided Cities? *Cambridge Journal of Regions, Economy and Society, 15*, 93–116. https://doi.org/10.1093/cjres/rsab031

UCT (University of Cape Town). (2022). *Statistics on children in South Africa*. http://childrencount.uct.ac.za/indicator.php?domain=2&indicator=10

Venter, A., Marais, L., Hoekstra, J., & Cloete, J. (2015). Reinterpreting South African housing policy through State Welfare Theory. *Housing, Theory and Society, 32*(2), 346–366. https://doi.org/10.1080/14036096.2015.1048895

World Bank. (1993). *Housing: Enabling Markets to Work.* World Bank.

PART III

Housing and welfare in the 21st century

This final part of the book builds on the empirical findings of the country and regional chapters in Part II to draw together the changes in housing and welfare systems. The analysis employs the framework of housing and welfare suggested in Part I.

Part III is divided into three chapters. The first chapter draws out the main changes in housing and welfare regimes based on the country chapters, supplemented by additional comparative data. The second chapter focuses on the nexus of housing and welfare in the case-study countries using the analytical framework, based on the critique of previous understandings provided in Part I and the empirical chapters of Part II. Based on our analysis, the third and concluding chapter summarizes the findings and discusses the main theoretical and empirical contribution of the book towards advancing the understanding of the relationship between housing and wider welfare regimes.

DOI: 10.1201/9781003212690-18

16

CHANGES IN HOUSING AND WELFARE – A GLOBAL ANALYSIS

Martin Grander and Mark Stephens

What conclusions can be drawn from the national examples outlined in the previous chapters in terms of changes? In this chapter, we will consider changes in the scope and depth of the components of the case study countries' "wider" welfare regimes, where we consider the distributive tendencies of the labour market, social security systems and taxation.

Regarding housing, we will draw out the main changes in housing policy discussed in the country chapters, with a special focus on changes in housing tenure and the role of homeownership in relation to rented housing, especially the cost-rental sector. We will discuss changes in the governmental policies that encourage different types of housing tenures and pay specific attention to the rise in mortgaged homeownership, and its subsequent reduction in some countries.

In summarizing and analysing the changes in welfare and housing systems in the countries, some contextual comparisons of interest could be made. We will therefore relate the examples from the case study chapters to a comparison of the countries and regions included in the case study chapters regarding changes in some indicators relevant to understand the housing and welfare nexus, namely income inequality, income poverty, housing tenure, and mortgage indebtedness.

As in part II, we have grouped the countries in relation to their welfare and housing systems, starting with the Anglo-sphere countries (the US, England/the UK, and Australia) in one group. Germany, Austria and Switzerland and the Nordics (Sweden, Norway, and Finland) constitute a second group. A third group comprises the southern European countries of Italy and Spain. The post-socialist countries of Russia, Hungary and China make a fourth group and a fifth and final group covers Chile and South Africa from the Global South.

Changes in welfare regimes

A prerequisite for understanding welfare regimes is their ability to mitigate poverty and inequalities through their redistributive approaches. As described in the first part of this book, welfare regimes comprise state entities and social institutions that protect and promote the economic and social well-being of its citizens. Different welfare regimes have had different approaches to this. Based on Esping-Andersen's (1990) classic categorization, we have said that, while the liberal regime is based on market mechanisms, the social democratic regimes have relied heavily on state

DOI: 10.1201/9781003212690-19

measures and conservative regimes on the civil society and the family. Moreover, we have drawn on Holliday's (2000) addition of a "productivist" regime in parts of Asia.

However, the development during recent decades has changed the basis for the ideas of divisions between welfare regimes. As we will show in the coming section, we see tendencies of both convergence and divergence in the different countries examined in the case study chapters, which results in a situation where, for example, the relation towards poverty and inequalities differs.

Diverging poverty and income inequalities

As housing could be regarded as a market commodity, income poverty and inequality set the scene for housing accessibility and affordability. In order to measure income inequality and poverty, we have used statistics collected and presented by the OECD on disposable income and its calculations of the Gini coefficient. The OECD data for income inequality (measured through the Gini coefficient) and poverty (using 50 per cent of the median income as the poverty line) are based on incomes after taxes have been paid and social security benefits have been received. They are also based on incomes that have been adjusted ("equivalized") to take account of household composition.

Starting with poverty rates, this rate is the percentage of people (in a given age group) whose income falls below the poverty line; taken as 50 per cent of median household income, after taxes and transfers, of the total population in the country. This means that two countries with the same poverty rates may differ in terms of the absolute income level of the poor (Figure 16.1).

Among the case study countries, by far the highest poverty rate can be found in South Africa and China, where 25 per cent or more of the population falls below the national poverty line. They are followed by the post-soviet states, Romania, Chile, and the US with rates ranging from 15 to

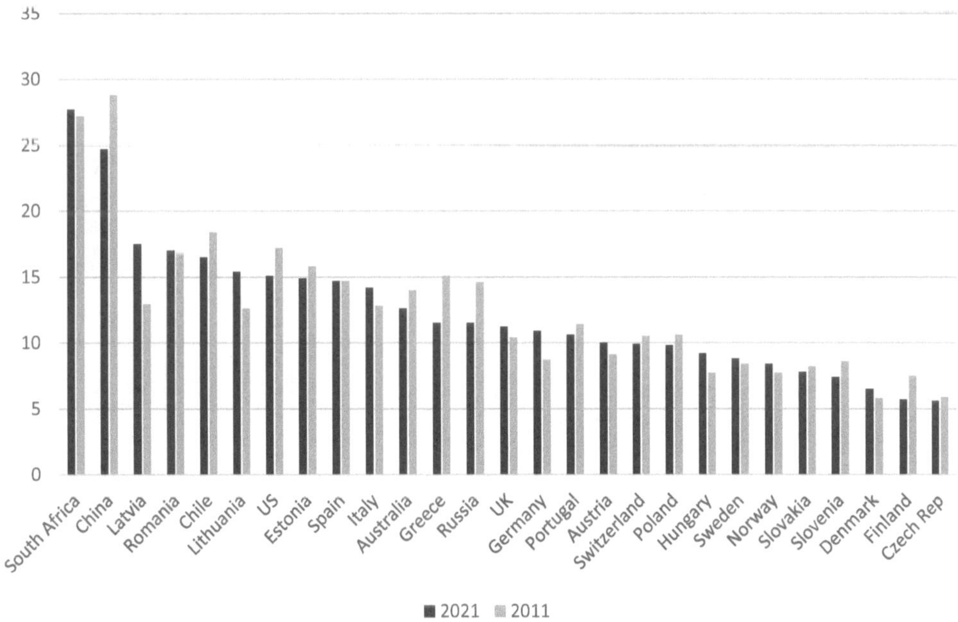

Figure 16.1 Poverty rates (% individuals) in case study and nearby countries (latest available data (2019–2021) compared to 2011).

Source: OECD (2023), Poverty rate (indicator), except for China 2019: World Bank (2023).

18 per cent. Among the Anglo-sphere countries, Australia and the UK have lower poverty rates than the US. The southern European countries Spain and Italy see a significantly larger share of poor people than western European countries. At the other end of the spectrum, we find the Nordic countries, Slovenia, Slovakia, and Chile with a rate of 6–9 per cent of the population.

Poverty has decreased in many of the poorest countries. Among countries with the largest reduction of poverty rate, we find China, Russia, Chile, and the US. Among the countries with increasing poverty, we find some of the post-communist states (Hungary, Lithuania, and Latvia) as well as Germany and Austria.

The Gini coefficient is the most commonly used indicator of income inequality. It is based on the comparison of cumulative proportions of the population. Income inequality among individuals is measured here by five indicators against cumulative proportions of income they receive. A value of 0 represents perfect equality, meaning all households had the same amount of income. A value of 1 indicates total inequality, meaning that one household had all the income.

From what we can see in Figure 16.2, income inequality is lowest in some of the Central-Eastern European countries (Slovenia, Slovakia, and Hungary), the Nordic countries, Austria, and Germany – all having a Gini coefficient below 0.3 in 2019. The most unequal countries by this measurement are South Africa and China, with a Gini well above 0.5, with the UK, the US, and Chile following with high rates of inequality.

Looking at the development of the Gini coefficient, the largest increases in inequality can be found in Western Countries with generally low levels of equality (Switzerland and the Nordics), but also Hungary, Lithuania, and the UK. In contrast, Russia, China, and Estonia, all countries with a high level of inequality have seen notable decreases in inequality.

***

What patterns could be found in the above? Starting with the liberal welfare regimes highlighted in this book, the Anglophone countries of Australia, the US and the UK are characterized by high rates of inequality. The high (and rising) inequality rates in the US and the UK are primarily explained by changes in the labour market and demography, exacerbated by changes to social security, including reductions in coverage and the increased use of conditionality for

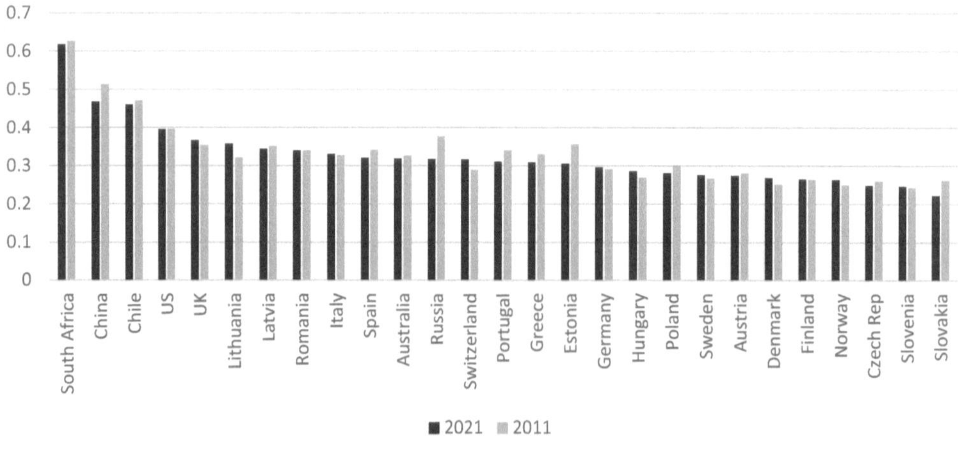

Figure 16.2 Gini coefficient of case study countries and nearby countries (latest available data (2019–2021), compared to 2011).

Source: OECD, except for China 2021: World Bank.

unemployment-related benefits. In the US, Tighe et al. (this volume), argue that the past 50 years of reduced welfare benefits have had a disproportionate effect on people of colour who have been "shut out of the wealth-building opportunities of previous generations".

In the UK, the "New" Labour governments of 1997–2010 emphasized raising employment levels as the best route out of poverty. Measures included expecting lone parents to be available for work sooner than before. To this end a statutory minimum wage was introduced and a system of in-work benefits (tax credits) was established. Child and pensioner poverty were also targeted through more generous benefits, but social assistance levels for childless working-age adults remained at low levels. The post-2010 austerity agenda promoted by the Coalition (2010–2015) and subsequent Conservative governments, involved a welfare reform program entailing a ramping up of conditionality (with sanctions), social security cuts such as benefit caps, reductions in local housing allowance limits, and declining additional benefits past a second child, has according to Marshall et al (this volume) fallen disproportionately on lower-income households. After the reforms, perhaps surprisingly, overall rates of poverty and income inequality in the UK have increased only slightly. However, destitution has become pronounced, and the use of food banks has risen as safety nets have become less effective. There are, according to Marshall et al. (this volume), significant group and spatial variations, i.e., growing inequalities. Young single men and migrants are particularly at risk of falling through the social security safety net and being frequently subject to sanctions. The authors summarize the post-2010 development with that "[t]he effect of welfare reform has been to widen the gaps in the social security safety net and exacerbate housing stress" (Marshall et al., this volume). Certainly, the strategy of promoting employment as the best route out of poverty has been shown to have its limits: most people who are poor now live in a household where at least one adult is at work.

The Australian welfare state, categorized by Esping-Andersen as a liberal welfare regime, has previously relied on a wage-earner/bread-winner principle. As discussed in previous chapters, the history of Australia poses a problem in welfare regime categorization as there has been a commitment to social-democratic principles of egalitarianism and social justice but leaning on redistribution of market incomes via wage regulation rather than tax-and-spend policy as in the social democratic nations. As the bread-winner model has been challenged in welfare policies from the 1990s and onward (and should be regarded as obsolete), an increase in contract, casual and part-time work and wage has been seen, resulting in wage increases significantly below the rate of house price increases (Hulse et al., this volume). The more recent development of the welfare regime has included tightening of social security measures, increased primacy of market competition and smaller government. Hulse et al. (this volume) describe how social protection for working-aged people is kept at levels well below the poverty line to encourage- people into paid work. Poverty and inequality rates are improving, but in sum, the authors claim, the past decade's changes have brought Australia increasingly similar to other liberal welfare regimes, and thus distanced itself from social democratic ones.

The Nordic countries, grouped by Esping-Andersen in the social democratic welfare regimes, have seen rapid development of the welfare state in line with what is described in many liberal regimes. A transition from the welfare systems (such as the distinct Rehn-Meidner model in Sweden) to more of workfare systems, characterized by labour market flexibility and increased conditionality in social security seems to be evident also in traditional bastions of social democracy such as Sweden and Norway. Poverty rates are for the first time in a long while increasing and, as described by Sørvoll et al. (this volume), income inequality has increased in all Nordic countries, most notably in Sweden, due to ideologically motivated cutbacks of welfare programmes. The financial crisis of the 1980s and 1990s struck Sweden harder than oil-rich Norway. The banking crisis in Sweden led to a decisive shift away from the social democratic economic and social model, involving both labour market and social security reforms, which together with extensive immigration from war-struck countries in the Middle East, North Africa and Asia has led to a sharp increase in unemployment in low-income groups. Sweden has also decreased the reach of its welfare systems more extensively through austerity measures than its neighbour, although Norway has not been able to "escape entirely from the ideological winds of welfare conditionality and workfare" (Sørvoll et al., this volume). While Sweden has maintained a high level of social investment in child care, parental leave, and other forms of public services boosting the level of participation and competence in the labour force, inequality has increased due to the fact that the number of recipients has increased due to immigration and that the levels of subsidies, such as the housing allowance, has not been in line with the general inflation (Grander, 2018). Furthermore, changes in the pension systems have increased generational income inequality, as it has less progressivity than previous models. On the other side of the spectrum, financial policies have concentrated asset-based incomes, not the least regarding housing, which we will come back to.

Turning to the welfare development in the conservative/corporate regimes, we find a similar trend as in the Nordic countries. As described in the chapter by Wetzstein (this volume), the regime has relied on an extensive corporatist social security system with access to cash benefits related to work position, financed via earnings-related contributions. After the "golden age" of steady growth from 1945 to the mid-1970s, the Bismarckian welfare states have struggled to respond to problems of unemployment, lower productivity, and an explosion of social cost, much related to international competitiveness and migration.

In Germany, the so-called Hartz reforms in the 2000s, characterized by increasing the flexibility of work arrangements by introducing part-time work models with lower or gradually rising taxes and insurance payments as well as providing start-up grants for aspiring entrepreneurs, can be seen as a combination of state retrenchment and privatization measures. This has been accompanied by changes in pension policies, unemployment protection/labour market policies and family policies and lowered social entitlements for unemployed people, where contributions are conditioned by legally binding obligations to improve their job situation. In sum, the reforms have added to national economic competitiveness (Germany is regarded as the economic motor of Europe) but also stigmatization and impoverishment. As we can see from the figures, both poverty and inequality are on the rise. Interestingly, as noted by Wetzstein, the 2019 ruling of the Federal Constitutional Court has prohibited harsh sanctions against benefit recipients, and in response, new measures have increased the standard unemployment benefit and reduced conditionality.

Meanwhile, the Swiss and Austrian welfare states are seemingly a bit more resilient and path dependent. Austria has, according to the author, emerged as the regional leader in universalism in social security, with regard to pensions and general universal welfare. Poverty and inequality are consequently decreasing. Switzerland has continuously increased social spending to become a leading welfare state in Europe, although we find that poverty decreases, inequality increases – arguably due to the concentration of corporate wealth.

Nevertheless, Wetzstein concludes that traditional institutions and instruments protecting the Bismarckian citizens from social risks have been weakening as a consequence of economic, demographic, social and political changes. Both Germany and Switzerland are seeing increasing economic inequality. The retrenchment in Bismarckian countries has been motivated by the view that welfare costs too much, and that the growing financial burden it imposes jeopardizes national competitiveness and discourages entrepreneurial activity. Consequently, the Bismarckian states have all developed a stronger focus on individual activation and female labour market participation in recent years in order to re-fuel the engines of economic prosperity (Wetzstein, this volume).

In the familial welfare regimes of Southern European countries, we find that both poverty rates and inequality have decreased, but are turning upwards again. While the global financial crisis and the following austerity plans initially became a "great equalizer" for Southern European welfare states (Guillén et al., 2022), the development thereafter has turned, and poverty and inequality seem to be exacerbated by the 2020 pandemic.

As in many other European countries migration has become an issue connected to the development of the welfare state. However, the southern European non-comprehensive welfare regimes have had less ability to provide welfare for migrants. As the Catalonian chapter (García-Lamarca, this volume) describes, devolved welfare-related capabilities over the last 40 years have gradually reduced the role of the state vis-à-vis the market, especially in housing. Combined with increasing immigration from outside the EU, this development has worsened stratification and polarization. Immigrants from Africa and the Middle East have in many cases become part of the informal economy, this being the only option for undocumented immigrants. Working in the informal economy means not being eligible for employment benefits or social security support, as they fall outside the non-comprehensive social insurance system in Spain. Indeed, the informal economy is widespread in southern European countries. Non-European immigrants are heavily impacted by unemployment due to racism in the, but also on the housing market.

Women and young people are seemingly other disadvantaged groups. In Italy, young adults are being severely affected by cuts in the welfare system. As noted above, both inequality and poverty have increased during recent decades and are – according to Leonardi (this volume) – especially affecting young adults in Italy, not the least after the COVID-19 pandemic. Among the age group 18–34, the frequency of poverty is at 11.1 per cent in 2021 and 13 per cent of young adults rank among the working poor. Since the welfare system is dependent on such a familial logic, the rate of young people between 18 and 34 living with their parents is highest in Europe, 62.5 per cent compared to a European average of 48.1 per cent.

The development of the Russian welfare regime after the fall of communism is well documented. Inequality has skyrocketed. Looking historically, the Gini coefficient in Russia almost doubled in the decade after the fall of communism – from 0.29 by 1989, to 0.64 in 1996. As discussed in the Russian chapter (Kalyukin & Puzanov, this volume) the welfare reforms have involved a shift from direct funding of welfare by the government to financing through a system of insurance funds. Such reforms have made access to benefits based on individual citizenship rather than group membership and replaced the universal provision of entitlements with targeted, categorical and means-tested social assistance. Moreover, the responsibilization of citizens is a common theme in Russia and the other post-soviet states. Social benefits such as the free supply of medicines and free public transport have been replaced with in-cash payments, much to public dismay. Nonetheless, the welfare system could still be argued to be comprehensive and according to the OECD data, income inequality is currently decreasing quickly reaching a Gini coefficient of 0.32 in 2019.

The central-eastern European countries, which have seen a similar ideological regime shift during the 1990s, have seen increasing inequalities. In Hungary, price control and full employment were abandoned after the fall of communism, and the new welfare system was built against a backdrop of ideological debates, with incremental changes having often unpredictable consequences. As noted in the Hungary chapter, the new welfare system appears as something of a hybrid, where measures are either approaching a neoliberal position or a social democratic one. The latest decade has seen the transformation to what is called a "populist authoritarian" or "illiberal" regime, where welfare provision is subordinate to the political aims of maintaining power. Hegedüs (this volume) exemplifies the development of the private pension funds, which in 2010 were integrated into the central government budget to counterbalance the losses from the GFC. While this de-privatization on the one hand can be seen as a movement away from the neoliberal welfare system, Hegedüs argues that the real reason was the state's need for budgetary resources.

The chapter on China, a regime that could be described as a productivist welfare regime (Holliday, 2000; Zhou & Ronald, 2017) due to its (re-emerged) embeddedness of social protection in a post-state-socialist system devoted to economic growth, describes how a state-organized social insurance system has developed to become the backbone of China's welfare state (Miao, this volume). In contrast to many other regimes, the socialistic regime in China has not been associated with liberal reforms, but rather the opposite. However, income inequality is considerably higher than in other countries in this study and rose considerably after the economic reforms in the late 1970s, but plateaued in 2010 and has since declined. The sharp increase in inequality in the closing decades of the 2^{0t}h century is arguably connected to the local variations and the divide between the urban and rural populations (Xie & Zhou, 2014), inferring that there is not one welfare model in China but many. Moreover, the large amount of migrant and irregular workers, close to 300 million in 2021, have notably inferior social rights compared to regular urban residents.

What about the other countries omitted from Esping-Andersen's classical typology? We find that these countries seem to be the ones that stand out the most. South Africa, which was not included in Esping-Andersens's characterization, but could arguably be seen as having a liberal welfare regime, has the largest rates of poverty and is considered the most unequal country in the world. Both poverty and inequality figures are rising. According to the World Bank, 10 per cent of the population owns 80 per cent of the country's wealth. The two most important contributors to this unflattering position are the apartheid legacy and the present government's governance style. As highlighted in the South African chapter in this volume, the high degree of poverty and inequality arguably stems from inequalities in the labour market due to discrimination. Seemingly, taxation policies and social security systems have played a small role in reducing such inequalities and the exclusion of black people under apartheid has resulted in that "intergenerational asset accumulation for these households has been difficult" (Mmbadi et al., this volume). The reliance on housing policy, more specifically on homeownership has seemingly not decreased poverty, despite such ambitious aims.

Chile, which saw a radical political transition from a corporatist and later socialist country in the 1930s–1960s, through Pinochet's' dictatorship 1973–1990 to the Concertación governments after that, has seen a similar development of a hybrid welfare state. During the 1960s the socialist rationale sought to advance social rights for previously excluded sectors of the population. The socialist government saw an expansion of the state's role in key areas of the economy and the start of the political incorporation and mobilization of previously excluded groups, such as the agricultural labourers and the urban poor. The dictatorship, which took power in 1973, drastically reversed this direction with neoliberal objectives, making structural adjustment policies, including deregulation of the economy and the privatization of state companies. It reduced the state's previous role as the key agent in fostering industrialization and production. Pinochet established a new contract between state and capital, where the state would play a subsidiary role in economic development – reducing the state's economic role to a minimum, i.e., to assure that individuals were free to make contracts between a variation of private parties. As discussed by Abufhele and

Angelcos (this volume), this principle gave the state power to "create markets where none existed but excluded it from participation in them, except when they were not functioning". Accordingly, the state withdrew from key sectors, which were privatized, including the education, health, social security, urban and housing sectors (Abufhele & Angelcos, this volume).

After the fall of the military dictatorship regime, the new Concertación coalition had to deal with very fragile political stability, not the least between the government and the armed forces on the one hand and disarticulating leftist groups on the other. Thus, it avoided engagement in confrontational politics and grassroots activists were mainly excluded from the political process. The subsequent Concertación governments have sought to combine free market and redistributive social policies – a strategy referred to as "growth with equality" (Abufhele & Angelcos, this volume). This strategy illustrates, according to the author, a combination of rationales: a neoliberal orientation under which the state's role was to create markets to make room for private profit and foster macroeconomic growth, accompanied by a social-democratic rationale to counteract the detrimental effects of neoliberal policies on the most disadvantaged sectors of the population. This hybrid form has included an increase in redistributive social spending and the expansion of paternalist assistance to the poor, why the Gini coefficient has gradually decreased. Still, as noted by the OECD, Chile's system of taxes and transfers does much less to reduce income inequality than in most other OECD countries.

Conclusions on the welfare development in the case study countries

From what we can see from the above, the wider welfare regimes (where we include labour market policies, taxation principles, and social security systems) seem to become increasingly conform, yet they still have some key differences in their distributional tendencies. Western economies have been clearly deindustrialized and shifting towards economies based on services. The distinctive ideologies of social democracy with universal welfare benefits have been eroded in the Nordic welfare regimes. This has increased the commodification of labour, and as the pension systems have become more dependent on financial investments from the state (for example in housing) and taxation systems seemingly work in a more stratifying manner – resulting in increases in both poverty and inequality. While the last decade indicates small changes, the historical development since the 1980s, when the Nordics were regarded as the most equal countries in the world, has been dramatic, which is exemplified by the abandonment of the Rehn-Meidner economic model in Sweden, which we will come back to in the next chapter.

In Western European conservative regimes, Christian democracy with its compensative measures from civic society is no longer so clearly reflected. In the above, this is exemplified by Germany, where the economic crisis of the 1990s led to the Social Democrat-led government introducing the Hartz IV labour market and social security reforms, that have accelerated the demise of the "male breadwinner" characteristic of the corporatist labour market. While the state has clearly taken a clearer role and Germany has become the motor of economic growth in the EU, income inequality is still on the rise.

The liberal systems have arguably become even more liberal, which could be exemplified by the UK emergency loan from the IMF in 1976 which signalled a dramatic change in politics, leading to the deregulation of the labour market, tax cuts, and a less solidaristic social security system. In parallel, demographics have changed, increasing the share of ageing people and immigrants. As we have seen, this has contributed to a rise in poverty and inequality. The changes in the minimum wage and the tightening of social security measures are examples of changes moving Australia towards other liberal welfare regimes.

In sum, we argue that the Anglophone, the Nordic, and the Western European countries all have reformed their wider welfare regimes in response to economic crises in the past decades in a way that it could be argued that all countries have strengthened the liberal tendencies in the welfare regime. Although reforms in Spain have developed the social security systems after the Franco regime, Spain, as well as other southern European regimes, is still heavily dependent on the family as welfare cuts and austerity policy have decreased the welfare state's comprehensive measures, which has increased polarization reinforced by the last decade of immigration and the COVID-19 pandemic.

In the eastern part of Europe, the variation is larger. The liberalization of Eastern economies such as Russia and Hungary has abruptly ended previous regimes, leading to the emergence of hybrid regimes with clear populist power characteristics. The socialistic regime in China has not been associated with liberal reforms, but rather the opposite, so it could be argued that communist China has strengthened its traditional welfare configuration.

Changes in housing regimes

As discussed in the first part of this book, it cannot be assumed that a welfare regime with a high degree of decommodification and a low degree of social stratification means a housing system with the same principles. We have argued that the question as to whether the housing system replicates, reinforces, or counters the distributional tendencies of the welfare regime, therefore, is an empirical one. In the following, we will point out the key changes in the housing regimes when it comes to their ability to affect the distributional aspects of the welfare regimes.

We will start with an overview of the tenure breakdown in the countries, highlighting the tendencies of the state to "steer" the population to a certain housing tenure. Next, we will analyse the government's eagerness to make homeownership more accessible for households in general – and low-income households in particular.

Housing tenures change: Increasing homeownership

As illustrated in Table 16.1 and Figure 16.3, the tenure composition differs largely. Homeownership rates in the case-study countries range from around 40 per cent in Switzerland and Germany to 80 per cent in Russia and the former Soviet states.

Regardless of the traditional categorization by Kemeny in "home-ownership" and "cost-rental" societies (see Part I), we find that homeownership has risen since 1990 in all countries included in the table except for Australia and the UK (see Table 16.1). While homeownership societies such as the US, the UK, Australia, and Norway have very high levels of homeownership, we also find that the cost-rental regimes are characterized by home ownership being the dominant and most rapidly increasing tenure. But as we see in the cases of Germany, Sweden, and Switzerland, the rental housing sector is still significantly larger than in the homeownership regimes.

Zooming in on the traditional homeownership regimes, the already ownership-dominated Norway has seen an increase in homeownership rate from 64 per cent in the 1960s to 77 per cent at the present date, while rental housing has decreased from 36 to 19 per cent. Subsidized rental housing is, and always has been, negligible, as described by Sørvoll et al (this volume).

The three Anglophone countries pictured in this book are usually categorized as homeownership societies with dualist housing systems, often characterized by a very large rate of owned detached housing, not the least in Australia and the US. Here, some authors claim that governments

Table 16.1 Tenure composition (%) in case study countries, 1960, 1990 and 2021

	Homeownership			*Market rented*			*Subsidized rental*			*Other (incl. free)*		
	1960	*1990*	*2021*	*1960*	*1990*	*2021*	*1960*	*1990*	*2021*	*1960*	*1990*	*2021*
Australia	70	69	67	23	21	27	4	5	4	2	4	2
Austria	38	46	54	35	22	30	16	21	16	11	11	0
Chile	43	73	68	57	18	18	0	0	0	n/d	5	14
China	10	33	69	0	0	22	90	66	4	0	0	5
Germany	29	36	45	n/d	36	50	16	17	4	n/d	11	1
Hungary	67	74	92	0	5	4	33	19	3	0	4	1
Italy	46	68	75	54	32	17	n/d	n/d	4	0	0	4
Norway	53	56	63	36	27	19	0	0	4	11	14	14
Russia	32	22	82	n/d	5	12	68	67	6.2	0	0	0
South Africa	n/d	n/d	67	n/d	n/d	18	n/d	n/d	0	n/d	n/d	15
Spain	52	73	77	43	21	15	2	3	2	0	0	6
Sweden	43	55	64	43	21	17	14	23	16	0	0	4
Switzerland	34	33	40	56	58	55	8	7	3	2	2	2
UK	32	69	64	19	11	17	18	21	18	0	0	0
US	62	64	65	38	36	35	1	1	1	0	0	0

Source: 2021: Housing Europe (2022) and OECD (2023) Affordable housing database, except for Russia and Chile: country chapters. 1960 & 1990: country chapters.
Note: UK, South Africa, and Chile: percentage of households. Russia, Italy, and Spain: percentage of individuals. Others: percentage of housing stock.

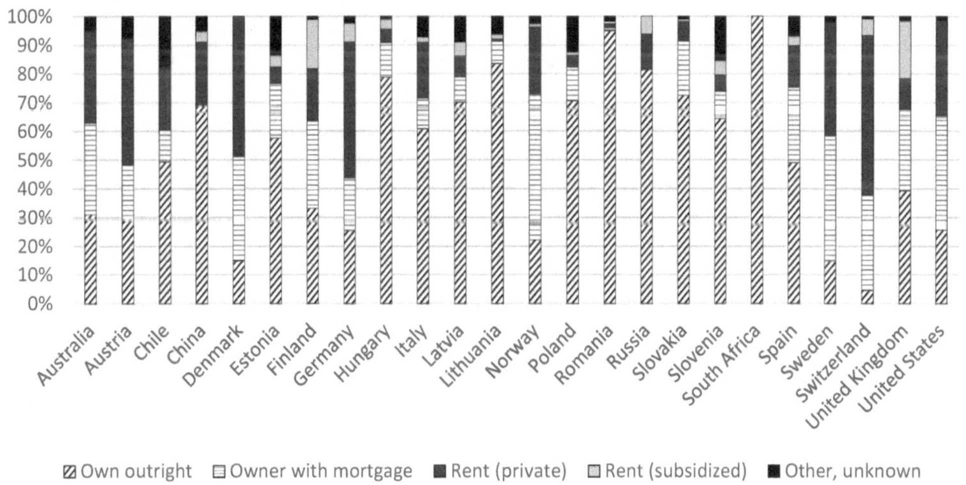

Figure 16.3 Tenure composition ((%) share of households) in case study countries and nearby countries, including mortgaged ownership, 2021 or the latest available data.

Source: OECD (2022) and country chapters.

have promoted house price growth as a "growth strategy", through generous tax reliefs on mortgage interest payments and land ownership.

In Australia, the rise of home ownership post World War II was generated by rapid domestic population growth, waves of migration from European countries, and favourable institutional settings, where financial institutions have been advancing mortgages with low-interest rates, sales of public housing to sitting tenants and offering capital grants to first-time buyers. Taken together, these measures resulted in an increase in homeownership rates, from 53 per cent in 1947 to 70 per cent in 1961. Thereafter, the overall homeownership rate in Australia has been rather stable, but the modest "headline" reduction disguises much larger falls among younger age groups who have been priced out. There was a "golden wave" of public housing investment during the 1960s–1970s, albeit the share of public rental housing never exceeded 6 per cent and has been in decline since then resulting in the current all-time low. Increasing numbers of households are dependent on insecure free-market rental housing.

The US appear to have had a remarkable stability of tenures. As illustrated by Table 16.1, homeownership levels have remained around 60 per cent over the decades, where it stabilized after the Second World War, since. the 1960s. The homeownership rate has increased slightly since the 1960s to the current rate of 65 per cent. The rental sector has seen a corresponding incremental decline from 38 per cent in 1960 to the current figure of 34 per cent. The public housing sector has failed to establish. Although its share increased from 1960 to 1990, it has since descended to an all-time low. Thus, public housing has remained a very minor sector in the US. As we will come back, the US housing system, like many others, has however changed radically in terms of affordability and accessibility.

Also in the UK, homeownership has grown over the decades. After the Second World War, the rate of homeownership increased from around one-third of the housing stock to 58 per cent in 1980. Accordingly, the share of the private rental sector decreased from 50 to 11 per cent during the same time span, as much of the stock was demolished in strategies of slum clearance, and landlords sold up in response to rent controls. In parallel, the share of social housing units increased from 18 to 31 per cent, increasingly serving as a supplement to homeownership. After the Thatcher regime took office in 1979, homeownership, not the least through the Right-to-Buy (RTB) of council housing, became even more prominent in housing policy, illustrating the central role of the state in driving commodification and financialization (Marshall et al., this volume). Another key factor facilitating the expansion of homeownership was financial deregulation which liberalized the mortgage markets, where credit towards homeownership became increasingly available. Accordingly, homeownership increased rapidly, reaching a share of 68 per cent in 1997, while social housing decreased steadily. Homeownership peaked around 2004, and since then, it has declined. As in Australia the seemingly modest reduction in the "headline" rate disguises larger falls among younger age cohorts. As social renting and homeownership have declined the private rental sector (PRS), deregulated since 1989, has increased and has overtaken social renting in England. Whilst the last Labour government (1997–2010) invested in improving the quality of existing social housing (the Decent Homes programme), it did not prioritize new social rented housing and was content to allow the liberalized private rented sector to grow as "buy-to-let" mortgages became available. Since around 2015, there has been something of a backlash against private renting, with less favourable tax policies intended to make it a less attractive investment and seeming to end its growth.

The most dramatic changes in housing tenure composition are found in the countries which have abandoned communist/socialist regimes. As is described in the chapter on Russia (Kalyukin & Puzanov, this volume), Russia's transition from a state-socialist plan-driven system to a market system premised on private homeownership rights began during perestroika in 1988 with the transfer of fully paid-off cooperative units into optional resident ownership. Many of those who chose to privatize their homes were low-income households unable to bear the full cost of homeownership. As a result of the mass giveaway privatization, the share of owner-occupied housing has dramatically increased since 1991, while the share of public/social rental housing subsequently decreased tenfold. Thus, Russia has changed from a clear example of a cost rental regime to a home ownership society.

From the country chapters, it is evident that countries often labelled as cost-rental regimes, with unitary rental markets, are subject to a change towards increased homeownership In Sweden, the core of the housing regime has been characterized by almost surprising stability, as pointed out by the authors of the Nordic chapter (Sørvoll et al., this volume). However, we find that while the ownership rate of detached houses has remained stable, the tenant ownership of apartments in multi-family housing has replaced public housing as the dominant tenure in such housing. The price regulation on tenant ownership was lifted in 1968, which paved the way for the tenant-ownership housing boom in the era of financial deregulation in the 1980s. The system of municipal public housing, which once was regarded as a bastion of the housing system, where a quarter of the population resided in the 1990s, has declined and is currently housing 15 per cent of the households. Simultaneously, private rental housing (companies or private personals owning multi-family housing estates) has increased. Such development is explained by several factors; abolished privileges and subsidies for public housing, new accountancy legislation which has decreased possibilities to construct new housing and massive privatization of public rented housing to tenant-ownership in larger cities during the last decades (Grander & Westerdahl, 2023).

Germany and the other DACH countries (Switzerland and Austria) were, like Sweden, categorized by Kemeny as cost-rental societies, which is reflected by the large share of rented housing, averaging 50 per cent according to the latest OECD statistics. Switzerland features the lowest homeownership rate in the OECD, followed by Germany. Over time, the tenure share shows a remarkable consistency in these countries. In Switzerland, high house prices and the taxation of credited rent are two key factors in limiting homeownership and keeping rented housing as a viable option for the bulk of the population (Wetzstein, this volume). While the social rental sector is small, private rental housing has remained over 55 per cent since the 1960s, while homeownership only increased at a slow pace. In Germany, national and local policies have explicitly favoured renting, which has led to the country having the lowest share of ownership in Europe. As noted by Wetzstein, there is currently no strong financial support for either owner-occupied housing or the rental sector in Germany, but the building-up of a large and stable rental housing sector has been made possible through a historical combination of regulation, housing subsidies and generous tax treatment. However, social housing has been characterized by temporary subsidies which have encouraged the transfer of social housing dwellings to the private market, leading to a large growth of the PRS and a decline in social rental housing. During recent decades, homeownership in Germany has increased. In Austria – often described as the promised land of social housing – homeownership has played a less prominent role than in most other countries over the post-war decades. There has been political consensus that housing should be seen as "a basic human right" (Wetzstein, this volume) that should be provided by the public sphere rather than mainly by market mechanisms. Hence, the Austrian share of social rental dwellings is among the highest in Europe. Yet, the last decade has seen a decrease in social housing and an increase in homeownership also in Austria.

China's system of housing provision could – like many other case study countries – be argued as going through a process of commodification and has become akin to the liberal interventionist system as portrayed in many Western housing regimes. Since the late 1990s, four main tenures: commercial housing; public housing ownership, economic and comfortable housing (ECH); and cheap rental housing (CHR) have emerged. The ECH and CHR sectors are argued to be the main form of welfare-based housing in China and have gradually replaced the public/social rented housing sector which had dominated the market for many decades. In 2020, the public housing sector accounts for only 4 per cent of all housing in China. However, the ECH and CHR sectors have, just like the commercial sector, seen a maturation of housing commodification (Miao, this volume) and thus lost their significance while homeownership has risen dramatically. Non-subsided homeownership (excluding self-built homes) went from non-existent in the 1990s to 47 per cent of households in 2020. Such a change has been underpinned by state policy. As described by Miao (this volume) the government has asserted control of commercial lending, planning and the speed and scale of land supply. The increasing homeownership trend in housing consumption in China is thus heavily influenced by the central and local governments.

South Africa has seen a large shift in housing policy due to the end of the apartheid regime. Homeownership has traditionally been a dominant form of tenure for the white and black population in both urban and rural areas, but during the apartheid regime, blacks were prohibited from owning housing, leading to black homeowners being forced to leave their homes in urban areas and were resettled to other locations, without homeownership. From the early 1950s to 1983, urban homeownership was available only to the white population. The post-apartheid regimes have searched to mend ethical disparities by once again providing homeownership to the black population. Overall, landownership linked to capital subsidies has been a prominent feature of politics encouraging homeownership, which has risen in urban areas. While data on housing tenure are fragmented and "sometimes contradictory" (Lochner & Gbadegesin, this volume), the statistics available reveal a country where 61 per cent of the urban population are homeowners, or 4.5–7.1 million households, depending on statistical sources. Private rental housing houses 1.9 million households, a number which is decreasing. This lack of growth of the PRS contrasts with the substantial growth in rent-free accommodation, i.e., households who live with families or in backyards of existing stands and do not pay rent. The social rented sector is marginal. The post-apartheid government initiated a social rental programme in 1994, but the construction of such units became very limited, and the social housing sector has thus failed to establish itself in South Africa. Some 35,000 households live in social rented housing, and even though this figure has tripled during the last two decades, the sector remains negligible.

Another country with a radical political regime shift is Chile, where the instalment, reforms, and subsequent abolishment of the military dictatorship under Pinochet (1973–1989) incrementally changed the housing sector. As described by Abufhele and Angelcos (this volume), the dictatorship made several marked-led institutional reforms of the planning and housing systems and introduced a social housing subsidy programme with demand-side subsidies. Such a shift also permitted the commodification of urban land and housing through the deregulation of the land market, privatization of social housing construction and general financialization of housing provision. This increased the share of homeownership from 54 per cent in the 1960s to 65 per cent in 1990. After the fall of the dictatorship, the new regime followed the same liberal housing policy, leading to a continuation of the rise of homeownership. but has fallen during the last decades, albeit still close to 70 per cent. Changes in tenure have thus been small, albeit housing construction and the availability of social housing – which in Chile consists of subsidized homeownership – have increased. As in the US, the system of housing provision has been criticized as the housing construction from the 1990s created homogeneously poor and segregated neighbourhoods, contributing to social exclusion, failure to address the needs of the urban poor, as the rise of homeownership did not necessarily improve life chances or quality of life, or integration into society for the population.

Summing up, most countries presented in this book, regardless of how they previously have been defined in terms of Kemeny's distinction of cost-rental vs homeownership societies with integrated vs dual rental markets, have seen a clear political direction of increasing homeownership at the expense of rental – especially social – housing, resulting in often radical changes in tenure composition. The clear exceptions are the DACH countries, where the rental housing sector still is large and stable, and where social housing remains a significant housing tenure for those with low – or as in the case of Austria – medium income. Sweden, which has traditionally been regarded as a cost-rental society with a similar political ideal of "tenure neutrality", where rental housing is politically regarded as an equal alternative to homeownership (Haffner, 2003), as in the DAHC, could, on the other hand, be argued to move towards a homeownership society, but not as a result of clear political or ideological directives, but as a result of a fiscal policy steering households into ownership as this has been regarded as a capital investment. In the next section, we will look further into this development and reflect on the implications of case study countries' endeavours for broad homeownership. Relating to Kemeny's theses, we could argue that many of the countries that previously could be categorized as cost-rental societies are turning towards home-ownership societies, where political ideologies and fiscal politics encourage the bulk of households, regardless of income, to get access to the owned housing market.

Increasing household indebtedness

In line with the increased focus on homeownership described above, the expansion of mortgage finance has risen rapidly in most countries in recent decades. Governments have overseen the expansion of mortgage finance to widen access to homeownership for a broader section of the population. The share of mortgaged versus outright ownership in the case study countries is illustrated in Figure 16.3. Among the countries with high rates of mortgaged ownership, we find the traditional homeownership countries of Norway, the US, and Canada. However, we also identify Sweden, Denmark, and Switzerland as exhibiting high levels of mortgaged ownership, even though they are traditionally categorized as cost-rental regimes with large rental sectors. However, there are other countries, such as Hungary and other post-communist countries, with very high ownership rates but low mortgage indebtedness.

This variation can be further analysed by investigating household debt in relation to disposable income and GDP. Household debt, in relation to disposable income, gives an indicator of the affordability of debt for households since it considers their income levels and thus provides insights into the financial strain on households and their ability to service their debts. Following the definition by OECD, household debt is defined as all liabilities of households (including non-profit institutions serving households) that require payments of interest or principal by households to the creditors at a fixed date in the future. Debt is calculated as the sum of the following liability categories: loans (primarily mortgage loans and consumer credit) and other accounts payable. The indicator is measured as a percentage of net household disposable income.

From Figure 16.4 we find a large divide between western and eastern economies. While households in the post-socialist economies are the least indebted, their counterparts in Australia., the Nordic countries, and Switzerland are highly indebted. Denmark and Norway lead the rankings in the European Union when it comes to the total level of household debt, while Sweden is placed fourth. All three countries display an indebtedness rate of more than 200 per cent of the disposable income. Thus, despite the differences in housing regime regarding homeownership policy outlined above, the homeowners in these countries are all highly indebted. Other cases of interest are the DACHs countries which, like Denmark and Sweden, have a high share of rented housing. Germany and Austria both have moderate levels of household indebtedness, while Switzerland has a very high rate of household indebtedness of more than 225 per cent of the disposable income. In the case of Switzerland, the perhaps surprisingly high level of debt may be explained by high

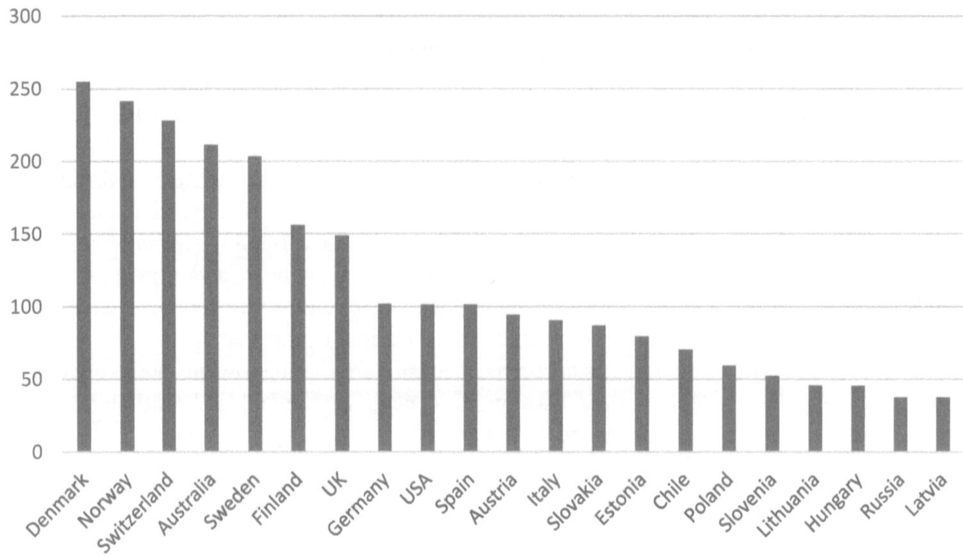

Figure 16.4 Household debt as a percentage of disposable income in case study countries and nearby countries (2021 or latest available data).

Source: EMF (2022).

Note: No data available for China and South Africa.

house prices, mortgages taken out by small-scale landlords to buy apartments and tax incentives to maintain high levels of debt (Schneider & Wagner, 2016).

Another indicator that can enhance our understanding of the role of mortgage indebtedness in the housing-welfare nexus is the household debt to GDP. This indicator helps us evaluate the systemic risks associated with household debt and provides an assessment of the overall vulnerability and potential impact on the economy as a whole; thus, it is an indicator of the macroeconomic implications of household indebtedness. The indicator is illustrated in Figure 16.5, which shows the mortgage debt in relation to GDP in the case study countries in the years 2000, 2010 and 2021.

From the figure, we find that the countries with high rates of household indebtedness do not necessarily mean that the rate of outstanding debt in relation to the country's GDP is high. It should be noted that the GDP measure is volatile and subject to continuous revisions. Here, Switzerland is something of an outlier with an incredibly high and increasing debt-to-GDP ratio, despite its low homeownership rate. The high level of indebtedness in Switzerland has historical reasons: high prices for real estate, interest-only mortgages (mandatory amortization was not introduced until 2012), and the fact that most rental apartments are owned by private individuals (Schneider & Wagner, 2016). Omitting Switzerland, the European countries having the highest household indebtedness (Norway, Sweden, and Denmark) also have high rates of debt-to-GDP. The same principle applies to Australia.

As illustrated by Figure 16.5, the rate of mortgaged housing to GDP has changed drastically over the last 20 years. Most countries have largely increased their household mortgages. Among the case study countries, the most notable case is Australia, which has gone from a rate of 19.6 per cent in 2000 to 100 per cent in 2021, an increase of 500 per cent. The other Anglophone countries, the US and UK, have had negative trends during the last ten years, after increasing quickly in the

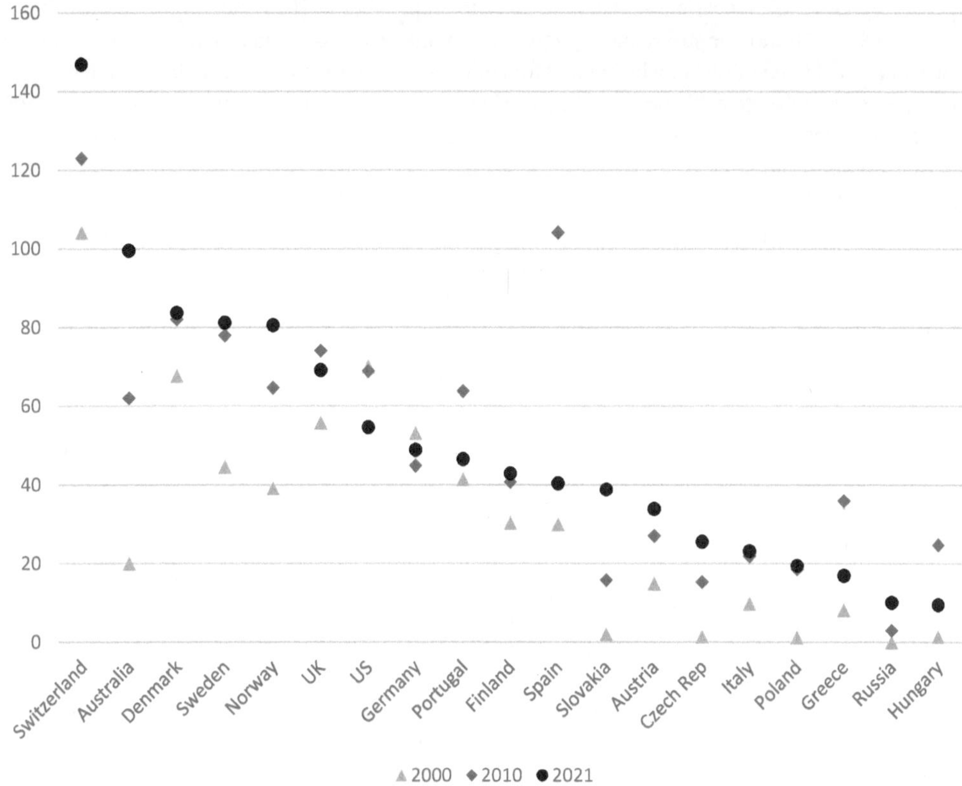

Figure 16.5 Total outstanding debt of households to banks and other financial institutions as per cent of GDP in case study countries (2000, 2010 & 2021 or latest data).

Source: EMF (2009, 2022).

Note: No data available for China and South Africa.

first decade of the 2000s. The development is clearly connected to the GFC and the real-estate crash in its aftermath, most notably in the US where the rate has dropped from 70 per cent in 2010 to just below 55 per cent in 2021.

Among the social democratic countries, we find that Sweden is increasing its debt-to-GDP ratio drastically, from 45 per cent in 2000 to 81 per cent in 2021. Another clear example is Norway, which we have described as a traditional homeownership society, which has seen a relative increase of 200 per cent over the 20 years. Finland exhibits a more moderate increase.

The corporatist countries of Germany and Austria are seeing diverse trends. While the mortgage-to-GDP ratio has increased by a factor of 2.5 in Austria, the households in Germany are less indebted in relation to the GDP in 2021 than in 2000. This is a clear example of the volatility of the GDP, where the German economy has been notably strong in the past decades, but the figure might also reflect the governmental policy of encouraging rental housing.

The development in the post-communist countries reflects the growth of the economies after the abandonment of previous regimes. The clearest example is Russia, where financing of housing production has shifted from the state budget to financing by private developers and household savings and mortgage loans have become increasingly important. Here, the mortgage debt as a share of GDP increased from near non-existent in 2000 to 10 per cent in 2021. According to Kalyukin & Puzanov (this volume), the still relatively low level of indebtedness could be explained by a cultural resistance to debt and the established perception of housing as a fundamental social right that should be guaranteed by the state, "hence creating a unique system of 'property without markets' in which most housing has been privatized but never fully commodified" (Kalyukin & Puzanov, this volume).

The southern European countries of Italy, Spain, Portugal, and Greece saw strong increases in debt during the first decade of the century but have since dropped. Like the US and the UK, they are illustrative examples of decreased indebtedness due to the GFC. The clearest examples of the real-estate crash in the wake of the GFC can be found in the (in)famous cases of Portugal and Spain. The latter has seen a drop in debt-to-GDP ratio from 104 per cent in 2010 to 40 per cent in 2021. Hungary follows the same principle, where the debt ratio increased from almost non-existent in 2000 to a rate of 25 per cent in 2010, only to drop to 9.5 per cent in 2021.

A conclusion from the above, which is supported by the country chapters, is that as the state is withdrawing from housing financing, households have stepped forward and taken on debt to finance housing production. These are clearly communicating vessels, which is illustrated in the Swedish example, where the national debt as a share of the GDP has fallen from 75 to 25 per cent since 1996, while the household indebtedness in relation to the disposable has risen from 95 per cent to 200 per cent over the same time span. The data also illustrate the significant risks for households with such a development, which has been painfully observable in the cases of the US, the UK and the Southern European housing markets.

It could be argued, in line with Aalbers (2008), that increased mortgage credit is a typical indicator of housing financialization, where increased and liberalized mortgage markets are allowing global capital to flow into national and real estate markets. Mortgage debt and house price increases tend to go hand in hand. Several studies (see e.g., Kholodilin et al., 2022) have connected the expansion of mortgage credit with increasing house prices. The ASR Global House Price Index (Absolute Strategy Research, n.d.) shows how house prices internationally have risen since the GFC, taking a small dip at the beginning of the COVID-19 pandemic, but then rising steeply to index values well above the 1980s booming market. The current financial setback has, however, led to a decrease since February 2022 when the Ukrainian war began and is projected to continue falling. The index fell to minus 1.5 per cent in March 2023, turning negative for the first time in over a decade. However, the data shows that while the Eurozone as a whole, and especially the Nordic countries are seeing large price drops, the UK and Germany are more resilient.

We have seen that house price increases and rises in mortgage credit tend to coincide. In many cases, national governments and their financial regulators have feared the emergence of a housing bubble. As a result, mortgage lending has been restricted, for example through stricter maximum mortgages and criteria regarding income and employment to take a mortgage. Although mortgage debt is high and often increasing, there are signs of stagnation in several countries. However, prices have continued to rise, at least until recently. Indeed, as stated by several chapter authors in Part II, the debt-driven model that has been observed among the countries is becoming saturated. García-Lamarca (this volume), for example, argues that the debt-driven model has become "fundamentally unsustainable" and has deepened social inequality. As Hochstenbach and Aalbers (2023) claim, housing wealth is arguably connected to accumulated profits among long-term homeowners and thus intergenerational reinvestments, as well as private landlordism (buy-to-let). Thus, the debt-driven model in many countries has arguably been replaced by wealth-driven housing markets.

Conclusions on the housing regime changes

From the preceding discussion, we can argue that most of the housing systems examined in this book are becoming increasingly commodified. The increased political quest for homeownership is a clear indicator of this. Over the past three decades, we find that most countries have enlarged their homeownership sectors at the expense of rental housing, especially social rental housing. Homeownership has all over the world been facilitated through mortgage credit and a sustained period of low interest rates, but in many cases also through subsidization of homeownership. However, it is doubtful that homeownership has become more accessible as house prices have risen accordingly – indeed in traditional "homeownership" societies such as Australia and the UK, the levels of homeownership have fallen, particularly among younger and even middle-aged households. While it might be argued that an expansion of the ownership sector, fuelled by subsidies, would increase the chances for households with lower incomes of lending money in order to gain access to homeownership, the examples from the case study countries seem to contradict the thesis that increased mortgage credits have increased broad accessibility. Housing prices have outpaced the subsidies, suggesting that the increased mortgage finance has become self-defeating. "Free money" has meant that middle-class households have become more indebted, but low-income households have still not been able to access homeownership. Moreover, "insiders" in the housing market have been able to capitalize on increased housing prices, without being heavily mortgaged. In parallel, the decline of affordable and social housing is putting strain on those households unable to join the homeownership race.

Overall, we find evidence of liberalization, commodification and financialization of housing markets in the case study countries. Many housing regimes have gradually gone from being rather national and welfare-oriented, to becoming a part of a globally financialized housing market, where national housing policy is increasingly detached from fiscal policies and international capital investments. However, these are very broad trends, and the evidence is not supportive of a simple convergence towards a uniform housing model.

Conclusions

In this chapter, we have described gradual changes in the welfare and housing systems across the world. We have shown that both housing and welfare systems – regardless of their earlier constitution – have changed drastically in terms of income inequality and housing accessibility and affordability. With the exception of China, most countries have reformed their wider welfare regimes through austerity policies and welfare cuts in a way which has strengthened the liberal tendencies and decreased the welfare state's comprehensive measures. This has strengthened the tendencies of stratification due to income polarization, which has been fuelled by housing wealth emerging among homeowners. On the other end, housing solutions for the least affluent are increasingly sparse as affordable housing is decreasing and increasingly means-tested, further contributing to stigmatization and social stratification.

The inequalities we observe in both housing and welfare have in many countries been reinforced by the last decade of immigration and the COVID-19 pandemic. The current economic crisis seems to continue to strengthen such inequalities as inflation and increasing interest rates tend to put an even harder strain on low-income households. Moreover, many newly mortgage-indebted middle-class households are facing financial difficulties as interest rates have tripled in 2023. Long-time homeowners with no or low mortgage debt are, however, faring well as the housing prices do not yet seem to drop significantly.

As this chapter has shown, the development of the national housing regimes is clearly detached from the national welfare state policies and simultaneously increasingly attached to fiscal policies. This development, we argue, calls for a more in-depth analysis of the interrelation between housing and the wider welfare state, which also includes the sphere of finance. We will commence such an analysis in the next chapter.

References

Aalbers, M. B. (2008). The Financialization of Home and the Mortgage Market Crisis. *Competition & Change, 12*(2), 148–166. https://doi.org/10.1179/102452908X289802

Absolute Strategy Research. (n.d). *ASR Global House Price Index*. https://www.absolute-strategy.com/asr-house-price-index. [Accessed: 25 April 2023].

Esping-Andersen, G. (1990). *The Three Worlds of Welfare Capitalism*. Princeton University Press.

EMF. (2009). *Hypostat 2008*.

EMF. (2022). *Hypostat 2022*.

Grander, M. (2018). *For the Benefit of Everyone. Explaining the Significance of Swedish Public Housing for Urban Housing Inequality* [PhD Thesis, Malmö University].

Grander, & Westerdahl, S. (2023). Financialization the Swedish way: A Case Study on how Public Housing is Replaced by the International Investment Portfolio of Vonovia. (Manuscript accepted for publication). *Journal of Urban Affairs*.

Guillén, A. M., Jessoula, M., Matsaganis, M., Branco, R., & Pavolini, E. (2022). Southern European Welfare Systems in Transition. In L. Burroni, E. Pavolini, & M. Regini (Eds.), *Mediterranean Capitalism Revisited: One Model, Different Trajectories* (pp. 149–171). Cornell University Press. https://doi.org/10.1515/9781501761096-009

Haffner, M. E. A. (2003). Tenure Neutrality, a Financial-Economic Interpretation. *Housing, Theory and Society, 20*(2), 72–85. https://doi.org/10.1080/14036090304262

Hochstenbach, C., & Aalbers, M. B. (2023) The Uncoupling of House Prices and Mortgage Debt: Towards Wealth-driven Housing Market Dynamics, *International Journal of Housing Policy*. https://doi.org/10.1080/19491247.2023.2170542

Holliday, I. (2000). Productivist Welfare Capitalism: Social Policy in East Asia. *Political Studies, 48*(4), 706–723. https://doi.org/10.1111/1467-9248.00279

Kholodilin, K. A., Kohl, S., Korzhenevych, A., & Pfeiffer, L. (2022). The Hidden Homeownership Welfare State: An International Long-term Perspective on the Tax Treatment of Homeowners. *Journal of Public Policy, 43*(1), 86–114. https://doi.org/10.1017/S0143814X2200023X

OECD. (2023a). *Affordable Housing Database*. http://oe.cd/ahd

OECD. (2023b). *Household Debt (indicator)*. https://doi.org/10.1787/f03b6469-en

OECD. (2023c). *Income Inequality (indicator)*. https://doi.org/10.1787/459aa7f1-en

OECD. (2023d). *Poverty Rate (indicator)*. https://doi.org/10.1787/0fe1315d-en

Schneider, M., & Wagner, K. (2016). Housing Markets in Austria, Germany and Switzerland. *The Narodowy Bank Polski Workshop: Recent Trends in the Real Estate Market and Its Analysis -2015 Edition*. https://ssrn.com/abstract=2841641

World Bank. (2023). *Poverty Headcount Ratio at National Poverty Lines (% of Population)*. https://data.worldbank.org/indicator/SI.POV.NAHC [Accessed: 12 May 2023].

Xie, Y., & Zhou, X. (2014). Income Inequality in Today's China. *Proceedings of the National Academy of Sciences, 111*(19), 6928–6933. https://doi.org/10.1073/pnas.1403158111

Zhou, J., & Ronald, R. (2017). Housing and Welfare Regimes: Examining the Changing Role of Public Housing in China. Housing, Theory and Society, *34*(3), 253–276. https://doi.org/10.1080/14036096.2016.1223165

17

THE EMBEDDEDNESS OF HOUSING IN THE WELFARE REGIME

Martin Grander and Mark Stephens

This chapter provides a detailed examination of the current state of housing and welfare in the five clusters of housing and welfare regimes we have identified: (1) The countries of the Anglo-sphere (US, the UK and Australia); (2) the Germanic (Germany, Austria and Switzerland) and Nordic countries (Sweden, Norway and Finland); (3) the southern European countries (Italy and Spain); (4) the post-socialist countries (Russia, Hungary and China) and finally; (5) the countries from the Global South (Chile and South Africa).

We draw (mainly) on the country chapters in this book and examine them within Stephens' (2020b) framework, which was outlined in Part I. We begin by examining these countries' wider welfare regimes, and then we examine the housing regimes, first within the context of Kemeny's typology of housing regimes, and then within the framework established by Stephens (2020b) where housing is examined through the spheres of production, consumption and exchange. Finally, we draw the two together.

Countries of the Anglo-sphere: Australia, the United Kingdom and the United States

The three Anglophone countries that are detailed in this book are each usually classified as having "liberal" welfare regimes, and "dualist" housing systems. They fall within Schwartz and Seabrooke's (2008) Liberal-Market grouping as a "variety" of residential capitalism, and Reisenbichler (2021) identifies the US and UK as examples of countries where governments have embraced financialization and promoted house price growth as a "growth strategy".

The wider welfare regime

The wider welfare regimes of the three Anglophone countries operated within the benign environment of the long post-World War II economic boom, at least until the oil price shocks of the 1970s. This facilitated something of an Indian summer for these late Fordist economies, allowing labour in the unionized industries to form the basis of the "male breadwinner" model – whereby the earnings of a man in a conventional nuclear family could support his wife and children. In all three countries, the labour market dominated the wider welfare regime, and the relative strength of

 DOI: 10.1201/9781003212690-20

labour allowed income inequality to be driven downwards, at least until the 1970s. The idea of the (male) "living wage" that could support a minimum standard of living – including housing – was particularly strongly embodied in the Australian ethos and became a central part of the notion of a "wage earners' welfare state". The notion of the male breadwinner was necessarily partial, both reflecting and reinforcing highly gendered societal norms.

Whilst the labour market may have dominated the wider welfare regimes, they were bolstered by progressive taxation, which in the two federal countries meant strengthening the powers of the federal government. The strongly centralized British state was able to introduce a strong, if parsimonious, system of social insurance after WWII, supported by a means-tested social assistance safety net that was meant to be residual but later became more dominant. Australia opted for a system of non-contributory means-tested benefits and is unusual in eschewing a formal social insurance system. Both a more strongly individualistic political culture and the federal system presented barriers towards the development of a comprehensive safety net in the US. Advances were made as part of Roosevelt's New Deal and Johnson's Great Society, but a comprehensive social assistance safety net for all adults was never developed, as well as the well-known absence of comprehensive health protection, even after the Obama reforms. The US' exceptionalism in this regard may also be attributed to the legacy of slavery which left a large minority subject to discrimination and a reluctance of white electors to support aid programmes that would benefit the Black population disproportionately.

The retreat from the high point of post-WWII wider welfare regimes therefore took place from rather different starting points. If deindustrialization undermined the male breadwinner model and weakened trade unions, economic restructuring combined with societal changes to support the growth in service sectors. In Australia, the removal of the family element from the minimum wage was a powerful symbol of societal change (Hulse et al., this volume). Whilst the expansion in higher education supported the growth in higher-paid jobs, each of these countries experienced a growth in part-time, less secure and lower-paid employment, albeit underpinned by statutory minimum wages. The growth in insecure and lower-paid employment was encouraged, especially in the US and the UK through the tightening of conditionality attached to the receipt of out-of-work social security benefits. The UK also followed the US in supporting low-paid employment through income supplements for lower-wage employees: low-income tax credits in the US and Working Tax Credits in the UK. Pensioners tend to be more generously treated than the work-age population. The term "social security" is reserved for pensions in the US, with the stigmatizing term "welfare" referring to working-age benefits. In Australia a compulsory employment-based pension system was established in the 1980s ("super"), while the UK made a concerted effort to tackle pension poverty first through more generous means-tested support and then a more generous (although flat-rate) state pension. The post-2010 "austerity" cuts, which included not uprating benefits by inflation, fell primarily on the working-age population and children.

These three Anglo-sphere countries form a cluster of relatively high-income inequality and relative income poverty regimes (Table 17.1).

The average Gini coefficient is 0.349, which is higher than the corporatist and social democratic country averages. Indeed, each one of the Anglo-sphere countries records a higher level of income inequality than every corporatist or social democratic country in this study. The same holds for the poverty rate. Nonetheless, it is notable that both income inequality and poverty are more pronounced in the US, compared to Australia and the UK. This is consistent with the institutional framework of the wider welfare regime described in detail in the regional chapters and summarized above (Table 17.2).

Table 17.1 Income inequality and poverty in the liberal welfare regimes

	Gini	*Poverty rate*
Australia	0.318	12.6
UK	0.355	11.2
US	0.375	15.1
Ave.	0.349	13.0

Source: OECD (2023) Income inequality (indicator); and OECD (2023) Poverty rate (indicator).

Note: % individuals. Poverty threshold: 50% median equivalized income.

Table 17.2 Housing tenure in Australia, the UK and the US (%)

	Australia	*UK*	*US*
Owner occupied (mortgaged)	35.1	28	
Owner occupied (unmortgaged)	31.6	36	
Owner occupied (total)	66.7	64	65.3
Private rented	27.4	17	34.7
Social rented	3.8	18	0.1
Other	2.1	na	na

Source: Australia, US: country chapters this book; UK: Family Resources Survey 2020/2021.

Note: Australia (2021); US (2020): stock; UK (2020/2021) Households.

Housing regimes

The role of cost rental housing

Australia, the UK and the US each conformed with Kemeny's notion of dualist rental systems, in the sense of sectors separated from the market rental sector. Historically, the social rented sector was almost exclusively owned and managed by the public sector – public housing authorities in Australia and the US, and local authorities in the UK. That profile has since changed in the UK where stock transfers and a switch to new build since 1988 have shifted the balance in favour of notionally independent housing associations. In Australia, too, federally funded housing programmes (1945–2003) have been supplemented with community housing programmes since the 1980s, although public housing remains dominant. Public housing in the US was conceived in 1937 and built in some volume with federal support mainly in the 1950s and 1960s, but federal funding dried up after 1972 and overall, the sector never grew to more than 1.5 per cent of the stock. "Maturation" enabled large amounts of the UK's local authority housing with positive values to be refinanced, transferred to housing associations and renovated without further subsidy, but some urban stocks in poorer condition and with negative values required additional subsidy. In contrast, the US stock was in very poor condition by the 1970s and federal programmes led to widespread demolition, partly to deconcentrate poverty. In Australia, too, relating individual rents to individual incomes has compromised financial viability rather than creating rental surpluses. Without these, the cost rental sectors in the US and Australia were faced with more of a downward slide than a critical juncture.

Institutionally, only the UK's local authority sector followed Kemeny's dualist pathway, in the sense of reaching scale (it housed one-third of households at its peak), and maturation, followed by the extraction of rental surpluses and forced discounted sales to tenants. It housed a reasonably broad section of the community, but the sector became more "residualized" first as it was opened up to poorer households, and then increasingly targeted as it became more scarce – although it still houses around 15 per cent of households in England. The Australian public sector houses only 4 per cent of households and the US around 1 per cent. Thus targeting is a question of degree, with the UK still able to provide a secure "safety net" for a broader range of households than Australia (where "[s]trict entry requirements (means tests)" were introduced in the early 1970s and is now reserved for those "with the most urgent and complex needs"– Hulse et al., this volume) and – especially – the US, where conditionality gives it the characteristic of being more of an "ambulance service".

An important question is how significant the "dualist" nature of these countries' rental systems actually is. In the UK the private rented sector was separate, but far from being protected as it was subject to rent controls almost until the last decade of the 20th century. The key dynamic for the latter half of the 20th century was the polarization between social renting and homeownership. The squeeze on private renting in Australia was less symmetrical with the growth in homeownership exerting more pressure than the growth in public renting.

Sphere of production

The three countries' housing systems are founded on market-based institutions, underpinned by support for property rights. Land is predominantly privately owned, and development is now dominated by the private sector. However, development rights are state-owned within the frameworks of planning systems that are zonal in Australia and the US and based on discretionary principles in the UK.

Whilst it is a very crude indicator, building rates adjusted for population give an idea of the relative levels of housing production. Notwithstanding differing population and household growth rates, building rates in Australia are much greater than in the US and the UK, which has consistently met estimated supply needs. However, the higher levels of supply in Australia and the US have not prevented rising house prices and deteriorating affordability (Table 17.3).

Planning systems have played important roles in shaping housing systems more broadly. In the US, zoning was instrumental in shaping the landscape of post-war urban America. Zoning exhibited a "hyper-protection of the single family [SF] home" (Tighe et al., this book) whilst discrimination in the finance system (see below) excluded most Blacks from suburbia. With more than 75 per cent of housing in US cities located in single-family zones, it is unsurprising that zoning contributes to class as well as racial segregation. Whilst some cities have followed Minneapolis

Table 17.3 Annual average building rates per 100,000 population in Australia, the UK and the US 2017–2021

	Building rate
Australia	776
UK	246
US	375
Ave.	466

Source: Calculated from EMF (2022) *Hypostat*, Table 17.14.

in eliminating SF zones, there is little evidence that this alone has affected prices. Moreover, the location of public housing built from the 1940s to 1960s in marginal locations also contributed to segregation.

In Australia, planning was instrumental in facilitating the development of low-density suburbs separating homes from the industry. Public housing tended to be located more centrally as part of slum clearance projects, at least until the 1970s when community campaigns pushed back. The state was a leader in planning during post-war reconstruction in the UK, but its ambitions are much more limited, and sometimes characterized as simply licencing individual developments rather than seeking to shape places.

However, the UK does make greater use of "planning gain" agreements with developers. Recognizing that the conferment of planning permission increases the value of land and that development creates externalities, developers agree that a proportion of units on new development will be set aside as social or affordable housing. This practice contributes towards around half of new social rented dwellings in England.

In the US, Low Income Housing Tax Credits (LIHTC), introduced in 1986 are the "principal mechanism" (Tighe et al., this book) for delivering affordable housing. The tax credits encourage private developers to build housing in lower-income neighbourhoods and to make it available at affordable rents to eligible households for a period of at least 15 years.

Both planning gain and LIHTC demonstrate relationships between state and market that are perhaps more blurred than Esping-Andersen might allow, and certainly outside Kemeny's maturation model.

Sphere of consumption

We have seen how cost rental systems in English-speaking countries developed along dualist lines, but the much larger – if diminished – sector in the UK allows it to fulfil a stronger and broader safety net function than in Australia or the US.

However, in these countries, support for low-income households is delivered in other ways, notably through demand-side interventions. The UK's housing allowance system (best known as Housing Benefit) applies across rental tenures and provides means-tested assistance with rent, in principle meeting all of it if incomes are no higher than social assistance levels. Most social tenants receive support in this way, but the cost of the scheme grew with the revival in private renting particularly since the turn of the century, leading the government to cut back on its generosity after 2010. Nonetheless, it remains the largest financial subsidy for housing.

The Australian Rent Assistance scheme is open to private (but not to public) tenants. Eligibility is more tightly defined than the British scheme, as the Australian scheme is limited among working-age tenants to those with dependent children. Nonetheless, it is the largest part of low-income housing support "and has been increasingly relied upon as the key plank of national housing policy" (Hulse et al., this volume).

The British and Australian rent allowance schemes are both demand-led – what in the US is called an "entitlement programme". They contrast with the US' closest equivalent, Housing Vouchers, which is a cash-limited programme, with the result that many households that meet eligibility criteria do not receive assistance. Those households who do receive support can have up to 75 per cent of their rent met, but rent ceilings act as a barrier to helping to create truly mixed-income communities – a feature it shares with the British system.

Homeownership is also supported through demand-side assistance. All three countries treat the tenure favourably in tax terms. The US provides generous tax reliefs on mortgage interest payments that exceed the value of support provided to low-income housing; Australia exempts owner-occupied properties from land tax; and the UK does not tax homeowners' capital gains. The UK and Australia have also provided other demand-side interventions ostensibly to assist first-time buyers from time to time. These include exemptions from transaction taxes (Australia and the UK), cash grants (Australia) and a shared equity scheme (UK).

Sphere of exchange

We have observed that in none of these countries does the *dualist* nature of the cost rental sector define the whole housing system. Rather, the defining – and uniting – feature of these countries is a cultural attachment to homeownership as the "ideal" tenure, captured in the ideas of "The American Dream", the "Great Australian Dream" and, in the UK, "the property-owning democracy".

Homeownership grew in all three countries during the 20th century assisted by reasonably efficient and risk-averse closed-circuit housing finance systems based on depository principles. The US system was, however, defined by the introduction of government insurance of mortgages through the Federal Housing Agency to help the country recover from the Great Depression. This, in turn, contributed to racial segregation as underwriting criteria were institutionally racist, excluding non-whites and contributing to inner-city decay through redlining – a form of stratification.

All three countries deregulated their housing finance systems from the 1980s, and, in the case of the US, it was mismanaged and resulted in the Savings & Loans crisis that necessitated a federal bailout. The development of complex financial instruments including the widespread use of securitization that supported the growth in sub-prime lending was an extreme example of the trade-off between risk and widening access to finance to support homeownership, resulting in the Global Financial Crisis. Less dramatically, in the UK and Australia, financial deregulation assisted the growth in mortgage lending and house prices and very much developed the "investment" character of housing. The irony is that in both these countries, rising house prices helped to reduce access to owner-occupied housing.

If financialization was initially manifested through the growth in mortgaged homeownership, it now takes more diverse forms. In the UK and Australia, access to mortgage finance helped to underpin the growth in small-scale "retail" landlords. Private renting grew, housing younger households priced out of homeownership (and to an extent unable to access socially rented housing). In all three countries, it is characterized by a lack of security (most acutely in the US), and a lack of rent control. This is changing in the UK, where Scotland has introduced a very high level of tenure security and is heading towards a national system of rent controls; and the UK government is committed to increasing security. In Australia, too, several states are seeking to follow Victoria in reducing the scope for "no-fault" evictions.

In all three countries, foreign investment, the growth of so-called "platform capitalism" (encapsulated by Airbnb) and simply holding properties empty for investment purposes have become manifestations of financialization. Within the social rented sector, the recent advent of equity investment in social rented housing by institutional investors can be seen as another example of financialization.

Housing and welfare regimes

The housing and welfare regimes of Australia, the UK and the US reflect some common underlying cultural values that have created similar organizing principles reflected in both the welfare and housing regimes. The US is nonetheless an "outlier" in the extent of its individualism and support for the protestant work ethic, combined with the unique impact of the legacy of slavery that has divided American society and inhibited the development of a welfare state.

The wider welfare regimes embrace relatively deregulated labour markets with decentralized wage bargaining and have experienced a growth in part-time and insecure jobs. None of these countries has sought to develop strong earnings-related social insurance systems seen in corporatist or social democratic states. There has, instead, been a greater – or, in the case of Australia, total – reliance on means-tested social assistance and other targeted benefits. Predictably, these institutional arrangements have distinctive distributional outcomes: the crude indicators of inequality and poverty suggest a higher degree of inequality than is the case in corporatist or social democratic counterparts. These crude indicators also suggest that the US is something of an outlier, with the highest level of poverty. The poverty figure alone cannot, however, capture more acute forms of deprivation that arise from the failure to seek to introduce a comprehensive income safety net in the US and the gaps in the safety net that have opened up in the UK.

We argue that the distributional outcomes of the wider welfare regime set the "boundaries of possibility" of the housing system. In largely market-based housing systems, we would expect housing consumption to mirror purchasing power, unless disrupted or amplified. We can make only a qualitative assessment of these disrupters and amplifiers. Clearly, the UK's greater (if diminished) stock of socially rented housing is one such disrupter compared to Australia and the US. The relatively high level of poverty in the UK means that a social market model is not feasible and instead, there is a high level of targeting (with needs tests rather than income means test), including homeless people, creating a secure safety net for those who benefit. The targeting is more acute in Australia and the US, and the sectors veer more towards the "ambulance service" model. However, the shallower subsidies associated with LIHTC housing mean that there is a broader decommodifying effect in the US system than is implied by public housing alone.

The level of housing consumption that private tenants in these countries are likely to reflect current incomes, and the growth of this tenure in the UK and Australia reflects a growing commodification in the system. The main proviso is that the decommodifying effects of housing allowances also play an important role. However, it is limited in the US due to the cash-limited nature of the voucher programme which leaves many of those who are eligible without support. In the UK it has become strikingly less generous primarily due to limits on eligible rents, and is inadequate in the high-cost cities in Australia.

Australia, the UK and the US are all "homeownership" societies, in that a cultural attachment is placed on this tenure that may be independent of narrow financial concerns. Governments have sought to boost homeownership through discounted sales of social rental housing (especially in the UK, but also in Australia), and generous tax and other demand-side subsidies. In the US, the state also supported homeownership through mortgage guarantees, but historically the institutional racism embedded within the underwriting rules limited homeownership among the Black population and exacerbated residential segregation – an important example of stratification. The planning system also facilitated the growth of single-family homeownership in low-density suburbs in Australia and the US, but zoning limiting areas to single-family houses in the US is another source of stratification.

Homeownership may be conceived as being hyper-commodified at the point of purchase (depending to an extent on savings reflecting past income or family support, and not only current but anticipated future income to secure a mortgage). However, it becomes a source of decommodification in retirement, with mortgages paid off and current income falling. The link with the wider welfare regime in all three countries is obvious, but its role as the "fourth pillar" of the pension system is most overtly acknowledged in Australia. The pricing out of younger households and their reliance on private renting will unavoidably become a growing source of commodification, meaning that housing systems amplify current and lifetime income inequalities.

Financialization flows through these housing systems, which deregulated their housing finance systems in the 1980s, ending the specialized circuits that underpinned the growth in homeownership in the post-war decades. The initial effect of deregulation was to widen access to mortgage finance, but to create more risk, and incentives for lenders to take risks. The Savings and Loan (S&L) crisis of the 1980s and the subprime crisis in 2007 are both exemplars of what can happen when incentives to take risks are built into the system. A longer-term consequence of the Global Financial Crisis was to establish a "long decade" of ultra-low interest rates, which lasted until the inflationary upsurge that followed the COVID-19 pandemic and the Russian invasion of Ukraine in 2022. Ultra-low interest rates in turn helped to fuel house prices, contributing to the locking out of younger households from homeownership. Other forms of financialization seen in Australia, the UK and the US include institutional investment in rental properties, the growth of platform capitalism (Airbnb) and equity investment in social renting (UK) (Figure 17.1).

Whether this constitutes a "growth regime", as posited by Reisenbichler (2021), is questionable. Certainly, policy (notably monetary policy) has supported house prices. However, central banks are responsible for consumer prices, not asset values, and are often critical of other government policies towards housing. We have seen that fiscal policies may favour housing, but that is often true in countries notionally less reliant on house prices. The suggestion that house price growth is intended to compensate households for parsimonious social security benefits or weak wage growth has some attraction but is clearly self-defeating as more people are forced into market renting and face much higher lifetime housing costs. If it is a growth regime, it is not a very smart one.

Two of the three countries are federations of states, which enjoy responsibility for housing policy. The UK has devolved responsibility for housing to Scotland, Wales and Northern Ireland. Yet, the two federal and the UK governments still define the housing regimes. The wider welfare regime is defined largely at the federal/nation-state level, which sets the "boundaries of possibilities". Many instruments that affect housing directly (such as tax treatment) are set at the federal/nation-state level, too. Financial institutions operate across countries and monetary policy is

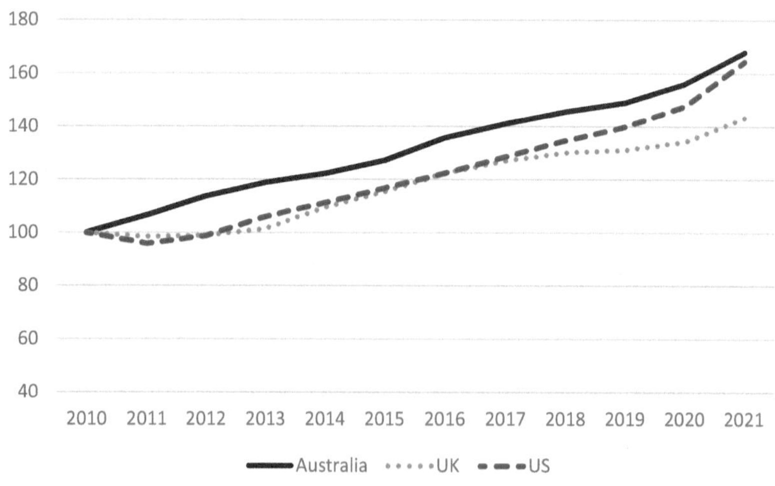

Figure 17.1 Nominal house prices in Australia, the UK and the US (2010–2021; 2010 = 100).
Source: Calculated from EMF (2020) *Hypostat*, Table 18.

uniform. The National Housing & Homelessness Agreement in Australia acts as a unifying force, and national/federal grants are also deployed in the US and UK. Important variations can occur through policy differences at the state (Australia, US) or national (UK) level, but they are unlikely to be sufficient to shift regime type. Market differences (between booming and depressed/post-industrial areas) also mean that similar institutions and policies may vary in their outcomes. Yet the organizing logic – the essence of a regime – remains broadly the same.

Conclusion

The Anglophone countries capture the essence of regime theory: how organizing logic derived from cultural dispositions shapes both wider welfare and housing regimes. This allows for variation in institutions and policy choices. The welfare and housing regimes may be more or less decommodifying, and more or less stratifying. On one level the three Anglophone countries can be grouped together as reflecting "liberal" organizing principles, yet on another they exhibit differences. The US, for example, *is* an outlier in the extent to which individualism and the legacy of slavery have inhibited the development of social protection in the wider welfare and housing regimes. Employing regimes in comparative research necessitates comparing *within* a cluster as much as *between* clusters.

The Nordic and Germanic countries

The Nordic countries of Denmark, Sweden, Norway and Finland tend to be grouped together due to their historical heritage of Social Democratic hegemonic history and their comprehensive welfare policies. This is at least what Esping-Andersen (1990) did in his analysis of welfare state regimes. With regards to housing provision, there are, however, large differences between the Nordic countries, which was highlighted by Bengtsson (2013) in a comparison of Nordic housing regimes entitled *Why so different?* (Swedish: *Varför så olika*?). Such dissimilarities were earlier observed by Kemeny (2006), who instead of focusing on the similarities regarding the social democratic influence of the welfare states, discussed the similarities between some of the Nordic countries (Sweden and Denmark) and the Germanic countries (Germany and Austria) when it comes to the degree of corporatism and the evolution of the cost-rental sector. In this section, we will therefore discuss the two clusters of countries together, focusing on the interrelationship between housing and welfare.

The wider welfare regime

The beginnings of the welfare model in the Nordic countries were already developed before the Second World War by a high level of taxation with the purpose of funding a social safety net. In the years after the war, social democracy increased its strength and the Nordic welfare models were further expanded to include modern health service and education systems.

Perhaps the most distinctive social and economic system was found in the Swedish Rehn-Meidner model. This model involved an industrial strategy that left industries in the private sector, but the state provided incentives to ensure high productivity growth, for example, by incentivizing the reinvestment of profits. The model also encouraged restructuring in unsuccessful industries with generous unemployment benefits and active labour market policies intended to ensure workers were well-matched when seeking employment. The system compressed wage differentials and for a period brought much economic success.

The economic upswing in the second half of the 20th century, thanks to rising exports and the availability of raw materials, strengthened these foundations. However, the countries have seen divergent economic development during the 20th century, where the Norwegian economy has continued to flourish, thanks largely to its considerable oil deposits.

The social democracy that emerged in the Nordic countries was the key marker of concern for Esping-Andersen's categorization of the countries as belonging to the social democratic welfare regime. The social democratic regimes have strongly exhibited corporatist tendencies, where labour and capital have been tightly knitted together under the state's orchestration. Such tendencies also emerged in the continental countries of Germany, Austria and Switzerland – the so-called DACH countries.

The DACH countries are bound together by the Bismarckian principles of the welfare organization, characterized by strong social insurance, introduced in the late 19th century. Such principles, which came to be dominant in the German Reich and the Habsburg Empire, had the intention of implementing the minimum aspects of the Social Democrats' welfare programmes, but without integrating any of their explicitly socialistic aspects. Thus, the intention was to find support for a socialistic welfare policy in a non-socialistic German government.

The period after Second World War is referred to as the "golden age" of the Bismarckian welfare states, characterized by their steady growth up until the mid-1970s. Like the social democratic welfare states, the welfare system that was developed provided universal health care and a strong system of social security, which included unemployment benefits, pensions and disability insurance. Like their northern counterparts, a progressive tax system was initiated in order to fund social welfare programs and support less fortunate households. The crisis-ridden 1970s and 1980s put hard strains on the conservative welfare states, leading up to large reform efforts across the countries, perhaps most notable was the Hartz IV labour market and social security reforms introduced by the German government to mitigate the 1990s crisis. Such reforms were highly corporatist in the way they searched to support and accommodate labour and confront the increasing burden of non-wage labour costs for employers. As Wetzstein (this volume) notes, the welfare regime of the "social insurance state" built up during this period was "much less concerned with reducing inequality than to provide workers with security, and to conserve status".

Esping-Andersen (1990) placed the Nordic countries in the social democratic group of welfare regimes, but Germany and its continental European neighbours in the conservative/corporatist regime family. Indeed, there are clear similarities between the nations, also in the way they have developed. As described in the previous chapter, the recent evolution of the social democratic welfare regime has seen rapid development in line with a transition from welfare to workfare characterized by labour market flexibility and increased conditionality in social security. In parallel, the Bismarckian welfare states have since the 1970s struggled to respond to problems of unemployment, lower productivity and an explosion of social cost, much related to international competitiveness and migration. Today, social democratic and corporatist countries display increasing income inequality and relative income poverty, but the rates remain significantly lower in the Nordic countries, as illustrated by Table 17.4.

While the social democratic countries are rather convergent in their numbers (with Sweden being somewhat an outlier with a higher rate than the others), the corporatist countries display larger dissimilarities. Austria has the lowest degree of inequality while Switzerland has by far the highest. Interestingly, the poverty rates show more consistency in the conservative regimes than in the social democratic. Finland, which has an inequality rate similar to its neighbours, has significantly lower proportions of households in relative poverty.

Table 17.4 Income inequality and relative poverty rate in the Nordic and Germanic countries

	Gini	*Poverty rate*
Sweden	0.276	8.8
Norway	0.263	8.4
Finland	0.265	5.7
Denmark	0.268	6.5
Ave.	0.268	7.4
Germany	0.296	10.9
Austria	0.274	10
Switzerland	0.316	9.9
Ave.	0.295	10.3

Source: OECD (2023) Income inequality (indicator); and OECD (2023) Poverty rate (indicator).
Note: see Table 17.1.

Housing regimes

When examining the housing regimes of the social democratic and corporatist countries using Kemeny's typology, we find large dissimilarities between the two groups of welfare state regimes. In social democratic countries, homeownership is now the dominant tenure, but as we will see, the state politics on tenure have varied greatly, leading to very different roles for and characteristics of their rental sectors (Table 17.5).

The role of cost rental housing

When examining the political trajectory of housing tenures, we find that the social democratic welfare regimes have markedly dissimilar strategies regarding cost-rental housing. In Norway, private rental housing in general is small and has been on the decrease for decades. Subsidized rental housing is, and always has been, negligible due to the political support for homeownership (Sørvoll et al., this volume). Where the homeownership rate is near 80 per cent, private rental housing stands at 18 per cent and social housing (strictly means-tested and subsidized) houses less than 2 per cent of households. The Finnish housing regime is close to the Norwegian case in its consideration of homeownership as the norm, and how government-subsidized rental housing is directed at disadvantaged low-income groups. Compared to the Norwegian case, Finnish social rented housing however accounts for a much greater share (11 per cent of the households). Finland has developed a clearly selective system of social housing to complement the otherwise dominant homeownership sector. The allocation of social housing apartments is generally conducted on a needs-tested basis, where the so-called *ARA* property owners (subsidized housing associations) select applicants using a procedure based on priority. As the Finnish cost rental market is subsidized and needs-tested, it is a clearly dualistic rental market. However, the size of this sector has diminished in recent years due to a general retrenchment of housing policy (Ruonavaara, 2017).

In Sweden and Denmark, which Kemeny labelled as cost-rental regimes with unitary rental markets, the role of cost-rental housing has increasingly begun to differ. While they both have stuck to a – in principle universal – model of public housing, open to everyone, Denmark has a major exception from the universal approach, while Sweden has a truly universal model – at least in discourse.

Table 17.5 Housing tenure in the Nordic and Germanic countries (%)

	Denmark	*Finland*	*Norway*	*Sweden*
Owner occupied (mortgaged)	43.1	35.3	57.3	49
Owner occupied (unmortgaged)	13.8	33.8	22.4	15
Owner occupied (total)	56.9	69.1	79.8	64
Market Rented	43	16.8	18.2	35.2
Reduced rent/free	0.1	14.8	2	0.8
Rented total	43.1	30.9	20.2	36
Social/public rented	21	11	4	17
	Austria	*Germany*	*Switzerland*	
Owner occupied (mortgaged)	21.1	19.1	38	
Owner occupied (unmortgaged)	32.8	25.8	4.5	
Owner occupied (total)	53.9	44.8	42.5	
Market rented	30.1	51.3	51.6	
Reduced rent/free	16	3.9	6	
Rented total	46.1	55.2	57.6	
Social/public rented	24	3	3	

Source: Rows 1–6, EU-SILC; Row 7, Housing Europe (2022).
Note: Year 2021, except for Norway (2020). First 6 Rows: % individuals. Seventh row: % housing stock year.

Due to the increasing adoption of business-like principles – originating from deregulation and marketization in the 1990s and strengthened with the EU's legislation on competition – the Swedish universal model of public housing (*allmännyttan)* has been stripped of state subsidizes and privileges and competes since 2011 with private actors in the rental housing market on a level playing field (Grander, 2017; Sørvoll et al., this volume). Thus, the Swedish rental market has become truly unitary, to use Kemenys' terminology, and might also be the only EU country clinging on to a "true" universal housing policy because it is the only regime that lacks a subsidized housing sector. Nonetheless, Eurostat (EU-SILC) treats the Netherlands in the same way, even though income limits have been introduced for social housing because of the government's endeavours to comply with the EU's legislation on competition.

Denmark appears as something in between a unitary and dualistic rental regime. Public housing (managed by social housing associations) is in principle universal, but the municipal state can allocate up to 25 per cent of all public housing associations *(almene boliger)* for households with limited income. Although the housing associations are formally private organizations, they are legally regulated and monitored by the government and municipalities and receive subsidies from the government to cover their operating costs and to keep rents affordable. The European legislation on competition has yet to affect the Danish rental market.

Germany and the other DACH countries (Switzerland and Austria) were like Sweden and Denmark identified by Kemeny as cost-rental societies, which is reflected by the large share of rented housing. All three countries were originally grouped by Kemeny into unitary rental markets but have clearly progressed towards dualization. According to Wetzstein (this volume), such a process towards more dualistic regimes with a greater degree of residualization of social rental housing has been a result of increasing attempts to regulate affordability and tenure security as well as a propensity to adopt business models that rely on cross-subsidization rather than state-subsidy. Today, all three countries have a clear means-testing principle for the distribution of social housing units by the application of income ceilings.

In Switzerland, the social housing sector is subsidized and means-tested by a national income ceiling; in Germany, it is means-tested according to a national income ceiling. However, local variations within the German states are large. Social housing units which are state-subsidized have a reduced rent for 30 years, but are then phased into market rent setting, while municipally and corporately owned subsidized rental housing have permanently reduced rent but are after 30 years released to the general public, without means-testing. Thus, there are elements of both dualist and unitary rental regimes in Germany. The same could be said to apply for Austria, where social housing is provided by means-testing according to an income ceiling. However, the income ceiling is remarkably high, so 60–80 per cent of the population is eligible for subsidized social rented housing (Bengtsson & Grander, 2023). Needless to say, the state expenses for subsidizing such a system are very high. Moreover, as in Sweden, the European Union's legislation on competition has made it more difficult to subsidize Austrian social housing, which has endangered the comprehensive system with its generous income ceilings in the social rental sector, which in practice excludes the richest part of the population (Mundt & Amann, 2010).

The extensive social housing system in Vienna has previously been argued to keep the rent in the PRS low and contribute to mitigating socioeconomic segregation. However, as shown by Kadi et al. (2022), income and housing polarization have increased in Vienna since the GFC, especially in the sense that "low-status areas" cannot keep up with the pace of upgrading observed in the city as a whole. As discussed by Wetzstein (this volume), Austria is increasingly facing challenges related to access and affordability, similar to those other European nations that also have a long history of social housing. In Germany, a severe division between existing and new PRS in urban growth regions has emerged according. Statistics show that, despite recent legislation on the "rent break" and a tendency for shifting to inflation-indexed rental contracts, the PRS rent levels have increased. This has also led to a more marked difference between social housing units and the PRS, contributing to a dualistic character.

Such tendencies can also be found in Sweden. Although the public housing sector has become more similar to the PRS, with the consequence that public housing has increased income barriers and thus excludes low-income earners (Grander, 2017), the median income is considerably lower, and decreasing, in the public housing sector compared to the PRS. As the different types of rental housing often are geographically separated, the income segmentation between the two connects to residential segregation – two forms of stratification not considered by Esping-Andersen.

The sphere of production

Regarding the production of housing, we find both similarities and differences between the clusters of countries. Building rates are generally higher in the Nordic than in the conservative countries. However, both Germany and Sweden have lower levels of construction than their neighbours.

The housing systems in Nordic countries are generally formally decentralized at the local (municipal) level. Planning is a municipal jurisdiction, and much land is owned by the local state (municipalities). In most cases, land is sold by competitive bidding. As housing construction is concentrated in urban areas, where demand is highest, the land prices are correspondingly high, making affordable housing construction hard to achieve. In Sweden, production of new public housing units has for many decades been significantly lower than the average market share, especially after subsidies were withdrawn in the 1990s. There are several reasons for this. European competition rules have made Swedish municipalities reluctant to favour municipal housing companies in land allocation. Combined with the abolished privileges for public housing and changes to accountancy rules (Westerdahl, 2021), public housing companies are unable to construct housing units unless they elicit a short-term profit. However, some companies are quietly protesting against this logic, arguing that a single housing project can be completed without short-term financial profit since the project might be cross-subsidized by other, more profitable, housing construction projects. Thus, there is a certain discretion in the regulation, which the public housing companies interpret in different ways (Grander, 2019). Although subsidies for non-means-tested public housing are no longer permitted due to EU legislation, the Swedish government have tried to find other ways of increasing the production of low-cost rental housing. Investment support in the form of financial subsidies was introduced in 2016, making it possible for all rental housing providers to get subsidies for rental housing construction with limited rent levels. This universal approach has however been criticized by conservative and liberal parties, but also (interestingly) by many rental housing companies who argue that the subsidy decreases the market values of their properties (Bengtsson & Grander, 2023). Subsequently, the right-wing government which took office in 2022 withdrew the subsidy. Consequently, there is no state support for the construction of rental housing, while the corresponding generosity of the tax treatment of homeownership has contributed to the decline in rental housing in general and in public housing, in particular (Bengtsson & Grander, 2023).

In Norway, housing construction is predominantly planned and financed by private actors. As described by Sørvoll et al. (this volume), homeownership is championed by the government through tax subsidies and state-backed mortgages for disadvantaged households). Thus, the subsidies are predominantly directed at households rather than housing production. As in Sweden, government subsidies for housing construction and comprehensive state regulation of housing markets were scaled back or abolished and replaced with modest economic support to disadvantaged groups and more liberal market regulations. For example, the State's Housing Bank, which previously had an important role in the mass construction of affordable housing, was transformed into an instrument primarily addressing homelessness and managing housing allowance. As noted by Sørvoll et al. (this volume), Norway spent 60 per cent more of the GDP on housing than Sweden, a difference partly explained by generous Norwegian tax and grant subsidies stimulating homeownership. Finland, in turn, spends 1.9 per cent of its GDP on housing, mostly consisting of housing allowances. The main support for housing construction consists of interest subsidy loans given to ARA and special housing.

In Denmark, support for housing construction in larger cities has in recent years been to a great extent funded by pension funds, but since land prices in greater Copenhagen have increased, the pension funds have been more reluctant to invest here (Tunström et al., 2020). Four strategies have been persistent in Denmark: (1) boosting the supply side by attracting private investors and developing new sites for residential building via land reclamation and on formerly protected areas; (2) increasing regulation in order to ease pressure on the demand side: restricting Airbnb and limiting the number of condominiums purchased by parents and rented out to their children; (3) making construction cheaper by reducing regulation and industrializing the building process, or through limiting the size of social housing units; and; (4) securing affordability by public regulation, by demanding that a certain proportion of residential property in new developments should be designated as social housing.

While constructing housing in urban areas is lucrative, albeit leading to high housing costs, construction in less populated areas has become increasingly challenging among the Nordic countries. As summarized by Eliasen et al. (2020), a common challenge in these countries is to build new housing in disadvantaged areas of cities and, even more evident, in rural locations because of low market values. As Denmark is densely populated, this issue is more evident in Norway, Sweden and Finland.

The current crisis of rising inflation, energy costs and high interest rates has led all housing constructors, also social and public housing, across the countries to reduce output. In Sweden, where the economy has been among the worst performers in the EU, construction has all but stopped as it has become unprofitable, and households have found mortgage availability to be more restrictive. Housing commentators are calling for state support, but the current government are not inclined to contravene market principles and urge consumers to wait until inflation drops.

In the conservative counties, we find a similar market-based principle for planning and housing construction, but with more evident market corrections in terms of state regulations. As illustrated by Table 17.6, construction rate performances are varied. Austria and Switzerland are seeing very high numbers, while Germany stands below average.

As discussed by Wetzstein (this volume), planning and housing are characterized by "growing and vexed contradictions" which are developing under the global market-oriented model. International finance institutions and central banks have increased their influence in all countries. Historically, the large rental housing stock in these countries has been facilitated by a combination of regulation, subsidies for construction and generous tax treatment. Responsibility for housing construction in Germany and Switzerland is divided between national and local governments. Social housing construction is often favoured, as in Switzerland, where the government offers tax credits to developers who build affordable housing for low-income households. In Austria, the government provides tax incentives to developers of affordable housing. In Germany, progressive movements, housing advocates and often municipal state factions have placed a renewed emphasis on decommodification, rent control, strategic land use policies and affordable urban housing solutions.

In Switzerland, settling foreign buyers and investors (for tax and quality-of-life benefits) has put sustained pressure on the housing market. The main housing-related policy issue in the recent past in Switzerland has been the limitation of urban sprawl and the preservation of a liveable countryside. An increasing political divide between urban populations – that are increasingly migrating towards rural areas – and rural regions that want to obtain a similar standard of living as the urban dwellers. Since then, this topic has become a key challenge not just for national cohesion but also for the effectiveness of direct democracy (Wetzstein, this volume).

Table 17.6 Annual average building rates per 100,000 population in the Nordic and Germanic countries 2017–2021 (except Switzerland, 2017–2020)

	Building rate
Denmark	537
Finland	706
Norway	563
Sweden	493
Ave.	575
Austria	624
Germany	352
Switzerland	577
Ave.	518

Source: EMF (2022) Hypostat, Table 17.14; World Bank populations, 2021.

In planning, liberal tendencies that have become more pronounced in recent decades are now – at least in Germany – being countered by more state regulation – moving towards the social democratic principles of state correction, while Esping-Andersen's ideas of corporatism do not seem to be applicable on the current development housing production. The current crisis is serious in the German-speaking countries. As business models for constructing housing are failing, and private developers can't add on extra costs to the residents, as inflation and increasing energy costs heavily affect their economic situation – a "cost-of-living crisis" has emerged. The public finance arrangements are not enough for public and social housing providers to build at a price that falls within the permitted threshold for social housing in Austria, and in Germany, the lack of state support has led to a mismatch between the government's construction targets and local outcomes. Wetzstein (this volume) notes that representatives from the German construction industry are calling for stamp duty land tax to be suspended, the goods and services/VAT tax for social housing to be reduced and a massive support package for affordable housing construction.

The sphere of consumption

While home ownership is on the rise across all social democratic countries, we have described a clear division of development towards the integrated cost-rental systems (Denmark and Sweden) and dualist homeownership systems (Norway and Finland).

This is also reflected in the demand-side support. All Nordic countries have some kind of housing allowance, but the scope and state expenses for those differ greatly. Sweden and Norway have allowances that apply for all housing tenures, but as Norway has a very small share of rental housing, demand-side support such as start-up mortgages to encourage low-income homeownership is the more common way of supporting demand. Housing allowance is targeted at the most vulnerable households, and the share of the population having this contribution is very small – almost neglectable among homeownership households.

Finland, with a dualist rental market, combines its extensive model of means-tested social housing with a housing allowance applicable for tenants in both social housing and the PRS. The state incurs very high costs for housing allowances, which are by far the largest state expense for housing (€2.14 billion in 2020, covering more than 20 per cent of the households. The system allows households with low income to reside in rental housing with market-based rents, but the system is criticized for becoming an unsustainable part of the state budget, and households are increasingly becoming dependent on contributions.

As Sweden currently lacks a social housing sector with subsidized rents (although all rents are below market due to its system of collectively negotiated use-value rent setting, see Enström Öst et al., 2014), its housing regime is even more dependent on housing allowances. The Swedish housing allowance runs across tenures, is means-tested and available to young adults, families with children and the elderly. In addition to generally helping households with low economic status, the grant aims to make these possible groups gain access to the rental market by strengthening purchasing power and thus the ability to pay the rent. However, state investigations and research have shown how the housing allowance has been watered down over the last decades. The cost of the scheme grew exponentially during the 1990s crisis, leading the government to cut back on its generosity after 2010. State expenditure on the housing allowance has more than halved during the last decades – which makes it in practice insufficient for easing access to the rental market or let alone helping households over the poverty line (Bengtsson & Grander, 2023). Only 157,000 households (3 per cent of the total number of households) were granted the allowance in 2021. Moreover, 182,000 households have social assistance from the authorities, where rent costs often are covered.

Denmark has the same principles for housing allowance as Sweden, and national statistics show that around 260,000 households, the lion's share in PRS, were granted this allowance in 2021 (10 per cent of all households), a number which is steadily increasing. In Sweden, social assistance can provide assistance for rent, and around 100,000 individuals receive such support.

Homeownership is also supported through demand-side assistance. Although there is a more explicit drive for homeownership in the national and local policies of Norway and Finland than in the other social democratic countries, owner-occupation is treated favourably by the tax system. Sweden provides a 30 per cent tax deduction on mortgage interest payments which greatly exceeds the value of support provided to low-income housing. Taxation on capital gains after selling owned housing is set at 30 per cent but is subject to roll-over relief. Finland has a similar figure of a 25 per cent tax deduction for home ownership and taxation on profit gains when selling. Here, however, mortgage interest on house purchases intended for renting is fully deductible. As mentioned, Norway has provided demand-side interventions ostensibly to assist first-time buyers. The Swedish government is currently looking at a similar policy.

The conservative countries have less dependency on housing allowances since demand-side expenses to a larger extent are focused on making housing affordable. Subsequently, there have been no significant policies on easing conditions for first-time buyers.

In Germany, the housing allowance is available across tenures to assist lower-income households with their housing costs, such as rent, mortgage interest and heating. There are also several other subsidies on the demand side. For example, the possibility for a rent reduction if the rented property has defects or deficiencies that affect the households' living conditions. Approximately 1 million households or 2.6 per cent of all households – most of them pensioners – in Germany received housing allowances in 2019. However, the share of households receiving housing allowance varies significantly across different states and regions of the country. Housing allowance is no longer available to social assistance recipients. Instead, 4 million households receive support for rent and heating through social assistance payments, which can cover all of the rent (Stephens, 2020a).

In Austria and Switzerland, the means-tested housing allowance is also an important demand-side support, sided by rental deposit guarantees. In both countries, means-tested allowance provides financial support to low-income households to help cover the cost of the rent. In addition to the housing allowance, there are also other forms of demand-side housing assistance available in Austria and Switzerland, such as interest rate subsidies for mortgages, which are available in some municipalities and cantons.

Approximately 9 per cent of households in Switzerland received some form of housing allowance or subsidy. This number, however, includes both demand-side subsidies as well as supply side subsidies, such as tax credits for developers of affordable housing. The corresponding number in Austria is 15 per cent.

The sphere of exchange

The booming markets for owner-occupied housing in various forms is a significant common feature of the Nordic housing regimes (Kohl & Sørvoll, 2021). A high level of mortgaged homeownership is a common feature of all Nordic housing regimes. The OECD, IMF and various policy experts have warned that the interrelated growth of house prices and debt constitute a major threat to the economies of the Nordic nations. While the current expansion of mortgage debt in the Nordic countries has been driven by financial market deregulation recently, history is also making an imprint. The deregulation of cooperative tenant ownership in recent decades, and earlier in Sweden than Norway, has "transformed highly state-regulated tenures into fully fledged examples of financialized homeownership" (Sørvoll et al., this volume).

All three countries deregulated their housing finance systems from the 1980s onwards– in the case of Sweden it was part and parcel of the financial crisis. Such deregulation assisted the growth in mortgage lending, and house prices and very much developed the "investment" character of housing. Consequently, Swedish tenant ownership has continued to grow in popularity in urban areas, whereas rented housing got the flipside of the deregulation (rents increased 50 per cent over two years in the 1990s).

Again, replicating the different attitudes towards homeownership in the countries examined in this section, we find – quite surprisingly – that the country with the greatest financial benefits for homeownership is Sweden. Here, and in the other Nordic countries, interest rates have been extraordinarily low during the past decade. Central banks held base rates around zero or even below, which has led to steeply rising house prices and increasing mortgage dependence. Both the quantity and type of Swedish mortgage debt have played a key role in this. As noted earlier, the amount of such debt relative to disposable household income is high compared with other countries, as is the proportion of such debt that is payable at variable rather than fixed rates. Therefore, rising interest rates have impacted immediately on Swedish households. This has placed pressure on household budgets at a time when they are experiencing the highest inflation rate in the EU. The central bank's policy of sub-zero interest rates (2015–2019) and the government's subsidies to homeownership through tax relief on mortgage interest together poured fuel on an already overheated ownership market.

The German-speaking countries have historically seen less of a homeownership price boom. However, house prices have risen over the last ten years and the current economic crisis has put pressure on governments to intervene in the market-based housing finance system which is regarded as insufficient in supporting housing construction and preventing homelessness. Wetzstein (this volume) describes the current situation in the Germanic countries as a "perfect storm for inflation", which is arguably the case for the Nordic countries as well. As inflation and interest rates are rising, and costs for materials and workers are increasing, housing construction is stalling across these countries. Indeed, the market-based business models for housing construction are challenged without greater state support.

As in the US, financialization was initially manifested through the growth in mortgaged homeownership but has now spread to the rental sector. In Denmark and Sweden, the rise of equity investment in social rental housing by institutional investors is the clearest example of financialization. Vonovia, a listed company based in Germany, is today one of the largest private landlords in Sweden. The until-recently low-interest rates, the housing shortage in urban areas and changes in public housing legislation which forces public housing companies to sell property in order to fund new housing construction have created a "unique cocktail" for housing financialization (Grander & Westerdahl, 2023). At the same time, the Swedish company Akelius (later purchased by Heimstaden) has become a major player in the rental market in Germany and elsewhere, taking advantage of rent regulation and renovation needs. For example, in Sweden, they can bypass rent regulations by increasing the standard of the apartments (thus renovating to a standard far above the actual needs). However, the rent-controlled nations in Europe now seem saturated for those large investors. Denmark recently passed legislation limiting the scope for international capital to invest in Danish multi-family housing. In 2021, Akelius declared that they would leave Sweden, Denmark, and Germany for more profitable and less regulated markets outside the EU.

Housing and welfare regimes

Overall, we find that the housing and welfare regimes of Nordics and the German-speaking countries do certainly reflect some common underlying cultural values that have created similar organizing principles reflected in both the welfare and housing regimes.

Both the social democratic and the Bismarckian regimes have catered for a rather comprehensive welfare state in the post-war decades. The wider welfare regimes have embraced highly regulated labour markets with strong unions and centralized wage bargaining and have encouraged high labour security and high welfare levels based on earnings-related social insurance systems if falling outside the labour market. The institutional arrangements have distinctive distributional outcomes: inequality and poverty rates have historically been low. Thus, even if the social democratic and conservative countries stem from different political ideas, it becomes evident how the Bismarckian welfare regime has roots in social democratic beliefs and logic. As we have argued that the distributional outcomes of the wider welfare regime define the "boundaries of possibility" (Stephens, 2020a) of housing, this cluster of countries would promote a housing system where the state corrects market limitations to the extent that housing consumption is not a complete replication of purchasing power, but that housing is more evenly distributed across households with diverse socioeconomic status.

However, as the recent development depicted in this book shows, the "poster child" countries of northern Europe seem to be losing their distinctive characteristics. Labour markets are increasingly deregulated, the gig economy is on the rise and wealth is accumulating among the richest. Combined with large numbers of unskilled immigrants, this has led to rising polarization where income inequality is increasing at a rate that is among the fastest in the world. As in the liberal regimes, it seems that housing systems amplify current and lifetime income inequalities, "both in terms of housing wealth and terms of housing affordability and quality" (Wetzstein, this volume).

The homeownership ideal, promoted by policies in housing and housing finance, has supported house prices and increased the role of mortgages, most notably in the Nordic countries. Arguably, these countries constitute a certain "growth regime". Reisenbichler (2021) argued that the growth regimes in the Nordic countries traditionally have not been reliant on house prices, but rather on high-tech exports and services. Thus, the financialization of housing markets would neither reinforce nor contradict the Nordic growth regimes. However, the recent development of prices in the homeownership sector, and the increasing interest from international investment capital actors, searching for investments in rental housing with quickly increasing market values in Sweden and Denmark, might lead us to the question of whether housing financialization is not becoming an important ingredient in the growth regimes of the Nordic countries (Figures 17.2 and 17.3).

One might conclude that there has been an attempt by governments to encourage ownership as a means to make households invest in their (and their children's) future lives, but as in many liberal regimes, the stratification has increased as younger people are forced into market renting and face much higher lifetime housing costs, creating a certain kind of "Generation Rent" (Hoolachan et al., 2017; Grander, 2023).

The commodification of housing is seemingly evident in all countries portrayed here. The divergence between the Nordic countries is, however, interesting. Kemeny classified Norway and Finland as homeownership societies, and Denmark and Sweden as cost-rental societies. However, in many countries in recent decades, governments have sought to boost homeownership through tax concessions, demand-side support and discounted sales of social rental housing (especially in Denmark and Sweden). The stratification resulting from such a strategy is evident – those who have not been able to compete in the housing-as-an-asset race have been left behind in public

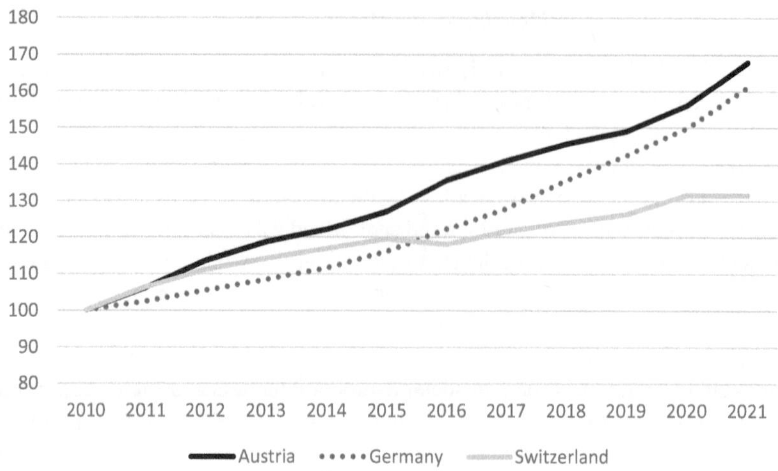

Figure 17.2 Nominal house prices in the Germanic countries (2010–2021; 2010 = 100).
Source: Calculated from EMF (2022) *Hypostat*, Table 18.

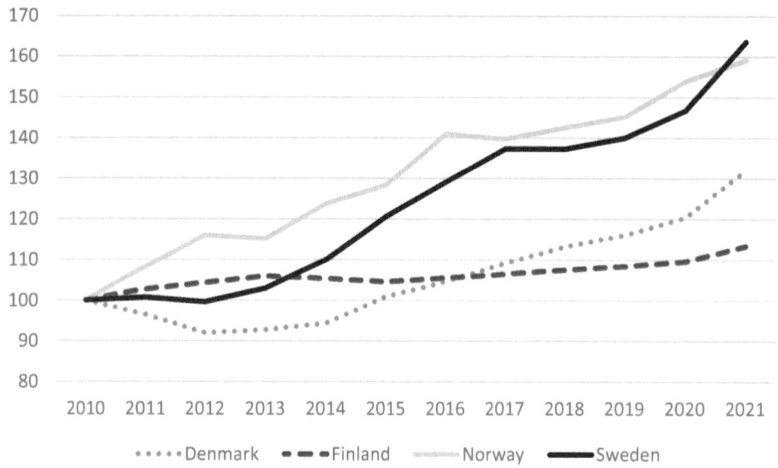

Figure 17.3 Nominal house prices in the Nordic countries (2010–2021; 2010 = 100).
Source: Calculated from EMF (2022) *Hypostat*, Table 18.

housing which is struggling between financial demands that conflict with social responsibility. The pricing out of many households and their reliance on renting at a considerably higher cost than owning clearly points to commodification.

As in many other contexts, homeownership must be seen as having become increasingly commodified as a result of central bank and governmental (intentional or unintentional) actions. However, the post-pandemic inflationary crisis caused house prices to fall by 20 per cent, leaving many households who invested recently with large mortgages, with high interest rates and property values well below the purchase prices. These households are also being squeezed by high general inflation and energy costs.

While Sweden has been clinging to the universal principles in the rental housing market, European legislation on competition has made it impossible for the cost-rental sector to maintain its cost-rental principles. Consequently, public housing is now governed and managed on strictly business terms, competing with the private rented sector. Public housing companies are thus implementing stricter rental policies with income barriers excluding low-income households from the tenure which in principle should be open to all. The decommodifying effects of housing allowances are very limited. Moreover, part of the public rental stock is transferred to international investors. Ultimately, the affordable housing stock diminishes while still, no means-tested social housing sector is on politicians' agendas. In parallel, the Swedish central bank – and the government – have directly and indirectly encouraged homeownership. This combination of regulations and financialization has created what Christophers (2013) called a "monstrous hybrid", the consequences of which are evident in the current financial crisis.

Denmark is faring batter in the current crisis but was on the other hand hit harder by the GFC than its neighbours, which arguably was the result of decades of increasing mortgages in the wake of tax reliefs in the 1990s, where the central bank also issued amortization-free housing loans.

Similarly, in Norway, the increasing market-based production, exchange and distribution of housing has contributed significantly to greater economic inequality and led to a reduction in the number of affordable homes for low- and medium-income households in urban areas. This explains why the central state and local governments have recently tried to respond to calls for more affordable homes in the owner-occupied and rental sector, where alternative models of affordable rental and owner-occupied housing are on the agenda of both political parties and civil society actors.

The link to the wider welfare regime in all three Nordic countries is obvious. The universal approach of the welfare states exists in principle, but in practice, it is clearly withering. The housing regimes still cling to the ideals of social democratic universalism – but operate under impossible circumstances dictated by a market-based system. Finland, with its combination of housing construction subsidies, well-established social housing system and high levels of expenditure on housing allowances, seems to better reflect a social democratic welfare regime, albeit being clearly dualist in its housing provision strategy.

The Germanic countries seem to fare slightly better in the transition, although there are clear indicators of commodification and financialization of both welfare and housing. The forms of financialization seen in many other contexts: institutional investment in rental properties, the growth of platform capitalism (Airbnb) and equity investment in social renting are evident also in Germany and Austria. Moreover, there are concurrent challenges in the potential overburdening of welfare commitments, demographic changes, an upcoming stall in housing construction, and the subsequent affordability/cost-of-living crisis. Consequently, there might be grounds for considering the "boundaries of possibility" to have actually widened. The Germanic countries indeed offer some reference points in this regard, not least the often-cited model of social housing in Vienna. Just like the Finnish model, this system of social housing was born in times of widespread poverty, acute housing need and the near absence of effective welfare. However, the model has survived and adapted to a well-developed welfare regime. The common thread is the large financial expenditure that both Vienna and Finland are directing to their housing systems in the form of social housing subsidies and housing allowances.

Formally, housing policy is clearly decentralized in this cluster of countries. Municipalities (in the Nordic countries and Austria) and states (in Germany and Switzerland) have responsibility for housing provision. However, as previously noted in this book, the wider welfare regime is defined largely at the nation-state level, which sets the "boundaries of possibilities". So too – in reality – is housing policy, as national governments set the general rules such as the legislative framework;

subsidies; and important aspects governing the tax treatment of different tenures. Central banks typically have the responsibility for meeting centrally set consumer price inflation targets, mainly through prevailing interest rates. Of course, within the Eurozone, the inflation targets are set across the whole currency area and the implementation of monetary policy is the responsibility of the supranational European Central Bank. Taken together, the distribution of responsibilities constrains the freedom of housing providers, municipal politicians and – most importantly – households. Thus, the hybridity in the housing-welfare regime type becomes evident. Yet the organizing logic – the essence of a regime – remains broadly the same.

Conclusions

Like the Anglosphere countries, the countries described here capture the essence of regime theory: how organizing logic derived from cultural dispositions shapes both wider welfare and housing regimes. However, recent developments seem to be moving these countries towards a more liberal regime type, regardless of their history. Between the Nordic countries, the welfare and housing regimes are diverging. The cost-rental social democratic countries of Sweden and Denmark seem, on the one hand, to be more commodified than the homeownership and dualistic rental regime of Finland as the ownership sector in Finland is less mortgaged and thus less financialized, and poorer households gain priority in accessing social housing. On the other hand, the Finnish dualist rental market with its needs-tested principle for housing the lower income groups is possibly more stratifying than the rental market in the neighbouring countries. Norway is an outlier with a very rich population. However, stratification is still evident as the small but increasing share of low-income households is directed into an increasingly stigmatized rental sector.

Germany appears to be following Sweden's example, whereby financialization has left indelible imprints on the housing system. In contrast, Austria, like Finland, has instead continued to rely on large state financing of the construction and management of social housing. Such state financing has arguably resulted in greater decommodification than in Germany and – especially – Sweden. This might give grounds for fearing a movement towards the kind of stratification often associated with the stigmatization and segregation of social housing estates in other contexts, such as the UK. However, lower levels of poverty in both Finland and Austria and the generous upper-income ceiling for social housing in Austria reduce this risk. On one level the countries can be grouped together as reflecting social democratic norms and principles in the welfare state. On another level, there are clear differences between financing, the discourse on homeownership and the role of cost rental housing. As stated in the Anglo-sphere section: employing regimes in comparative research necessitates comparing within a cluster as much as between clusters.

Southern European Countries: Italy and Spain

The Southern European countries have tended to be grouped together, as they share more than merely geographical proximity. Three of the countries – Italy, Spain and Portugal – share Catholic heritages and three – Spain, Portugal and Greece – were governed by authoritarian dictatorships until the 1970s, whilst Italy also had a recent history of fascist government. Economically, the southern European countries urbanized later than their counterparts in northern Europe. Some commentators (e.g., Allen, 2006) note that urbanization was often prompted by the collapse of agriculture rather than the growth of industry. Whilst this is doubtless a simplification, a shorter and less intense Fordist era helps to explain why these countries fit less easily into the logic of the welfare and housing regimes mainly developed in northern Europe.

Whilst Italy was a founding member of the European Economic Community (now Union) the establishment of democracy in Spain, Portugal and Greece was followed by accession in the 1980s. EU membership served as both a symbolic and material boost for modernization – not least through generous financial transfers through structural and cohesion funds that helped to modernise infrastructure. Italy, Spain and Portugal were founder members of the European Single Currency (Euro) in 1999 with Greece – controversially – being admitted in 2001. Euro membership in theory at least committed these countries to fiscal conservatism and they experienced falls in interest rates by joining a currency area dominated by countries such as Germany with historically low inflation. This meant that adjustment to economic shocks could no longer be absorbed through devaluation against fellow Euro countries, placing huge deflationary pressures on labour markets.

The Global Financial Crisis, therefore, had a disproportionate impact on the southern European countries. Greece, Portugal and Spain each became embroiled in the sovereign debt crisis that at times appeared to endanger the currency. The crisis was felt most acutely in Greece, but all three of these countries required financial support from the "troika" of the European Commission, European Central Bank and the International Monetary Fund as they struggled to borrow from the markets. Funding was conditional on austerity and structural reforms, for example, to labour markets. Although all three countries have left the support mechanisms, they – and Italy – have been tested by the pandemic. Such macro-level drivers, which include the ultra-low interest rates that prevailed between the GFC and the inflationary upswing in 2022, have played an important role in shaping the housing systems of southern Europe.

The wider welfare regime

Esping-Andersen (1990) placed Italy in his (large) group of conservative/corporatist countries in which market efficiency is eschewed in favour of the preservation of status, and with very limited redistribution. However, reflecting on the post-war period during which the Christian Democrats were the dominant party, he also noted the importance of the Church in shaping the conservative social values that shaped the welfare regime and the family as a central provider of welfare. These tendencies were also prominent in the welfare regime in Spain that emerged under Franco's dictatorship, characterized as being "corporatist authoritarian" (García-Lamarca, this volume). Both countries developed regulated labour markets that afforded relatively high levels of employment protection for the core (male) labour force, which also contributed to the development of informal labour markets without these protections. Core groups also benefited from social insurance systems including pensions available with relatively early retirement. The core labour market/social insurance nexus left those outside these core markets (notably women, and in recent decades non-European immigrants) with only weak levels of social protection – with a lack of comprehensive social assistance and dependence on charities often related to the Church. With family benefits encouraging non-employed motherhood, the notion of subsidiarity left the state with a notionally residual role.

Ferrera (1996) highlighted the role of clientelism (the receipt of state support in exchange for political support) which is often attributed to relatively "soft" states where support schemes are vulnerable to non-meritocratic practices. Notably in Greece, access to state jobs has been a key feature of the system, highlighted during the sovereign debt crisis. Allen (2006) outlines the links between access to core employment and a wider system of family kinship. She emphasizes the importance of having a family member in core employment who not only has a secure income and can access state support but can also assist the wider family through the contacts they enjoy. She also posits that the system of welfare derived from kinship and family may serve to break the link between income poverty and social exclusion.

Table 17.7 Income inequality and poverty in the southern European welfare regimes

	Gini	*Poverty*
Italy	0.33	14.2
Spain	0.32	14.7
Portugal	0.31	10.6
Greece	0.31	11.5
Ave.	0.32	12.8

Source: OECD (2023) Income inequality (indicator); and OECD (2023) Poverty rate (indicator).

Both countries struggled with unemployment and regional disparities, although safety nets have been expanded within the context of the EU which requires Members to develop social assistance safety nets. Nonetheless, the "familial" nature of welfare in these countries prevails and is called upon during both the GFC and the pandemic, particularly when younger adults cannot access secure jobs or the housing market.

Unsurprisingly, the southern European countries exhibit relatively high levels of inequality after taxation and cash transfers are taken into account. The Gini coefficient average of 0.32 is slightly lower than in the Anglo-sphere countries where it is 0.35, and there is little variation between the southern European countries. Relative poverty is also similar to the Anglo-sphere countries, but it is notably higher in Italy and Spain than in Portugal and Greece (see Table 17.7). But as we have seen, assessing the effectiveness of welfare on cash incomes alone may be misleading.

Housing regimes

Spain and Italy, along with other southern European countries, exhibit high levels of homeownership in comparison to those in the Anglo-sphere and northern Europe. However, these countries are not simply defined by the dominance of home ownership, but by its nature. Both countries experienced appalling housing conditions at the end of the Spanish Civil and Second World Wars, as the countries urbanized. These countries demonstrate how broadly similar welfare regimes can embrace housing systems that have evolved in distinctive ways.

The role of cost rental housing

Cost rental housing has never played a significant role in Italy and Spain, or indeed in the other southern European countries. Whilst EU-SILC records between 5.7 per cent (Greece) and 10.5 per cent (Portugal) of people living in "reduced rent" or "free" housing, this categorization is problematic in that it does not reliably reflect "cost rental" or "social rented" housing. We have therefore added figures for social/public housing from Housing Europe's (2022) country profiles in its *State of Housing in Europe* report. These are expressed as percentages of the (occupied) stock of housing, rather than individuals or households, but confirm the marginal role played by cost rental housing.

Both Spain and Portugal had large state-subsidized building programmes during the post-WW II decades. However, these focused on promoting homeownership. The *case populari* scheme launched in Italy in the 1970s promoted social rented housing. Consistent with its small scale and high levels of poverty and inequality, it can be characterized as performing an "ambulance service"

Table 17.8 Housing tenure in southern Europe

	Italy	*Spain*	*Greece*	*Portugal*
Owner occupied (mortgaged)	13.8	27.9	11.2	34.2
Owner occupied (unmortgaged)	61.1	49.4	64.8	44.4
Owner occupied (total)	74.9	77.3	75.9	78.7
Market Rented	16.8	14.6	18.4	10.8
Reduced rent/free	8.3	8.1	5.7	10.5
Rented total	25.1	22.7	24.1	21.3
Social/Public rented	3.8	1.1	n.a.	2.0

Source: Rows 1–6, EU-SILC; Row 7, Housing Europe (2022).
Note: First six rows: % individuals (2020). Seventh row (Social/ public housing): % housing stock. Italy (2011), Spain (2020), Portugal (2019).

function in that it was targeted at people on low incomes, and tenants had to leave if their incomes rose above qualifying levels. The Spanish social rented stock was largely built in the 1950s and 1960s, often at low standards and with cheap materials. It was later subject to sales to tenants and its revival did not feature as the country transitioned to democracy (McCrone & Stephens, 1995).

Alves (2017, p. 244) has suggested that Portugal represents "an 'exotic variation' of the dualist rental system". Yet the system is characterized by "an overwhelming share of owner occupancy…" (ibid.) set beside a very small cost rental sector. Allen (2006, p. 274) argued that marginalized rental sectors can, in part, be attributed to "the absence of a significant Fordist stage in economic development in the south…" This distinctive political economy means that the marginal role of cost/ social rented housing means that Kemeny's unitary/dualist schema is of limited relevance to explaining southern European housing systems.

The sphere of production

Both Spain and Italy launched subsidized housebuilding programmes in the post-World War II decades. As we have noted, these primarily supported private-sector housebuilding and homeownership. Both programmes exhibited tenets of corporatism and were weakly decommodifying, but strongly stratifying in terms of linking access to status.

The Italian INA-Casa Plan of 1949, as well as other housebuilding programmes, were based on social insurance principles and hence funded by contributions from workers' employers' and the government which were paid into a national insurance institute (INA). The structure of support ensured that the segmented/dualist nature of the labour market was replicated in the housing market. In the 1970s, public sector programmes featured, but houses were primarily sold at fixed or regulated prices. Rates of subsidized housebuilding fell with the first oil crisis and later support became more targeted on areas requiring renewal and on individuals.

Those people who were unable to access the subsidized housing schemes were left with a choice between free market rentals or self-building. The latter became a defining feature of housing production in southern Europe (with the exception of Spain) and provided one of the key linkages between the housing system and the wider welfare regimes with its reliance on familialsm and clientelism. Illegal building became commonplace to the extent that some 800,000 people were estimated to inhabit irregular self-built dwellings in Rome in the 1980s (Leonardi, this volume). As is typically the case in countries with large amounts of informal housebuilding, regularization programmes played an important, if erratic, role in shaping the system.

Table 17.9 Annual average building rates per 100,000 population in southern Europe 2017–2021

	Building rate
Italy	138
Spain	158
Portugal	138
Greece	87
Ave.	130

Source: Calculated from EMF (2022) *Hypostat*, Table 17.14.
Note: There is no data for Italy for 2017–2021. Italy's figure is for the single year 2016.

In Spain, self-building/promotion was more marginal after "Franco brokered a deal with large contractors, based on a strong town planning system, which reduced risk and increased profits in return for their political support" (Allen, 2006, p. 252). Large-scale subsidized housebuilding programmes from the 1950s lasted into the 1980s as the country began to turn towards more market-based solutions. During the period of subsidized housebuilding, state banks were used to create privileged sources of finance for construction and mortgage lending. Although subsidized housebuilding in Italy was part of a broadly Keynesian approach to the economy, It formed part of a developmental economic system in Spain, connected to the opening of the Spanish economy to foreign investment in the 1960s. As García-Lamarca (this volume) notes, "… it was during the later part of the Franco dictatorship (1960s–1970s) that the highest levels of state housing support existed". The main schemes under successive Housing Plans were known as VPO (officially protected housing) from the late 1970s, with VPT (housing at controlled prices) added in the 1990s. Access to housing and the support available depended on income, so stratification was baked into the system, and poorer households unable to access homeownership relied on the private rented sector which was heavily regulated until the mid-1980s after which it became highly liberalized.

A key difference between Italy and Spain was the way in which Spain transitioned to a market-orientated system but with housebuilding as a key economic driver. Helped by generous tax reliefs on the consumption of housing, major deregulation of land use in 1998 and financial deregulation, Spain achieved extraordinary levels of housebuilding in the years leading up to the Global Financial Crisis. According to García-Lamarca (this volume), some 6 million units were built between 1997 and 2008. The crash created so-called "ghost towns" of unsellable dwellings. Subsequently, the levels of housebuilding in Spain have fallen and although the highest of the four southern European countries in Table 17.9 above, these are internationally low.

The sphere of consumption

Housing in Southern Europe is consumed primarily through homeownership. In both countries, but especially in Spain, the state subsidized housebuilding, but from the 1980s onwards support began to shift towards the demand side.

The VPO/VPT system in Spain became more targeted over time, with subsidies directed more at purchasers through favourable mortgage interest rates and grants rather than simply at developers. In Italy, Leonardi (this volume) notes, that "the neoliberal turn has not been 'simply' a question of cutting back" and suggests that whilst access to owner-occupation is still supported, housing allowances (*fondo sociale per l'affitto*) have been introduced for tenants.

These shifts away from the high point of corporatist-influenced support for housebuilding probably create systems that are less formally stratified by weakening the link between core employment and state support. Arguably, they have also enhanced commodification in the sense that labour market income has become more closely aligned with access to housing that is secured through the mortgage market and market rentals. In Spain, the private rented sector was extensively deregulated in the mid-1980s by the Boyer decree, with both tenure security diminishing and rents being set by the market. The recent tightening of rent regulation in Catalonia marked only a brief reversal as after two years this regional law was overturned by the Spanish constitutional court.

As we noted at the outset of this section, the nature of homeownership is key to unravelling southern European housing systems. Outright (unmortgaged) homeownership is the largest single tenure in the four countries in Table 17.8 with more than 60 per cent of the populations of Italy and Greece living in this sub-tenure. Particularly given these countries' histories of self-build and underdeveloped mortgage systems these levels of outright ownership are likely to represent a high degree of decommodification, with the important caveat that mortgage payments are not the only cost associated with ownership, and repairs and maintenance remain. Spain and Portugal have rather higher levels of mortgaged ownership, which leads us to the role of finance.

The sphere of exchange

Historically, southern Europe was associated with underdeveloped mortgage markets, and to an extent that remains true of Italy and Greece, which both exhibit relatively low levels of mortgaged homeownership and outstanding mortgage debt. Yet as we have seen, an organized mortgage system was developed in Spain under Franco to create a privileged and closed circuit of finance for housing and liberalized in the 1980s after which the state mortgage bank (BHE) was privatized (Table 17.10).

The housing building and price boom that took place in Spain from the 1990s until the financial crash was closely allied to the expansion of mortgage credit within the broader context of falling interest rates globally and as a result of membership of the Euro. The process of financialization goes hand-in-hand with the commodification of housing in that income to a large extent determines how much can be borrowed. Yet wider access to finance also implies greater access to housing, although the macro-level feedback of higher prices means that over time it becomes more restrictive for future would-be purchasers. García-Lamarca (this volume) characterizes the era of mortgaged ownership and rising house prices as a form of "debtfare", but it has come to an end, at least in Spain.

Table 17.10 Mortgage debt: GDP (%) in southern Europe

	2010	*2021*
Italy	21.9	23.1
Spain	104.1	40.4
Portugal	63.8	46.5
Greece	35.9	16.9
Ave.	56.4	31.7

Source: EMF *Hypostat*, Table 17.8.
Note: No breaks in time series noted.

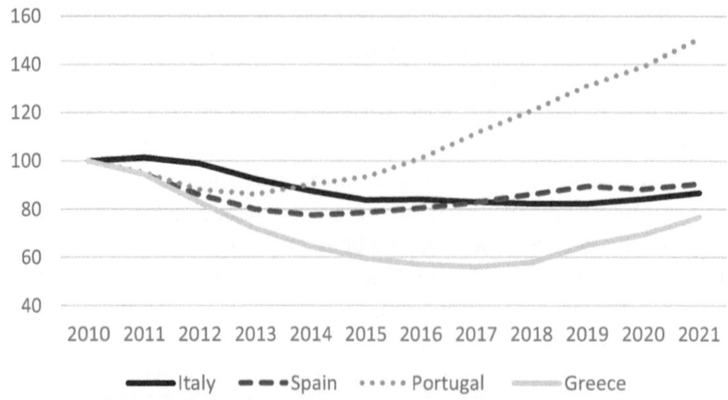

Figure 17.4 Nominal house price trends in southern Europe (2010–2021; 2010 = 100).
Source: Calculated from EMF (2022) *Hypostat*, Table 18.

Mortgage debt as a share of GDP, according to the European Mortgage Federation, has fallen in Spain from more than 100 per cent of GDP in 2010 to 40 per cent in 2021, and as we have seen, levels of housebuilding have become subdued (Figure 17.4).

Even nominal house prices in Spain as in Italy and Greece were lower in 2021 than in 2010 (with the rises in Portugal making it something of an outlier), suggesting something of a definancialization of owner-occupied housing since the GFC. However, the rental market has exhibited increased financialization with the development of Real Estate Investment Trusts, for example.

Housing and welfare regimes

Italy was the only southern European country to be included in Esping-Andersen's (1990) study, where it was categorized as being "corporatist" in nature. Subsequent scholarship identified distinctive characteristics among southern European countries which are now generally accepted to form a regime in their own right. Yet the geographical tag does not capture the regime element and the term "rudimentary" does not capture the dynamics of welfare in these countries. Their defining characteristics lie in underlying cultural values allied to the (Catholic or Orthodox) Church which place store on the patriarchal family, alongside subsidiarity that limits (or perhaps more accurately defines) the role of the state in welfare provision. The state helps to mould a segmented labour market with a protected inner core, which can access the patchwork of social insurance benefits that are only structurally similar to those in northern Europe – their limited coverage setting them functionally apart. Historically weak social assistance systems make interdependence between extended family members an important element in this system that only weakly decommodifies labour market income through the state, and is highly stratified in state provision.

The housing systems that emerged within the familial regimes reflected the same values. The states in Italy and Spain promoted homeownership under the same slogan, "a country for homeowners, not proletarians". State-subsidized housebuilding programmes were focused on expanding homeownership. A funding system based on social insurance in Italy and the structuring of support in Spain meant that core labour market workers were privileged and those outside it found it harder to access housing. Gaps in the system in Italy were filled by self-build – itself dependent on family networks and links between core and peripheral labour market members – much of it illegal and subject to regularization later. In Spain, the fascist government's clientelist links with developers and a stronger planning system limited the levels of self-build with outsiders reliant on a regulated private rented sector.

In both countries, governments turned away from subsidizing housebuilding and targeted subsidies including tax relief more on purchasers. Again, we see divergent approaches, as newly democratic Spain also embraced economic liberalism, deregulating the financial and planning systems, and, within the context of falling interest rates, facilitated a huge building boom that was a core part of Spain's economic system until the Global Financial Crisis.

As in other countries, there are now less clearly delineated housing strategies. The experience of Spain makes its appearance in Schwartz and Seabrooke's (2008) "familial" quadrant of countries that combined high levels of owner-occupation with low levels of mortgage debt curious. In terms of "growth models" Spain had clearly embraced one that depended on finance and house prices, but crucially – and in contrast to some countries in the Anglo-sphere – on the construction sector.

Conclusion

Strategically clear models are more difficult to detect as cultural values change, and political systems evolve to occupy less clearly ideological positions. Path dependence is evident in tenure structures where outright homeownership still provides the most obvious source of decommodification in housing, along with intra-family support. Spain sought to turbocharge its economy through a rapid transition to a financialized housing system, but is now experiencing a form of definancialization an alternative to which has yet to be devised. Italy might be thankful that its housing system – perhaps also reflecting a population less enthusiastic about embracing debt – avoided the boom/bust scenario experienced in Spain, but it is still to define a housing strategy that takes it beyond the legacy of familialsm.

The post-socialist economies

In this section, we have grouped together the countries with post-socialist housing systems: Russia, Hungary (with reference to other central European countries in the Visegrad 4) and China. This grouping includes both "upper middle" income (China, Russia) and "upper" income (Hungary/Visegrad 4) countries according to the Worlds Bank's classification. However, they are united in having developed socialist economies and housing systems on the Soviet model, and subsequently to have transitioned away from them. The grouping, therefore, provides a case study in the intra-regime comparisons and evolution within the context of the convergence/divergence debate.

The wider welfare regime

The wider welfare regimes in the socialist economies were part of an economic strategy of industrialization and urbanization, especially in the Soviet Union/Russia and China which were essentially agrarian economies at the time of the revolutions. This involved the suppression of consumption to support industrial (and military) investment – which contributed to the persistent shortages of consumption goods. It also involved the creation of a distinctive and coherent wider welfare regime, the core elements of which were created in the Soviet Union and exported to central and eastern European countries such as Hungary, and China.

The wider welfare regime was centred to an even greater extent than in capitalist models on employment and the workplace, with an assumption of full employment (unemployment not being recognized). Wages were low (sometimes called the "individual" – as opposed to the family – wage system) and there was a greater dependency on the provision of goods and services either at very low prices (food, utilities) or in-kind. "Social security" benefits derived from payroll deductions did play a role, but due to the centrality of the workplace pensions were the most significant.

These systems created very low levels of income inequality, but there was greater scope for differential access to the highly subsidized services that could form part of a "merit" system that allowed for management responsibility, party loyalty and military service to be rewarded (Hegedüs, this volume). There were also gaps in the system, which left room for private actors and family support – but whether the overall effect of these was to reduce or exacerbate inequality is a moot point (ibid.). The role of the control of population movement in both China and Russia/Soviet Union through registration requirements (*hukou* in China, *propiska* in the Soviet Union) was also important as a line of stratification, especially in China where different systems operate for urban and rural residents.

Transformation came rapidly in Russia and Hungary, due to the collapse of political Communism; in China, reform of the welfare regime within the context of economic transformation has occurred alongside continuity in the political system.

China's remarkable emergence as the world's second-largest economy was dependent on its rapid urbanization associated with raising productivity. This transformation from a centrally planned economy emerged as part of the wider globalization of the world economy, symbolized by China's membership of the World Trade Organisation in 2001. However, whilst the Chinese economy was liberalized in some ways, the transformation was gradual compared to Russia's shock therapy – it began with industrialization in rural areas which exerted competition on the state-owned enterprises before a cities-based strategy was adopted in the 1990s (Ren, 2013). Further, a school of research emphasizes the foundations laid by Chinese institutions that, for example, had already created high levels of literacy and life expectancy. With the state retaining control over large parts of economic activity, China's economic model is evidently not a simple case of "neoliberalism". Deng Xiaoping's characterization of "socialism with Chinese characteristics" differs only in emphasis from David Harvey's notion of "neoliberalism with Chinese characteristics" (quoted by Ren, 2013, p. 5). The rapid growth of the Chinese economy and associated rapid urbanization was founded on exports to global markets. However, during the Global Financial Crisis, the growth model switched towards an investment model in which local governments sold land-use rights to property developers and established Local Government Financing Vehicles (LGFV), which themselves issued bonds, and invested in infrastructure. Property development became the key driver of the economy, but the Chinese government's fears of collapsing property companies have led to controls being introduced that have contributed to slowing construction and it is widely believed that the economic model will need to change (Kynge et al., 2022).

As Miao (this volume) records, the evolution of the Chinese wider welfare regime began with the shift of responsibility for welfare from work units in the 1970s alongside economic change during which pension costs soared as early retirement was used to mitigate the impacts of redundancies. Indeed, the opening up of the Chinese economy and the attraction of international investment has entailed a departure from the Soviet-style system. Since 2000, social insurance schemes have been established to cover the usual "risks" of earnings loss or interruption (old age, ill-health, injury, unemployment, maternity), but are not universal, reflecting different rights arising from urban and rural and migrant residency. A social assistance safety net – also not universal – is determined locally. Although China has developed "an extensive system of public welfare", Miao (this volume) notes that "family, kinship and network fill important and substantial gaps".

Russia emerged from the collapse of the Soviet Union during which hardship was experienced broadly in society as price controls were abandoned and the guarantee of full employment was lost.

The country also adopted the neoliberal policies advocated by the World Bank and IMF and did so as part of a "shock therapy" strategy, which hastened the collapse of the socialist system. The economic system based on state-owned enterprises collapsed with enterprises – many of which represented major state assets – privatized and helping to create a class of super-rich "oligarchs". Living standards fell and as soon as growth had returned the country experienced a currency and sovereign debt crisis in 1998. The subsequent development of the economy has depended on the exploitation of natural resources, notably oil and gas, and their export.

In this chaotic transformation responsibility for welfare was necessarily shifted from the work units/state-owned enterprises to create a more mixed economy of welfare with social insurance systems being created to address the usual risks, although with private provision providing top-ups. Given Russia's demographic trends, it is unsurprising that pro-natalist policies have been introduced (e.g., maternity grants and subsidies for daycare), although pensions dominate social security spending. Crucially, as Kalyukin and Puzanov (this volume) observe, the "family remains an institution of welfare in Russia", with large numbers dependent on support from others and a high proportion of multi-generational families.

Hungary, along with the other formerly socialist states, experienced rapid economic transformation, although not so precipitously as did Russia/former USSR. Its evolution has involved its integration into the European Union and its Single Market (though not as yet, membership of the Euro). The chapter by Hegedüs (this book) suggests that initially, ideological debates produced reforms that could be thought of as being "social democratic" whilst others appeared "neoliberal" whilst the role of the family (e.g., multigenerational households) is clearly important. Since 2010, the country has been governed by a form of "populist authoritarianism" under which the state "mobilises welfare provision in service of legitimising and maintaining political power" (Hegedüs, this volume). This has created some "unorthodox" policies such as family benefits that are designed to be regressive, the apparent renationalization of pension funds as a means of gaining budgetary control, and the squeezing out of NGOs and local authorities from social care provision in favour of aligned religious organizations. Whilst the new system does not readily match any of Esping-Andersen's regime types, populist authoritarianism aimed at protecting an insecure middle class to cement political power, in our view, does represent an organizing principle with some parallels to Franco's Spain (Table 17.11).

Interestingly, levels of relative poverty in Hungary, the other Visegrad 4 countries and Russia are below 10 per cent, so lower than in the Anglo-sphere, Southern Europe and indeed the

Table 17.11 Income inequality and poverty in the post-socialist economies

	Gini	*Poverty rate*
Hungary	0.286	9.2
Czech Rep	0.248	5.6
Poland	0.281	9.8
Slovakia (P)	0.222	7.8
Visegrad 4 Ave.	0.259	8.1
Russia	0.317	11.5
China	0.514	28.8

Source: OECD (2023) Income inequality (indicator); and OECD (2023) Poverty rate (indicator); Russia and China = OECD Income distribution database.

Note: (P = Provisional). China = 2011; Russia = 2017. See Table 17.1.

"corporatist" countries. Levels of income inequality are similarly low in the Visegrad 4, although somewhat higher in Russia. China is an outlier with very high levels both of income inequality and poverty – a long way above any of the other groupings in this book, apart from South Africa and Chile. However, some care should be taken with the application of relative poverty figures to countries with very different income levels. Whilst China's levels of income inequality have risen steeply under economic reform, rapid and sustained economic growth has also taken almost 800 million people out of "extreme poverty" (defined by the World Bank (2022) as US$1.90 per day).

Housing regimes

Housing systems in Russia/USSR, China and Hungary were organized under the same organizing principles during the high-point of the Soviet period, by which we mean the period after 1956 in the cases of the USSR and Hungary when mass housebuilding programmes began. Before then, production was constrained by the political priority attached to industrial and military investment. A key change was the end of private property. Expropriation of existing properties in urban areas and shared (communal) apartments – in part a consequence of urbanization under acute scarcity – as well as an expression of collectivism. Interestingly "personal" property was permitted, and indeed in the Soviet Union the state-supported self-build. But "personal" property did not carry the same rights as "private" property so "ownership" essentially removed the right to buy and sell property on the market. In a controlled market, the asset element was removed so ownership rights were really reduced to a form of occupancy rights.

In the "high" Soviet period after 1956, production was driven in the Soviet Union by large enterprises and local Soviets and self-build became illegal in larger cities. Much of the same logic was followed in China. However, in Hungary, private housing was permitted alongside the large-scale rental programmes – but materials were nonetheless controlled by the state as part of the planned economy. A finance system was required and was provided by a state savings bank at very low interest rates. However, finance was not provided on the risk-based model that prevails in capitalist systems. With weakening economies from the 1970s, the "high Soviet" model came under increasing pressure, which in the case of Hungary led to a more mixed economy of provision. Hegedüs suggests that in the early 1970s, only 53 per cent of investment in state-controlled housing came in the form of subsidy. More than one-third came from loans and 12 per cent from household savings. Production was therefore dominated by the state in Russia, and led by the state in Hungary.

In socialist countries, the state also determines the distribution and hence the consumption. Access was in part determined by "merit" which meant privileging party membership, military service and managerial responsibility. Ironically, with a decommodified production process, in the sphere of consumption, a degree of *recommodification* occurred when managerial responsibility – which was barely rewarded in the wage system – could be rewarded through the housing system. Thus, the role of "public" housing was inverted compared to practices in capitalist economies. Even in Hungary where needs-based allocations based on income were introduced after 1971, state-owned enterprises and political organizations could purchase allocations, maintaining the role of "merit". Tenants enjoyed very high levels of security and could pass on tenancies to children. Moreover, informal sub-letting meant that markets of a sort operated. Such were the levels of security that renting in the socialist system has been characterized as being a form of "quasi-homeownership". Hegedüs (this volume) characterizes the Hungarian system as being a "dual market", which in a sense it was. However, Stephens et al. (2015) have argued that quasi-homeownership for renters combined with the stripping out of many property rights for owners led to a form of "unitary" housing system (though clearly not through the mechanisms that Kemeny described).

The notion of a uniform system of state housing is therefore somewhat misleading, especially in Hungary. The apparent disjuncture between urban housing systems in Russia/USSR and China dominated by public rental housing whilst Hungary exhibited an ownership rate of 72 per cent, is less real in reality – the attributes of tenure are crucial in understanding its "meaning". As Miao (this volume) argues, housing was "a key component of the urban welfare system".

Transformation in Russia and Central Europe: Spheres of consumption, production and exchange

The early years of transformation in Russia and the formerly socialist countries in Europe such as Hungary were dominated by a desire to establish market mechanisms along the lines advocated by the World Bank and other international agencies. This entailed the reintroduction of property rights, the "giveaway" privatization of municipal and work-unit housing and efforts to establish risk-based mortgage finance systems.

Privatization helped to move the post-socialist countries towards being "superhomeownership" states, with outright ownership being the dominant sub-tenure. A common effect of privatization was to carry over socialist-era inequalities (relating to allocation by "merit") into the post-socialist era. This also weakens the link between current income and housing consumption, although the effect must be diminishing over time. Public or social rented sectors are generally small and marginal, whilst market renting is often small (Table 17.12).

State support is directed towards subsidizing consumption, and there are some similarities between Russia and Hungary in this respect. Subsidies are not targeted at the poorest but fall within the parameters of the wider authoritarian populist welfare regime. Pro-natal policies regarding housing support operate in both countries, and subsidies often favour the middle classes. The design of relief for households trapped with foreign exchange mortgages also favoured the better off (see below). Gaps in state support, also reinforced by inadequately operating market institutions, are partly filled by family support, including multi-generational households. As Kalyukin and Puzanov (this volume) comment, for the least well-off households, "Intergenerational transfers remain the primary means of improving housing conditions… while the depopulation trends in the majority of Russian regions only emphasises the role of such transfers".

With the state having largely withdrawn from construction, supply is largely through the private sector. The statistics suggest varying levels of housebuilding, including apparently high levels in Russia, but information in this area is sparse (Table 17.13).

Table 17.12 Housing tenure in the Visegrad 4 and Russia (%)

	Hungary	*Czech Rep*	*Poland*	*Slovakia*	*Russia*
Owner occupied (mortgaged)	13.4	17.8	11.4	19.8	–
Owner occupied (unmortgaged)	78.7	61.1	75.4	73.1	–
Owner occupied (total)	92.2	78.8	86.9	92.9	81.5
Market rented	4.1	15.4	3.2	5.7	12.3
Reduced rent/free	3.7	5.7	10	1.4	6.2
Rented total	7.8	21.2	13.1	7.1	18.5

Source: EU-SILC 2021, except Russia = Kalyukin & Puzanov, this volume.
Note: % individuals; 2021, except Slovakia; 2020 and Russia; % of urban floor area.

Table 17.13 Annual average building rates per 100,000 population in the Visegrad 4 and Russia, 2017–2021

	Building rate
Hungary	209
Czech Rep.	320
Poland	544
Slovakia	361
Ave.	358
Russia	789

Source: Calculated from EMF (2022) *Hypostat*, Table 17.14.

Table 17.14 Mortgage debt: GDP (%) in the Visegrad 4 and Russia

	2010	*2021*
Hungary	24.7	9.5
Czech Rep	15.3	25.5
Poland	18.6	19.4
Slovakia	15.8	38.8
Ave.	18.6	23.3
Russia	3.0	10.1

Source: EMF (2022) *Hypostat*, Table 17.8.

One of the most intriguing aspects of the post-socialist countries is the subdued size of the mortgage markets, which reflects the relatively small size of the mortgage ownership sector with no more than one in five people living in households with a mortgage in the Visegrad 4. In these "superhomeownership" states, it is outright ownership that dominates.

Levels of mortgage debt are also low by international standards – the average in the Visegrad 4 is lower than in any other set of country groupings for which comparable data is available, and lower still in Russia. There is variation, however, with mortgage debt reaching nearly 40 per cent of GDP in Slovakia (Table 17.14).

Although much effort went into the establishment of mortgage systems, there are good reasons for them to have developed slowly. One is that relatively few people needed a mortgage due to the "giveaway" privatization of public rental housing. Another reason is the context of relatively weak demographic demand. Zavisca's (2012) study of Russia identified cultural resistance to mortgage debt as another factor contributing to what she termed "property without markets". As Kalyukin and Puzanov (this volume) comment, "most housing has been privatised, but never fully commodified". However, even in Russia, mortgage debt has risen over the past decade, and Kalyukin and Puzanov (this volume) suggest that more than half of transactions are supported by a mortgage.

The reduction in the level of mortgage debt in Hungary is likely to have arisen from the salutatory experience of foreign exchange loans (often denominated in Swiss francs) which became disastrously expensive when the forint devalued. The government's populist response was to force the banks to take the hit by allowing borrowers to prepay mortgages at advantageous exchange rates (Hegedüs, this volume) (Figure 17.5).

Nominal house prices have recovered following the GFC, although in most of these countries, the recovery did not begin until around 2014.

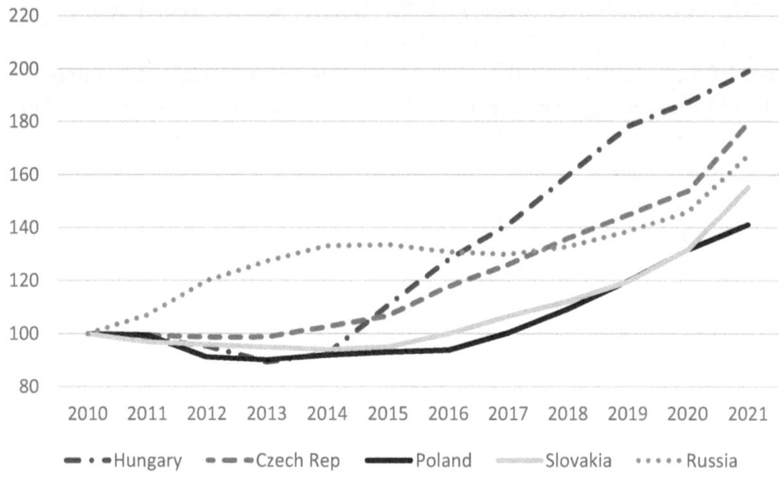

Figure 17.5 Nominal house prices in the Visegrad 4 and Russia (2010–2021; 2010 = 100).
Source: Calculated from EMF (2022) *Hypostat*, Table 18.

Transformation in China: Spheres of production, consumption and exchange

As in Russia, Hungary and other formerly socialist countries, China largely privatized its stock of public housing through sales to tenants at very low prices. Work units continued to build housing until this was stopped in the late 1990s. However, China's housing regime has evolved in ways that set it apart from the formerly socialist countries in Europe. This is partly explained by the country's rapid urbanization within the context of economic models founded on the export of manufacturing goods and since the GFC on investment, with property playing a key role.

Land in urban China (as in other southeast Asian countries) remains in public ownership, which places the state (in this case local governments) in a pivotal position in the sphere of production. It also accounts for why aspirations for forms of subsidized housing (e.g., Economic and Comfortable Housing, ECH) were limited. Local governments derive at least one-third of their revenue from the sale of land use rights, which made them reluctant to forgo revenue by reserving land for ECH.

The state retains a large role in housing production, especially with the ambitious public house-building targets adopted after 2010, but the emergence of private property developers has become a central part of the Chinese economic model, especially after the GFC. The housing construction sector through its contribution to employment and GDP plays a much larger role than in Western countries (Kynge et al., 2022). The model of development is heavily debt-driven, with loans raised from both domestic sources and international investors (including Blackrock and HSBC). The Chinese banking system remains under state ownership. Properties are sold to individual purchasers before they are completed. Concerns about over-leveraged property companies caused the government to introduce "three red lines" (debt ratios) that limited the ability of property companies to borrow domestically. At the time, more than half of companies did not meet these criteria and this resulted in a slow-down in the sector, causing many projects to be abandoned before completion – leading to protests from people who had purchased them in advance, often depending on family savings. There are high levels of vacant dwellings (between 10 and 40 per cent in different areas) and some 90 million dwellings in "ghost cities". The precarious nature of this property-driven economic model was highlighted when in 2021 Evergrande, the second largest developer, defaulted on offshore debt, and (in 2023) was subject to restructuring.

The legacy of the socialist system means there is a legacy of largely decommodified housing that was subject to "giveaway" privatization. However, the extent of urbanization in China since the socialist housing model was ended means that the system must contain a higher proportion of commodified housing than is the case in the formerly socialist European countries. Affordability ratios almost defy comprehension (an estimate suggested that "average" house prices in Beijing were 50 times the annual median household income, compared to 15 in London and 10 in New York (Financial Times, 2022)). Whilst owner-occupation remains the largest tenure, it is not clear how large the private rented sector is, partly because it is largely uncontrolled.

Conclusions

Russia, Hungary and China exhibited housing regimes governed by the same organizing principles during the socialist period. Then these housing systems were located firmly within the wider welfare regime to the extent that one can say they were part of it. It is more difficult to interpret the post-socialist period. Beyond the common practice of "giveaway" privatization, and the establishment of market institutions in Hungary and Russia, there is clear divergence with China. However, a common effect was to remove housing from being a central part of the welfare system.

In Russia and Hungary, housing might still be seen as existing with the same broad wider regime framework of populist authoritarianism whereby state support seeks to enhance the state's legitimacy by supporting a precarious middle class. This is exhibited in the way in which state subsidy and support in housing is directed towards such households, whilst less emphasis is placed on supporting the poorer households. Inadequately functioning markets and gaps in state support also encourage reliance on the family as a source of welfare, including housing welfare.

China's development has been quite different, and this is attributable to the way in which the country has undergone rapid urbanization. As Miao (this volume) argues housing has been separated from general welfare provision, it has become a central part of economic strategy, especially since the financial crisis and an investment-orientated policy adopted. This may be interpreted as exemplifying China's "developmental orientation" (Miao, this volume), but it is surely on a different scale than has been experienced in other southeast Asian economies. Certainly, the state is a leading actor, but the extent of reliance on debt accumulated by property developers that is now seen as a risk by the government makes it rather different from Singapore or Korea. It is also seen as being an economic strategy that has run its course, so further change can be expected.

South Africa and Chile

As discussed in the previous chapters, South Africa and Chile do not fit easily into the hitherto existing templates of welfare or housing regimes.

South Africa was not included in Esping-Andersens's characterization of welfare regimes but could arguably be seen as having liberal tendencies due to its policy of means-tested welfare subsidies and a large degree of stratification, harkening back to the apartheid regime. Chile has seen a radical political transition from corporatism via socialism, through Pinochet's dictatorship from 1973 to 1990, to the Concertación governments after that. Overall, this has led to the development of a welfare state characterized by hybridity, with clear elements of both (neo)liberal and social-democratic regimes. With regards to housing, though, we argue that both regimes have clear and increasing tendencies of Kemeny's homeownership societies although there are some key differences which we will discuss using the analytical framework of spheres of production, consumption, and exchange.

The wider welfare regime

The Chilean welfare state has had clear roots in corporatist principles, developed in the 1930s and after. The welfare regime transitioned gradually towards a social democratic model during the 1960s, as the socialist government expanded the state's role in key areas of the economy and set out to better include previously excluded groups in the welfare system. The state's role in economic development and the politically controlled incorporation of private actors advanced further from the mid-1960s under a social democratic moniker. The military dictatorship led by Pinochet took a complete reverse direction with the deregulation of the economy and neoliberal adjustment policies. When the dictator left power in 1990 after a referendum in 1988 where the Chileans voted no to an extension of his presidential term, the centre-left Concertación took power and searched for a return to social democracy, albeit based on a free market – a strategy referred to as "growth with equality" (Abufhele & Angelcos, this volume).

The welfare regime of today is thereby characterized by hybridity. Social-democratic principles of substantial welfare coverage, although directed at the most disadvantaged sectors of the population, try to balance neoliberal policies that encourage markets and private profit. According to Abufhele and Angelcos (this volume), the current rationale of the welfare regime includes a redistributive social principle and paternalist assistance to the poor. This combination of social democratic and liberal rationales has created hybrid practices that have been the base for social policies in all areas, housing included. The hybridity has resulted in an increase in redistributive social spending and the expansion of paternalist assistance to the poor. As discussed in the previous chapter, Chile's wider welfare regime does, however, seemingly mitigate income inequality to a lesser extent than most other OECD countries.

South Africa's welfare regime is, as described above, intrinsically connected to the apartheid legacy. Even if South Africa has – by far – the most extensive welfare system in Africa, the welfare benefits are minor and underdeveloped compared to other OECD countries. Inequalities in the labour market due to discrimination are evident and the wider welfare regime has a small impact on reducing such inequalities. The exclusion of black people is a historical burden for the welfare regime. As pointed out in the country chapter (Mmbadi et al., this volume), there are substantial degrees of stratification and commodification in the South African welfare system. All welfare benefits are means-tested and only those individuals or households with low incomes qualify for such grants. The health sector is heavily commodified – costly health insurance from private providers guarantees high-quality care for those who can afford it, while public services provide less care of poorer quality. Thus, there are reasons for placing South Africa in the group of liberal welfare regimes. As noted by Mmbadi et al. (his volume), though, there are elements of universalism in welfare distribution.

Evidently, South Africa is considered one of the most unequal countries in the world, and poverty and inequality figures are rising. The major cause is the inequality arising from the labour market, not the least because of the past exclusion of black people under apartheid and the heritage of this in the current labour market. Seemingly the wider welfare regime plays only a small role in reducing such consequences (Table 17.15).

While South Africa has a significantly higher Gini and Poverty rate than Chile, both countries have higher rates than the other countries in this book and the OECD as a whole. With an average Gini of 0.446, the two countries are closer to the liberal welfare regime than the conservative and social-democratic countries.

Table 17.15 Income inequality and poverty in Chile and South Africa

	Gini	*Poverty rate*
South Africa	0.618	27.7
Chile	0.460	16.5
Ave.	0.446	17.4

Source: OECD (2023) Income inequality (indicator); and OECD (2023) Poverty rate (indicator).

Note: see Table 17.1.

Table 17.16 Housing tenure in Chile and South Africa

	Chile	*South Africa*
Owner occupied (mortgaged)	15.4	n.d.
Owner occupied (unmortgaged)	52.8	n.d.
Owner occupied (total)	68.2	67
Private rented	17.9	18
Social rented/granted	7	–
Other	7	15

Source: Country chapters this volume, OECD.

Notes: Chile (2002): households, South Africa 2020: households. Chile has no social rental sector; "granted" = subsidized means-tested owned housing.

The housing regime

The current housing regime in Chile bears clear marks of the past 50 years of history, where the changes introduced during the military dictatorship (1973–1989) have made the largest imprints (Table 17.16). Before that, the Christian Democrat government (1964–1970) had established the Ministry of Housing and Urbanization, still present today, and became famous for launching a "basic needs" housing programme consisting of the provision of a serviced plot with a bathroom and kitchen.

The Pinochet military regime made several market-based institutional reforms to the planning and housing systems. The trajectory of market provision has continued under subsequent regimes, albeit with the ambition to balance liberal reforms with the return of social democratic ideas. If the welfare regime has become somewhat hybrid, as described above, the housing regime seems to have developed with clear dualist tendencies and a focus on homeownership through state subsidies mainly focusing on the poorest households.

The liberal tendencies in the national housing policy include a social housing subsidy programme which included the privatization of social housing construction, the commodification of urban land through deregulation of the land market, and what is referred to as a general financialization of housing provision (Abufhele & Angelcos, this volume). The reforms over the decades have clearly increased the focus on homeownership. Such long-term processes in the Chilean housing policy have during the last decade coincided with increased migration in Latin America, a political crisis in Chilean institutions and economic constraints arising from the COVID-19 pandemic.

The South African housing regime has seen a considerable shift in housing policy due to race-based regimes. As described, homeownership was during the apartheid regime limited for the black population until 1983. While the post-apartheid regimes have encouraged homeownership among blacks, the housing regime is still clearly segmented when it comes to housing tenure and race. Homeownership policies have had a small success in increasing the wealth of people of colour.

61 per cent of the urban population in South Africa are homeowners, while the clear majority of the remaining urban population rents privately. The social rented sector is marginal and there is a substantial number of households who live with families or in the backyards of existing housing and do not pay rent.

Kemeny's categorization of dualist versus integrated systems is difficult to apply to Chile and South Africa as the cost rental housing sector is non-existent in Chile and small – indeed almost negligible – in South Africa. Instead, both countries are clear examples of homeownership societies where the government steers all households – also the poorest – to homeownership. The social housing sector is therefore mainly built upon the idea of subsidizing households to buy or self-build owned housing.

In Chile, a homeownership-based social housing system with a differentiated subsidy programme provides basic security for around 7 per cent of households. The subsidy system helps households to – according to their relative needs – get access to housing, but also determine the location and typology of the housing to which they are entitled. Accordingly, the model is contributing to the stratification of low- and middle-income households (Abufhele & Angelcos, this volume).

The South African cost rental sector has been built up by the post-apartheid governments starting with the introduction of a social housing programme in 1994. Independent third-sector social housing institutions provide rental housing by means-testing, where local governments determine the rents. While the system is clearly dualist, the use of third-sector institutions for social housing provision has been inspired by the Dutch approach to public housing where the government supported independent social housing institutions (Mmbadi et al., this volume). Indeed, due to the historical ties between the Netherlands and South Africa (the Dutch colonized the Cape of Good Hope in 1652, which then oscillated between Dutch and UK rule until the independence of South Africa in 1931), Dutch consultants have for a long period provided support to the South African government in housing issues. The establishment of the cost-rental sector has however been very slow. As noted in the South African country chapter, only about 35,000 houses have been constructed in the last 25 years and the social housing programme remains smaller than its subsidized homeownership, where the government has facilitated ownership of between 3 and 4 million houses.

The sphere of production

Housing production in the two countries is based extensively on market principles. State support for housing construction in Chile is aimed predominantly at promoting social housing for ownership. In South Africa support is provided in a more universal way through general subsidies for housing construction.

There is no reliable data on housing construction in the two countries, but they both display interesting examples of state and market interplay. In Chile, the post-dictatorship decade (1990–2000) became famous worldwide and is still regarded as a role model in developing countries for its improvements in housing, not the least regarding housing production for the poor. As described by Abufhele and Angelcos (this volume), Chilean housing policy was in the 1990s regarded as "best practice" and "showpiece of social policy" in many parts of the world. The country might be the clearest example of the international housing programs emerging in the developing world, often dubbed the "enabling markets housing approach", which was promoted by international organizations such as the World Bank and the U.S. Agency for International Development (Gilbert, 2002). Indeed, because the Chilean housing strategy was in many cases adopted by other Latin

American countries, it can be argued that the Chilean housing model was a precursor of the enabling markets approach to housing across the region (Navarro, 2005).

Since the regime shift in the 1990s, governments have further strengthened the model's capacity, increased housing production, and sought to diversify the housing programmes to better target the urban poor. Compared to the state-led models of housing present before the dictatorship regime, there was a strong emphasis on economic efficiency, which also was reflected in housing production. Social housing construction increased, and the housing shortage was reduced considerably. Despite its international reputation, the system of housing construction eventually became subject to criticism, not the least for its "peripheral location of housing projects, the resulting segregation, the deficient quality of housing and the state-regulated channels of participation in the design and construction of housing projects" (Abufhele & Angelcos, this volume). More recently, increasing rates of migration, the 2019 social uprising in Chile and the economic crisis generated by the COVID-19 pandemic have further stressed the capacity of the system of housing production. Nonetheless, the overarching principles for housing production introduced in the 1990s have remained the main standard until today.

In South Africa, housing production is based on state financing through a fixed capital subsidy for housing construction. The state is a major landowner, although there is a large variation between urban and rural housing. Overall, state landownership linked to such capital subsidies has been a prominent feature of successful efforts to promote homeownership in urban areas. While urban land is mostly owned by the state, land ownership in rural areas varies. Commercial farmers have land ownership, but generally, few farmworkers enjoy property rights (Mmbadi et al., this volume). In most cases, households have occupancy rights but do not own the land. Thus, despite the national policy focus on homeownership, many people living in houses built with the support of the state housing subsidy have yet to acquire ownership of their homes, as the deeds have not been registered resulting in the houses remaining the government's property. Housing production is financed in four ways, by mortgage finance to middle- and high-income households; by micro-lending (microfinance); and by government funds aimed at the least affluent households. Finally, many households finance their own self-built houses.

Recent decades have seen a decline in housing production due to decreasing funds available for housing coupled with inflation. The state does not intervene in any price setting for housing. The private sector involvement in construction has varied over time and as of today, the focus is on large-scale developers and mega-projects. As noted by the chapter authors, the application of the fixed amount of capital subsidy has resulted in the location of housing projects on the urban fringes, where land values are the lowest. There is thus a certain degree of commodification and stratification within the housing production system, as the fixed amount of subsidy together with a focus on housing size has "reinforced historical urban patterns" and thus "perpetuated the racial structure of the apartheid city" (Mmbadi et al., this volume).

The sphere of consumption

The demand-side support of housing in Chile is connected to the social housing system through subsidies for the acquisition of housing by eligible applicants. According to Navarro (2005), 4 per cent of housing subsidies in 2004 were aimed at construction while 96 per cent of the resources were targeted at subsidy programs.

According to the "enabling markets" approach, the state has encouraged a private housing market for the poor population to facilitate the supply side in form of expansion of the private sector in housing provision and the demand side through subsidies for the acquisition of housing by eligible applicants. The demand-side support is financed through a three-tier system: a capital subsidy provided by the government, the beneficiary's own savings and a mortgage taken out by the beneficiary. These subsidies are targeted at people living below the poverty line and cannot access the private housing market by their own means. For the extremely poor, "Debt-free social housing" exists, meaning that no mortgages are required, although this model has required a small amount of savings. There have also been subsidy programs designed for low-income people above the poverty line, where private banks have agreed to deliver mortgage loans for housing valued with low or moderate costs, as well as a subsidy for slightly more expensive houses to promote mixed-income units in private housing projects. Those programs have been less extensive.

The system of targeted subsidies proved effective in meeting the housing needs of low-income groups. It was regarded as a major welfare improvement for the extremely poor. Despite criticisms, the housing construction system has, therefore, remained in place.

The criticism has focused on the system's ability to segment and stratify access to housing. As discussed by Abufhele and Angelcos (this volume), it has on the one hand reduced public spending and increased effectiveness as the model has served to secure a housing demand that provided incentives for construction companies to cater for low-income groups. On the other hand, the programme was supposed to enable low-income families to participate in the market, but those households have not been able to choose the location of the house but are assigned a housing unit based on the availability of housing projects, mainly on peripheral low-cost land.

Thus, despite its social-democratic ambitions, the housing policy has remained weak in counteracting the model's segregating and stratifying effects. As Abufhele and Angelcos (this volume) observe, several improvements have during the last decade tried to integrate social housing projects into the urban fabric through the implementation of a conditional subsidy based on the location of housing projects, thereby creating incentives for developments with a better location within the city. Yet, this effect of the subsidy has been weak, since the costs of building on better-located land proved incompatible with the profitability for construction companies required for the model to function. Thus, the model is still contributing to socio-economic segregation.

In South Africa, demand-side support is – like Chile – limited to state subsidies for housing construction as described above. Only a very small percentage of state housing funds goes to rental housing. A housing allowance exists but is limited to public officials. The *Government Employees' Housing Scheme* provides a housing allowance which is paid as a monthly contribution to qualifying employees to assist with recurring housing costs. The allowance is applicable for homeowners and tenants but with a lesser amount for households renting their dwellings (Government of South Africa, 2022).

The sphere of exchange

The housing financing production system in Chile during the past decade has reflected the government's social democratic orientation of a liberal system, emerging since the fall of Pinochet. According to longstanding tradition, housing policies have encouraged "the dream of home ownership" through the combination of subsidies for social housing construction for low- and middle-income households.

As described, Chile is an interesting example of the "Enabling markets housing approach" emerging in the developing world during the 1970s. As part of this approach, the international development institutions advocate a "basic needs" strategy in developing countries, which includes the provision of sites and services, slum upgrading, and core housing and slum upgrading. As described, there are no government-backed efforts to develop, construct, and sell houses utilizing loans featuring subsidized interest rates. Rather, Chile started several reforms in the financial sector and in social housing programs, among which was the creation of the first program in the world to subsidize the demand to buy housing. The current housing finance policy has origins in reforms in the early 2000s, which stopped giving mortgage loans and gave up the direct construction of houses, with the notable exception of the social housing subsidy discussed above.

In South Africa, the large-scale housing subsidy programme, which has provided about 3.2 million houses over the past 25 years shows, according to Mmbadi et al. (this volume) how dominant the government has been in the process of housing financing. The private sector and individuals are directly involved in the construction process, which involves state financing in three ways. First, mortgage finance from the state-owned National Housing Finance Corporation financial institution, based on affordability criteria. The government established the institution in the 1990s to make housing affordable to middle-class households. Here, the state does not intervene in any price setting for housing. Second, micro-lending is a common way of financing housing for less affluent households outside the urban areas. The Rural Housing Loan Fund provides micro-finance for small-scale housing extensions. Third, the government funds housing solutions for the poorest households through a capital subsidy which entitles the target group with a small home of about $40m^2$, or through a social housing programme that runs through social housing institutions.

Housing and welfare regimes

Both countries display interesting examples of the tandem of housing and welfare regimes. As the Chilean welfare regime has shifted from a conservative/corporatist through a neoliberal to a hybrid regime with social democratic influences, the housing system has followed suit. While the housing system built during the decades has gained much international attention, it still is heavily commodified and seems to create certain stratification, also through the segregating effects of the state subsidies.

Abufhele and Angelcos (this volume) identify three phases in the relationship between housing and welfare that have formed the current housing system. The corporatist era was characterized by housing policy being seen as a key area for promoting both economic and social development – housing was understood to be a social right to which every family should have access regardless of socioeconomic status (in practice, though, most housing units built by state subsidies during this period were for the middle classes). The dictatorship regime era transformed the housing regime in line with a market-oriented blueprint, reducing the state's participation to a subsidiary role according to the principles of supply and demand, where the state took on the task of contracting out the construction of low-income housing to private companies and allocating it to eligible beneficiaries. As described by Abufhele and Angelcos (this volume), the dictatorship regime's interest in tackling poverty was mostly driven by a need to provide incentives for the functioning of low-income markets and housing wealth, in line with other "growth regimes" pictured in this book.

The current phase of housing in Chile reflects the hybrid welfare regime that has emerged. It has remained neoliberal in its basic tenets, particularly as regards the state's subsidiary role. Housing production remains private and continues to be regarded as a key activity for the economy. Housing is also still regarded as a commodity and social groups are channelled into socioeconomically segmented and segregated markets. State housing subsidies are distributed selectively, not universally, in contrast to the education and health systems that have begun to move towards "universal" rights of access.

Social democratic principles have although made a comeback in the later stages of thc post-dictatorship regimes. Public expenditure on housing has increased. The welfare regime's orientation has shifted from a "poverty approach" to a vulnerability and rights-based approach, which also applies to the housing regime. As the poverty rate has dropped, the focus of the welfare state has shifted to a social protectionist approach to guard against families falling back into poverty. As discussed by Abufhele and Angelcos (this volume), the social housing subsidy has functioned as a form of asset-based welfare as it funds a durable good. The subsidy counts as an asset and reduces monthly expenditure on housing, freeing up income for other necessities.

The current Chilean housing regime and its crisis-ridden tendencies are described by Abufhele and Angelcos (this volume) as "[the] exhaustion of a neoliberal-oriented model of housing production, provision, and distribution". Housing production in the last three decades has been high, but simultaneously created homogeneously poor and segregated neighbourhoods contributing to social exclusion. The social housing system with targeted subsidies has reinforced the stratification of middle and low-income households. As discussed, changes in tenure have thus been small, albeit housing construction and the availability of social housing have increased. The critique against the housing policy has been consistent since the 1990s and in the 2000s the national authorities acknowledged the problems. However, the housing deficit peaked in 2017 and housing informality has also increased steadily since the mid-2000s. Land prices have increased, and the value of subsidies is regarded as insufficient to build low-income housing units in urban areas. Thus, although the production of new housing has been comparatively high, the current housing deficit and informality rates, reinforced by the COVID-19 pandemic and its subsequent economic crisis have put the escalating housing crisis on the agenda, also thanks to the pressing demands of social movements.

The South African development of housing and welfare is not as evident as in the Chilean case but is heavily dependent on the aftermaths of the Apartheid regime. While the stratification between the white and black population was evident for many decades, the current system has searched to mitigate such inequalities through an extensive welfare system unknown to the rest of Africa. The emerging welfare regime is clearly liberal in its way of targeting the poor by means-testing, continuing the model's history of stratification. The housing regime has seen a similar transition by liberal means, as housing subsidies have declined rapidly – and focus on housing as a tradeable asset in a market where households should actively participate in order to create wealth. Thus, government housing subsidies are meant to provide a platform for poverty alleviation and wealth creation, but the results are still enigmatic and mostly provide evidence of a highly commodified housing regime, which Mmbadi et al. (this volume) note "continues the historical social stratification created under apartheid and colonial rule".

To conclude, while the Chilean housing production system was thought to articulate both neoliberal and social democratic principles effectively, the housing regime appears to mirror such hybrid characteristics, but with a clearly liberal bias as the "dream of home ownership" increasingly gains ground. Despite its social-democratic ambitions, the current housing regime has remained weak in counteracting segregating and stratifying effects. The development after COVID-19 has however further stressed the capacity of the system of housing production, and rising rents, leading some renters to seek lower-cost housing solutions in the informal market, revealing what the authors call an "exhaustion" of the market-led housing model, which has failed to mitigate the inequalities. Similarly, the current precarious position of South Africa's public finances paints a negative outlook for the housing regime. The authors argue that the former high level of subsidization is unlikely to resume and that the first signs of cost-cutting measures are already visible. One potential outcome is less funding for the starter homes and a return to previous models that were despised by the first post-apartheid government.

Conclusions

In this section, we have scrutinised the relationship between housing and welfare in Chile and South Africa. Both countries display an interesting history of radical governmental regimes, in South Africa the apartheid regime and in Chile the military dictatorship of Pinochet. The subsequent governments have searched for strategies for combining liberal market ideas with social democratic principles, with mixed results in housing provision. Historically and present, the two countries do not fit very well in Esping-Andersen's categorization of welfare regimes or Kemeny's categorization of rental markets. As argued in the South African chapter (Mmbadi et al., this volume), the relationship between welfare and the housing regime in South Africa is too complex to label neoliberal or government-driven. The same applies to the Chilean case., where the authors argue that "welfare regimes and housing policies do not respond to uniform political and economic orientations, but to rationales of government between which there is tension" (Abufhele & Angelcos, this volume).

Using Stephens' (2020b) framework of production, consumption, and finance, we have found that the two countries share a similar strategy for promoting homeownership to all kinds of households – also the poorest – through extensive governmental funding, both with an aim of a "growth strategy" and as a means for social housing for the poorest population. The rental sector is almost entirely market-based. A social rental sector does not exist in Chile and is very marginal in South Africa. Thus, the governmental focus has been on subsidies for constructing (supply side) or purchasing (demand side) rather than housing allowances. In Chile, the governmental funding for homeownership is limited to households with low income, constituting a social housing sector of subsidized homeownership – for the least affluent mortgage free – while the South African homeownership subsidies have covered a larger part of the population.

While both countries periodically have been successful in increasing housing production and mitigating the limited access to housing for low-income households, the historical heritage of segmentation, stratification and segregation still applies. The Chilean authors argue that

> Although the state continues to advance towards better-quality social housing and has striven to expand access and improve the policy's efficiency, it continues to treat housing as a commodity governed by market forces and to provide stratified and selective access to housing subsidies for middle and low-income groups.
>
> (Abufhele & Angelcos, this volume)

In the South African chapter, Mmbadi et al. (this volume) argue that the apartheid heritage is a hard burden which is manifested in large differences in homeownership and landownership between the black and white population, also connected to a clear. rural-urban divides. They pose a pessimistic outlook when they ask, "[i]n one of the more unequal countries in the world, the question is whether it is possible to find alternatives".

References

Allen, J. (2006). Welfare Regimes, Welfare Systems and Housing in Southern Europe. *European Journal of Housing Policy, 6(*3), 251–277. https://doi.org/10.1080/14616710600973102

Alves, S. (2017) Poles Apart? A Comparative Study of Housing Policies and Outcomes in Portugal and Denmark, *Housing, Theory and Society, 34*(2), 221–248, https://doi.org/10.1080/14036096.2016.1236036

Bengtsson, B. (Ed.) (2013). *Varför så Olika? Nordisk Bostadspolitik i Jämförande Historiskt Ljus*. Égalité.

Bengtsson, B., & Grander, M. (2023). *Bostadsfrågan som Politik och Intressekamp*. Egalité.

Christophers, B. (2013). A Monstrous Hybrid: The Political Economy of Housing in Early Twenty-first Century Sweden. *New Political Economy, 18*(6), 885–911. https://doi.org/10.1080/13563467.2012.753521

Esping-Andersen, G. (1990). *The Three Worlds of Welfare Capitalism*. Princeton University Press.

Eliasen, S. Q., Vestergaard, L., Sigurjonsdottir, H., Turunen, E., & Penje, O. (2020). *Rural housing challenges in the Nordic Region* (Report 2020:7). Nordregio. Available at: https://vbn.aau.dk/ws/portalfiles/portal/402017903/rural_housing.pdf [Accessed: 12 April 2023].

EMF. (2022). Hypostat 2022.

Enström Öst, C., Söderberg, B., & Wilhelmsson, M. (2014). Household Allocation and Spatial Distribution in a Market Under ("Soft") Rent Control. *Journal of Policy Modeling, 36*(2), 353–372. https://doi.org/10.1016/j.jpolmod.2014.01.002

EU-SILC. (2021). *Population by Tenure Status, Level of Activity Limitation and Poverty Status* (product code hlth_dhc010). https://ec.europa.eu/eurostat/web/products-datasets/-/hlth_dhc010

Ferrera, M. (1996). The 'Southern Model' of Welfare in Social Europe. *Journal of European Social Policy, 6*(1), 17–37. https://doi.org/10.1177/095892879600600102

Financial Times. (2022). *Evergrande: The End of China's Property Boom, video*, 4 October. Available at: https://www.ft.com/video/2fce1e2e-52a1-4882-868e-c18802793e8a [Accessed: 12 May 2023].

Gilbert, A. 2002. Power, Ideology and the Washington Consensus: The Development and Spread of the Chilean Housing Policy. *Housing Studies, 17*(2), 305–324. https://doi.org/10.1080/02673030220123243

Government of South Africa. (n.d.) *Government Employees Housing Scheme*. http://www.gehs.gov.za/?p=Housing%20Finance%20Access%20Service&q=Housing%20Allowance [Accessed: 12 April 2023].

Grander, M. (2017). New Public Housing: A Selective Model Disguised as Universal? – Implications of the Market Adaption of Swedish Public Housing. *International Journal of Housing Policy, 17*(3), 335–352. https://doi.org/10.1080/19491247.2016.1265266

Grander, M. (2019). Off the Beaten Track? Selectivity, Discretion and Path-Shaping in Swedish Public Housing. *Housing, Theory and Society, 36*(4), 385–400. https://doi.org/10.1080/14036096.2018.1513065

Grander, M. (2023). The Inbetweeners of the Housing Markets – Young Adults Facing Housing Inequality in Malmö, Sweden. *Housing Studies, 38*(3), 505–522. https://doi.org: 10.1080/02673037.2021.1893278

Grander, M., & Westerdahl, S. (2023). Financialization the Swedish Way: A Case Study on How Public Housing is Replaced by the International Investment Portfolio of Vonovia. Manuscript Accepted for Publication. *Journal of Urban Affairs*.

Hoolachan, J., McKee, K., Moore, T., & Soaita, A. M. (2017). 'Generation Rent' and the Ability to 'Settle Down': Economic and Geographical Variation in Young People's Housing Transitions. *Journal of Youth Studies, 20*(1), 63–78. https://doi.org/10.1080/13676261.2016.1184241

Housing Europe. (2022). *State of Housing in Europe 2021*.

Kadi, J., Banabak, S., & Schneider, A. (2022). Widening gaps? Socio-spatial Inequality in the "Very" European City of Vienna Since the Financial Crisis. *Cities, 131*, 103887. https://doi.org/10.1016/j.cities.2022.103887

Kemeny, J. (2006). Corporatism and Housing Regimes. *Housing, Theory and Society, 23*(1), 1–18. https://doi.org/10.1080/14036090500375423

Kohl, S., & Sørvoll, J. (2021). Varieties of Social Democracy and Cooperativism: Explaining the Historical Divergence between Housing Regimes in the Nordic and German-Speaking countries. *Social Science History, 45*(3), 561–587. https://doi.org/10.1017/ssh.2021.16

Kynge, J., Yu, S., & Hale, T. (2022) China's Property Crash: 'A Slow-Motion Financial Crisis'. *Financial Times*, 4 October.

McCrone, G., & Stephens, M. (1995). *Housing Policy in Britain and Europe*. UCL Press.

Mundt, A., & Amann, W. (2010). Indicators of an Integrated Rental Market in Austria. *Housing Finance International, 25*(2), 35–44.

Navarro, M. (2005). Housing Finance Policy in Chile the Last 30 Years. *Land Lines*. The Lincoln Institute of Land Policy.

OECD. (2023a). *Household Debt (indicator)*. https://doi.org/10.1787/f03b6469-en

OECD. (2023b). *Income Inequality (indicator)*. https://doi.org/doi: 10.1787/459aa7f1-en

OECD. (2023c). *Poverty Rate (indicator)*. https://doi.org/10.1787/0fe1315d-en

Reisenbichler, A. (2021). Housing Finance between Social Welfare and Growth Strategies. In A. Hassel & B. Palier (Eds.), *Growth and Welfare in Advanced Capitalist Economies* (pp. 320–347). Oxford University Press.

Ren, X. (2013). *Urban China*. Polity Press.

Ruonavaara, H. (2017). Retrenchment and Social Housing: The Case of Finland. *Critical Housing Analysis, 4*(2), 8–18. https://doi.org/10.13060/23362839.2017.4.2.382

Schwartz, H., & Seabrooke. L. (2008). Varieties of Residential Capitalism in the International Political Economy: Old Welfare States and the New Politics of Housing. *Comparative European Politics, 6*, 237–261. https://doi.org/10.1057/cep.2008.10

Stephens, M. (2020a). How Housing Systems are Changing and Why: A Critique of Kemeny's Theory of Housing Regimes, *Housing, Theory and Society, 37*(5), 521–547. https://doi.org/10.1080/14036096.2020.1814404

Stephens, M. (2020b). Towards a Multi-layered Housing Regime Framework: Responses to Commentators. *Housing, Theory and Society*, *37*(5), 584–596. https://doi.org/10.1080/14036096.2020.1816573

Stephens, M., Lux, M., & Sunega, P. (2015). Post-Socialist Housing Systems in Europe: Housing Welfare Regimes by Default? *Housing Studies, 30*(8), 1210–1234. https://doi.org/10.1080/02673037.2015.1013090

Tunström, M., Liliegreen, C., Granath Hansson, A., Kurvinen, A., Nordahl, B. I., & Sveinsson, J. R. (2020). *Building affordable homes: Challenges and solutions in the Nordic Region* (Report 2020:2). Nordregio. https://doi.org/10.6027/R2020:2.1403-2503

Westerdahl, S. (2021). Yield and the City: Swedish Public Housing and the Political Significance of Changed Accounting Practices. *Critical Perspectives on Accounting, 80*(October 2021). https://doi.org/10.1016/j.cpa.2020.102161

World Bank. (2022). *Lifting 800 Million People Out of Poverty – New Report Looks at Lessons from China's Experience*. Press release, 1 April 2022. https://www.worldbank.org/en/news/press-release/2022/04/01/lifting-800-million-people-out-of-poverty-new-report-looks-at-lessons-from-china-s-experience [Accessed: 15 May 2023].

World Bank. (n.d.) *Population, Total– OECD Members*. https://data.worldbank.org/indicator/SP.POP.TOTL?locations=OE [Accessed 12 May 2023].

Zavisca, J. R. (2012). *Housing the New Russia*. Cornell University Press. https://doi.org/10.7591/9780801464300

18

CONCLUSIONS

Welfare regimes in the 21st century: From labelling to explaining

Martin Grander and Mark Stephens

This book has examined housing systems from across the world. Its principal focus has been on the relationship between housing and welfare. While it undeniably focuses primarily on countries in the Global North, it has sought to include some countries from the Global South. The geographical scope is somewhat broader than in most studies, which allows us to consider whether frameworks developed in the Global North have a broader application.

The analysis of the country chapters has been conducted within the extended welfare regime framework proposed by Stephens (2020b). This seeks to move beyond the static identification of the roles of markets, states and households/families as sources of welfare, and to capture dynamic relationships.

Within this framework "housing" refers to the "housing system" by which we mean the middle range of institutions (market institutions, households, governments, etc.) that influence the way in which people are housed and how housing is used. These may be thought of as operating in three spheres: production (supply), consumption and exchange (finance). "Welfare" refers to the broader frameworks in which housing systems operate. The housing system enjoys a two-way relationship with a "wider welfare regime" which refers to labour market institutions, the levels and distribution of income these produce and how this income is redistributed through taxation, social security and services. The way in which wider welfare regimes are organized, therefore, has distinctive distributional tendencies that influence how housing is consumed and distributed. We refer to these as representing the "boundaries of possibilities". The boundaries of possibility capture both the limits to the options available to lower-income households in the housing market and to the ways in which social or cost rental housing can operate. For example, it is difficult to operate social markets (where tenants can exercise market-like choices within the context of below-market rents) when poverty is high because "choices" or so constrained and trade-offs involve the sacrifice of other necessities. Moreover, the role of the cost/social rented sector becomes more constrained when poverty rises. It is relatively easy to operate "housing for all" allocation policies when there is relatively little acute poverty-related stress; the choices become more acute when income poverty is high. Then it seems more likely that social housing will be targeted, either to provide a "safety net" (secure affordable housing targeted on those in need) or in even more constrained systems, an "ambulance service" (targeted on those on low incomes and with additional vulnerabilities on a temporary basis) (Stephens, 2009).

 DOI: 10.1201/9781003212690-21

Above these middle-range institutions there are macro-level drivers, which include both the broad forces of globalization (and within it financialization) and macro-economic policies, both fiscal and monetary. Beneath the middle-range institutions sit regional and local housing systems below the level of the sovereign state. Housing systems may operate differently for reasons such as policy differences and different market pressures (for example in capital cities).

Welfare regimes of course have a long antecedent, as was outlined in Part I of this book. Often authors (among whom we include ourselves) use the terms "regime" and "system" synonymously. Yet it is useful to maintain a conceptual distinction between them. A regime may be thought of as representing the *cause* of the system, while the system represents the design and operation of the institutions. An analysis focused primarily on systems might be used to categorize a country (for example whether its institutions are primarily market or state-focused), whereas one that also considers regimes is seeking to identify the causes of system design and operation. This step is needed to move beyond categorization and into the territory of explanation, or theory.

Esping-Andersen's (1990) choice of "liberal", "corporatist" (used synonymously with "conservative") and "social democratic" identifies the often neglected theoretical ambitions of his work. They may also be subject to reverse causality: for example, if the market dominates, then it must be "liberal". However, as Kemeny (1995) noted, Esping-Andersen was employing "working class mobilisation theory" in which the relative power (strength or weakness) of the labour movement explains the nature of the regime. This seems overly crude, and perhaps antediluvian in a post-Fordist world. Yet Esping-Andersen's labels also reflect different political ideologies, that is views about how society should be organized and the institutions that need to be designed and built to create it. Each regime, therefore, provides a *logic* or a set of organizing principles that explains why systems are designed in particular ways. A broader conceptualization of *why* such ideologies prevail might be attributed to societal norms. Perhaps confusingly, competing political parties that might even be named after competing ideologies may in practice when in government adhere to the societal norms that define a regime in a different way. (German Social Democrats largely adhered to the Christian Democrats' social market inspired by ordo-liberalism.) But whether class conflict or societal norms are the shapers, they are also mediated through political and other institutions, something that Kemeny emphasized when he differentiated between countries that adopt adversarial political, legal and social institutions, and those that seek to balance competing interests through corporatist ones. The emergence of alternative regime frameworks – the Varieties of Residential Capitalism (Schwartz and Seabrooke, 2008) and more recently Growth Regimes (Reisenbichler, 2021) – whose groupings bear a striking congruence with Esping-Andersen's, also point to the salience of regimes reflecting underlying values, and of the social and economic world being linked.

Nevertheless, the ideological nature of Esping-Andersen's schema is inherent in his choice of decommodification and stratification as the defining outcomes of a regime. Decommodification identifies the ability to enjoy a level of welfare independently of labour market position; stratification refers to the way in which welfare regimes maintain social distinctions between groups. That people with different ideological preferences might prefer to focus on efficiency (for example) is worthy of note, but a broader point is that welfare and housing regimes should not be characterized as being exclusively *social* in nature. Just as welfare systems are embedded in models of economic management, so too are housing systems, the division between the social and economic spheres is somewhat artificial. The conflation of the social and economic spheres was most evident in the socialist regimes that were constructed in the Soviet Union and its satellite states in central and Eastern Europe, and in Maoist China.

The choice of labels for regimes added to Esping-Andersen's three worlds is instructive. When geographical labels, such as "Southern European" are employed, classification appears to be the limit of ambitions. Although often used synonymously, the label "familial" reveals much more, not just that the family performs an elevated role in the production of welfare, but that this reflects an ideology that places a high value on the family as an institution, contrasting to a form of individualism associated with both liberal and (ironically) social democratic regimes. Similarly, the notion of a "productivist" regime, with social institutions serving overriding economic development objectives provides a clear set of organizing principles.

The macro-level regime

Esping-Andersen's classification of regime types was formulated at the end of the post-Second World War recovery period that was ended by the inflationary pressures brought about by the Vietnam War and oil price shocks. During this period, western governments had the need (reconstruction), inclination (popular pressure) and means (economic growth) to seek to shape societies. Although there were exceptions, a broadly "Keynesian" macro-economic framework existed in which governments employed fiscal policy to secure growth and full employment. It was also a time when the world economy was regulated, and sovereign political states enjoyed economic sovereignty to a degree that is unimaginable today. It was in these circumstances that coherent social and economic models took shape in the west, with a non-capitalist alternative developed behind Europe's Iron Curtain.

This economic – and geopolitical – world has now gone. Geopolitically, communism collapsed in Europe and China opened its economy and joined the WTO in 2001. A global economic system emerged founded on a belief in the benefits of free trade and economic interdependence, and a degree of economic sovereignty was lost. A new era of low interest rates and plentiful finance (what Aalbers (2016) refers to as the "global wall of money"), combined with low consumer price inflation. A fashion for placing monetary policy under the control of "independent" central banks charged with targeting consumer price inflation took hold.

The dangers of financial interdependence in this new world economic order were demonstrated when the US sub-prime mortgage morphed into the Global Financial Crisis, which in turn led to sovereign debt crises in southern European countries resulting in international intervention and acute austerity programmes. Governments' – or more precisely central banks' – response was to cut interest rates and to adopt the "unconventional" policy of quantitative easing (QE), in its first iteration mainly focusing on the purchase of bonds from banks to free up capital for lending on assets, which included housing. The world moved to ultra-low interest rates in the "long" decade between the GFC and the COVID-19 pandemic, when central banks intensified QE, this time often purchasing government debt to support huge government support packages during lockdowns. The ultra-low interest rate environment, therefore, continued, until inflation rose as post-pandemic supply chains were slow to reopen and energy prices rose in response to Russia's invasion of Ukraine. QE has, in some cases, given way to quantitative tightening.

These developments in the macro-level economy have set a new context in which housing systems operate. It is broadly the same global economic regime which sets the context for both wider welfare and housing regimes. Most directly the delivery of plentiful and cheap finance has impacted house prices by cheapening the cost of credit to those who want it and who are able to access it. It has contributed to the rise in house prices recorded in almost every welfare regime for which data was available, apart from in the Familial countries of southern Europe most of which

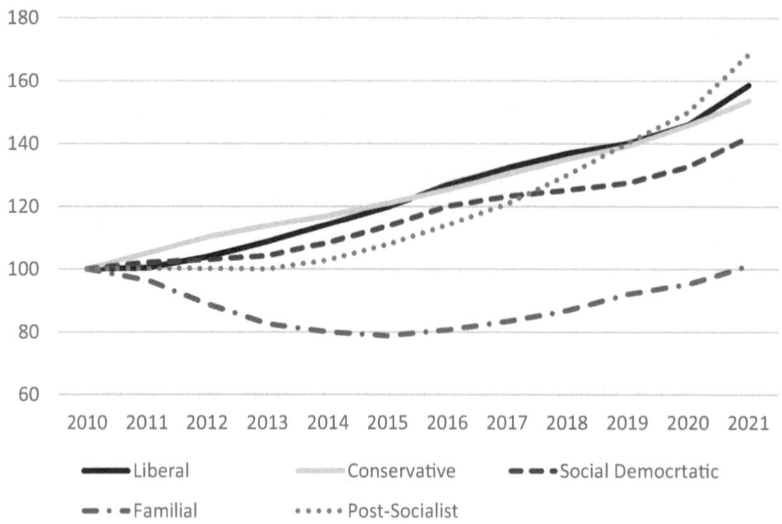

Figure 18.1 Nominal house prices 2010–2021 in the regimes identified in this book.
Source: Calculated from EMF, *Hypostat*, Table 18.1

experienced sovereign debt crises and endured austerity programmes. Even in these countries, house prices rose from around 2015.

The resurgence in inflation in 2022 has caused central banks to raise interest rates. Whether this is a temporary interruption to the era of ultra-low interest rates or marks a return to more historically "normal" interest rates will impact every welfare and housing regime.

Wider welfare regimes

The "classic" wider welfare regimes, by which we mean labour market institutions and the tax and social security systems that redistribute income, have undergone change since they were established in their modern forms during the post-World War II period.

Labour markets in the liberal regimes have experienced reductions in collective wage bargaining and have become more deregulated since the 1970s. Labour markets have also changed and become feminized. Rises in inequality occurred in the 1980s, and social security has tended to become more conditional. The UK's relative poverty rate,[1] for example, doubled between the mid-1970s and 1990 according to OECD statistics. The stagnation of living standards experienced by lower-income groups in the US since the 1970s has been replicated in the UK since the GFC. As a group of countries, the key organizing principles remain distinctly liberal, with their reliance on markets with the state providing income safety nets that are often inadequate, and sometimes with gaps in coverage. The US, however, remains an outlier even within this group due to the huge gaps in the coverage of the social assistance safety net.

Declining economic fortunes, rather than a full-scale crisis, led to reforms to Germany's welfare regime exemplified by the Hartz reforms that deregulated labour markets and reformed social security. Some of the defining institutional features of the conservative/corporatist welfare regime, such as national corporatist wage setting, social security arrangements defined by employment status, and a socially conservative attitude to female employment have largely disappeared. There has, for example, been some rise in relative poverty and inequality in Germany since the mid-1980s (when OECD statistics began). However, the belief in social insurance as a concept and a mechanism remains strong, and both poverty and income inequality levels remain below those experienced in the liberal regimes. Indeed Austria has been able to maintain a stable level of poverty, which is roughly at the same level as in Germany.

The welfare regimes in the social democratic countries have also undergone change. The especially distinctive Rehn-Meidner Swedish system has undergone an especially radical change, but all of these countries faced a severe economic shock in the form of the Nordic banking crisis of the early 1990s. As in the corporatist countries, the extent of collective wage bargaining – that tends to place a floor under the bottom end of the market and a ceiling on the top – has diminished, and labour markets are less regulated, with Denmark adopting a dual approach known as "flexicurity". Coverage of social insurance benefits which (under the Ghent system) is voluntary has shrunk, especially among younger people, as contribution rates have been increased. Levels of employment remain high, the focus on skills and education is strong, and public services remain generous by international standards. Relative poverty rates have risen markedly in Sweden – from just 3.8 per cent in the mid-1970s to 8.8 per cent in 2020, and in Norway since the mid-1980s. However, poverty rates in Finland and Denmark are much the same now as they were in the mid-1980s.

The familial countries of southern Europe have also evolved as they have become more integrated into the European Union. The historic cultural influence of the Church that helped to define these welfare regimes has left a strong legacy evidenced by continued high incidences of multi-generational families. The legacy of distinctively dualist core/periphery labour markets continues, leaving room for a significant informal economy and contributing to high levels of youth unemployment. Although social assistance mechanisms have improved over time, the impact of the GFC on these countries was pronounced as all but Italy experienced sovereign debt crises and austerity programmes imposed by the "Troika" of international agencies. The OECD's relative poverty statistics only go back to the mid-1980s, but they suggest that poverty rates over 10 per cent have been maintained consistently, and in Italy and Spain approach 15 per cent.

The review of the European post-socialist states focussed on Hungary and Russia. These countries experienced the socialist economic model in which welfare was integrated into the workplace. Full employment guarantees a basic level of income supplemented by services and goods provided free or at very low prices often through the workplace. This highly decommodified system collapsed and it is not possible to see replacement systems defined by ideologies in the way of the "old" worlds of welfare capitalism. There are some commonalities, however. Where markets are not fully functional and state support lacking, a form of familialism seeks to fill the gap. This tendency can be discerned across central/eastern Europe through data on multi-generational households. Moreover, in Russia and Hungary, forms of populist authoritarianism have emerged, whereby political leaders devise policies that are aimed at assisting the precarious middle class to cement support for the government. In many of the central European countries, poverty and inequality are perhaps surprisingly low. These are higher in Russia, though still below the liberal regimes.

China has also transitioned from a socialist economic model, but its welfare regime is different in both context and design from its European counterparts. The Chinese growth model has been predicated on urbanization on a grand scale, and an economy based on exports and more recently on highly leveraged property development. Responsibility for welfare has shifted away from work units to a shared responsibility of employers, state and households. Social insurance and assistance schemes have been developed to protect parts of the urban population from the usual risks. There is much local variation in social assistance schemes and stratification based on household registration status is a distinctive part of the system. The Chinese welfare regime, which can be characterized as being developmental, produces distinctive distributional outcomes: urbanization and growth have reduced extreme/absolute poverty, while inequality has soared to way above the highest levels seen in the Global North, and this is also reflected in very high relative poverty rates.

South Africa and Chile point to the importance of path dependence effects of previous political regimes. South Africa has developed a welfare system based on means-testing which is extensive compared to other African countries. However, the legacy of colonialism and apartheid continues to produce huge labour market income disparities, leaving the country with one of the highest levels of income inequality in the world. The legacy of the military dictatorship in Chile also continues to influence distributional outcomes. Although there have been moves towards the provision of more universal services, these follow the "neoliberal" experiment under the former dictatorship. Levels of inequality and relative poverty are high.

Conclusion: Wider welfare regimes

Table 18.1 seeks to draw together the preceding analysis by drawing out distinctive institutional features of different welfare regimes and linking them to distributional outcomes.

The classic welfare regimes have undergone change. Many have lost what have been regarded as being their historic institutional features, while arguably retaining underlying organizational principles. The retention of distinctive distributional tendencies – which are consistently held within and between regime types – suggests that welfare regimes adapt to the changed macro-level context, as well as to more progressive social values (notably regarding the role of women). Where poverty and inequality have risen – for example in Sweden and Germany – we would expect the "boundaries of possibilities" to have shrunk, increasing pressure on lower-income households in the housing market, and making it more difficult to operate "housing for all" cost rental sectors. Relative poverty in social democratic Sweden (8.8 per cent) is now higher than it was in the liberal UK in the mid-1970s (6.2 per cent). While clusters show consistency, there is also variation within them, and the gaping holes in the US social protection system are a case in point.

To reflect the dilution in institutional distinctiveness which combines with the retention of distinctive distributional tendencies, we refer to the classic regimes as "legacy" regimes. Nonetheless, while underlying organizing principles have been maintained the distinctive distributional tendencies of different regime types are generally operating a higher level of poverty than was the case during the high point of the classic regimes. (There are exceptions: for example, in Denmark, Finland and Austria poverty rates have remained low.)

Outside the countries included in the "classic" regime types, the picture is less clear. Distinctive socialist welfare regimes once operated outwith the worlds of welfare capitalism. These collapsed in the European post-socialist states. Social protection mechanisms have been developed while family and kinship play important roles where markets are dysfunctional or there are gaps in social protection. However, the crucial question is whether an organizational logic is discernible. The cases of Russia and Hungary are suggestive of authoritarian populism as an underlying explanatory cause of state interventions designed to support the precarious middle class for political support. But this may not be applicable to other post-socialist countries not detailed here. China, as we have seen, has developed its own distinctive form of developmentalist welfare regime, with social protection as part of a wider system designed to promote economic growth.

Finally, it is difficult to establish clear organizing principles for the two other countries from Global South, South Africa and Chile. However, the powerful influence of the previous "radical" (apartheid; military dictatorship) seems to exert a powerful legacy. These may not be organizing principles, but they provide some explanation for the persistence of very high levels of inequality and relative poverty.

Table 18.1 Institutional features and distributional outcomes in wider welfare regimes

	"Old" regime	*Legacy regime*	*Gini (income inequality)*	*Poverty (%)*
Liberal	Emphasis on markets; social assistance	Emphasis on markets intensified	0.35	13.0
Conservative	Emphasis on social insurance, collective wage agreements; and male breadwinner	Deregulation of labour markets weakens collective wage agreements; female employment increases; and social insurance remains central	0.30	10.3
Social democratic	High levels of employment; collective wage agreements; generous social insurance; and universal services	High level of employment; Deregulation of labour markets; weakened wage bargaining; and reductions in coverage of social insurance.	0.27	7.4
Familial	Core/periphery labour markets; partial social insurance coverage; and importance of family	Development of social assistance coverage; and impacts of GFC and sovereign debt crises.	0.32	12.8
Socialist and post-socialist (Europe)	Welfare centred on work units; full employment; and basic wages supported by subsidized goods and services.	Abandonment of socialist system. Hungary and Russia characterized by authoritarian populism: interventions aimed at a precarious middle class to gain political support for the regime.	0.27	8.8
Socialist and developmental (China)	As above.	Highly stratified system of social insurance and local social assistance, within the growth model. Developmental in orientation.	0.51	28.8
The path-dependent Global South	System of apartheid in South Africa; and market-orientated military dictatorship in Chile	Mechanisms developed for social protection, but systems remain dominated by legacies of former regimes.	0.54	22.1

Source: author analysis; OECD (Gini and poverty statistics)

Housing regimes

The wider welfare regime, combined with the macro-level regime set the context for housing regimes to operate. This section provides the opportunity to explore the relationships between the wider welfare regime and the housing regimes in a comparative context. The framework we have adopted posits that the wider welfare regime sets "boundaries of possibilities" around the housing

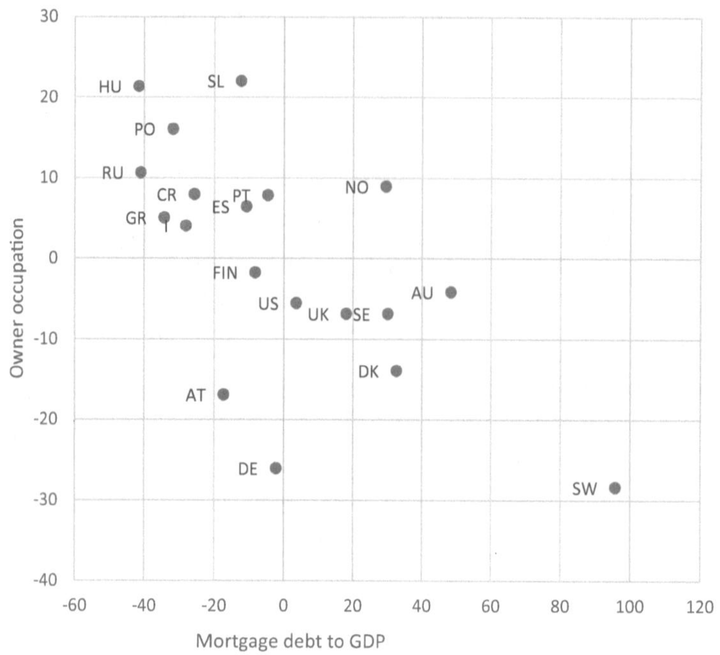

Figure 18.2 Relative deviation from average levels of mortgage debt to GDP (2021, or most recent) and owner occupation (ca. 2020) (percentage points).

Key: AT (Austria), AU (Australia), CR (Czech Republic), DE (Germany), DK (Denmark), ES (Spain), FIN (Finland), GR (Greece), HU (Hungary), I (Italy), NO (Norway), PO (Poland), PT (Portugal), RU (Russia), SL (Slovakia), Sweden (SE) SW (Switzerland), UK (United Kingdom), US (United States)

Sources: Authors' calculations from EMF (2022) (Mortgage debt to GDP); OECD, Housing Europe (2022) and country chapters (Owner occupation rates).

Note: This figure follows a similar logic to Figure 18.1 in Schwartz and Seabrooke (2008), but the benchmarks (averages) are based on different groups of countries and the data relate to a different timeframe.

regime, with levels of income inequality and poverty limiting the choices within the design of the housing system. This, however, is only one aspect of the question. Another is whether housing regimes can differ from welfare regimes. If regimes reflect societally determined organizing principles, then one would expect them to reflect these. But Kemeny (2006) suggested that "sector regimes" exist, and these may differ due to different power relationships within them. As we have suggested, in principle a housing system may exacerbate, counter or merely replicate inequalities in the wider welfare regime. This is something that we explore here, too.

Housing in the liberal welfare regimes

The same organizational principles behind liberal welfare regimes of Australia, the UK and the US, are easily identified as key drivers in their housing equivalents. A qualification might be that although the rhetoric of free markets might be prevalent, they sit alongside a cultural – and political – value being attached to homeownership. Historically, the US housing finance system was supported by the state to extend finance and hence ownership. Support for purchasers in Australia and the discounted sales of public housing particularly in the UK helped to raise ownership levels. The embrace of financial deregulation in the 1980s changed the nature of ownership by making it more liquid, hence blurring the lines between wealth and income. As some welfare regimes have ceased to deliver sustained rises in living standards, particularly at the bottom end, some academics posit that financialized homeownership has been adopted as a strategy to compensate for low-income growth; and may even form the basis of a model of economic growth. The former is self-defeating as it causes the pricing out of younger households, while unequal access to cheap finance leads to growing inequality as equity-rich households acquire more property with cheap finance from which many derive an income stream from the same people who have been priced out. The labour market distortions arising from such a dysfunctional housing system seem most likely to undermine any growth strategy.

In Figure 18.2, which employs the average (among the countries included) level of homeownership and the average level of mortgage debt as a share of GDP as benchmarks, all three liberal countries can be seen to occupy the bottom right quadrant, indicating that they have *below average* levels of homeownership and *above average* levels of mortgage debt. This contrasts with Schwartz and Seabrooke's (2008) diagram which was benchmarked against the OECD averages over the period 1992–2002.

While (thus far) similarly liberal approaches to private renting[2] have been adopted, the UK's housing allowance is clearly more extensive than the Australian and (especially) US equivalents. This leaves the role of social renting within the regime. Although identified as being organized on "dualist" principles by Kemeny, the division between social and market renting does not define any of these housing systems. The high levels of poverty and inequality, however, make "housing for all" approach to social housing deeply problematic, as there are few secure and affordable housing alternatives for low-income groups. Yet, policy matters and the UK's still much larger social rented sector allows "safety net" principles to be applied: it is targeted but provides secure housing for those selected. In contrast, both Australia and the US operate "ambulance service" models, in which access is strictly means-tested, linked to another vulnerability such as disability, and may be withdrawn should income rise.

Social democratic and conservative countries

The social democratic and conservative countries are especially interesting cases in housing regime studies, due to their different treatments by Esping-Andersen and Kemeny. The latter argued that Germany and Switzerland shared with Sweden commonly "corporatist" regimes that created "unitary" housing systems, while Finland, for example, had developed a "dualist" rental system.

To an extent, grouping depends on how much one emphasizes similarity or difference. Perhaps the more one moves from a European analysis to a global one, the more similar the welfare and housing regimes of these countries appear. We have seen how wider welfare regimes have changed, both losing some of their distinctive institutional features. Although poverty and inequality have risen in some of these countries, the patterns remain the same. From a global perspective, these countries enjoy high levels of social protection and relatively low levels of poverty and inequality.

However, one wishes to group these countries, they demonstrate intra-regime variation in housing, in that Norway and Finland both promoted homeownership, while Sweden, Germany, Austria and Switzerland promoted (cost) renting within unitary rental markets. These have evidently come under strain, partly because "boundaries of possibility" have narrowed, even though they are wider than in most other countries. The Swedish example demonstrates the need for subsidy to maintain supply – a situation complicated by the impact of EU competition rules. Scarcity and lack of subsidy have at once undermined the "housing for all" approach, while also making municipal housing companies more wary of letting people on lower incomes.

Diminished supply is also a defining feature of Germany's cost-rental system, and its integrated rental sector in which cost-rental housing exerted a competing influence on the market sector that rendered crude rent controls unnecessary. However, the affordability crisis has led to a reversion to forms of rent control, and the social market is retreating towards a more regulated one. Austria – or more precisely Vienna – is held out as an example of how cost rental models can continue to operate on unitary principles, but even here there is some narrowing of scope. Austria is one country where low poverty rates have been maintained along with cost rental supply. Austria and Finland (another country that has maintained low poverty rates) now appear to have the strongest commitment to cost rental housing. Equity investment on scale in cost rental housing is a relatively new phenomenon, which represents a form of financialization, and falls outside Kemeny's model which is predicated on debt finance and its maturation.

The crucial divide has opened up between the social democratic countries which embraced financial deregulation in the 1980s, and the conservative regime countries that have remained resistant to mortgage debt (see Figure 18.2). Austria and Germany are distinguished by combining below-average levels of homeownership with below-average levels of mortgage debt. In this sense, they remain rental societies. Finland also reflects this combination, although its owner occupation level is only slightly below average, and is much higher than in Austria or Germany. In contrast, Sweden and Denmark have below-average levels of owner-occupation (though again notably higher than in Austria and Germany) combined with high levels of mortgage debt. Norway is unique in our selection of countries by combining above-average ownership rates with an above-average level of mortgage debt. Given the growth in the power of the macro-level regime, the social democratic countries are likely to be more exposed to international developments and changes in monetary policy.

Given the way in which these countries' housing systems have evolved, as shifting organizing principles lead to some institutions changing while others exhibit inertia, distributional patterns are likely to be mixed. Particularly in Sweden, those inside the still highly regulated and secure rental sector can expect to remain protected, while those outside it are more vulnerable. With parts of these housing systems remaining decommodified, and other parts commodified and subject to financialization, stratification becomes of elevated significance. As noted earlier the intergenerational aspect to this is obvious.

Familial countries

The political commitment to homeownership in the southern European countries can be interpreted as being reflective of the defining familial characteristic in the wider welfare regimes, combined with the value of "subsidiarity".

The familial element is reflected in the tradition of self-build housing, which also links to the core-periphery labour market and the informal economy. In Italy, those excluded from the social housing programmes tended to resort to self-build, often illegally, with regularization occurring, sometimes on a massive scale, later. Spain under fascism provides an example of how housing systems, informed by the same underlying values, can vary within a regime type. Franco's clientelistic relationship with the housebuilding industry led to subsidized building programmes rather than self-build, but with stratification ensuring that the beneficiaries were not the poorest.

The divergence between housing systems opened up with the deregulation of mortgage markets in Spain and Portugal, with apparently far less appetite for debt in Italy and Greece. The legacy of the centrality of housebuilding to Franco's model was perhaps reflected in the housebuilding and debt boom that defined Spain's economic model in the years leading up to the GFC. Thereafter a degree of definancialization has occurred as the level of mortgage debt has fallen – an important illustration that the common post-GFC environment of ultra-low interest rates did not have a uniform impact. All four southern European countries fall into the top left quadrant in Figure 18.2 indicating that they combine a below-average level of mortgage debt with an above-average level of homeownership. With consistently high levels of poverty and inequality, the role of informality/self-build, familialism and unmortgaged homeownership are likely to remain features of the housing systems, reflecting the welfare-housing regime nexus.

The post-socialist countries

The post-socialist countries have all moved away from a model in which housing was thoroughly embedded within the wider welfare regime, which itself merged social and economic functions. In abandoning the socialist wider welfare regime, these countries detached housing from welfare. Privatization of public and work-unit housing at "giveaway" prices acted as a shock absorber against rapid economic change in Europe, and in all cases provided a form of "state legacy welfare" (Stephens et al., 2015) – that is outright ownership made possible by the investment made by the state during socialism. However, commodification is limited because markets have not fully developed, for example, take-up of mortgages remains low internationally, although it has risen over the past decade. In these circumstances, the family remains important, for example with multi-generational households and inter-generational transfers. The reduced state seems to intervene in Hungary and Russia based on populist authoritarianism, meaning a somewhat erratic approach to the role of the state and market. Like the familial countries of southern Europe, the European post-socialist countries lie within the top left quadrant (Figure 18.2) indicating above-average homeownership rates and below-average levels of mortgage debt. However, the post-socialist countries are more extreme: homeownership rates are higher (hence "super-homeownership") and mortgage debt levels are almost always even lower.

China's trajectory has been quite different from its European counterparts. Privatization of work-unit housing has protected benefiting households from the consequences of economic transformation and rapid urbanization. New generations are more exposed to the market and at the household level, family support has become very important for younger households seeking to secure homeownership. However, the housing system has become defined by its extreme developmental nature arising from the role of highly leveraged developers and the purchase of land-use rights from local authorities the revenues from which are used to fund investment in infrastructure. The state's concerns about the risks surrounding developers suggest that the model, which helps to drive the world as well as China's economy, is vulnerable to shocks.

Countries in the Global South

This book covers three countries from the Global South. China has been considered with other post-socialist countries. This discussion relates to Chile and South Africa, both of which have very high levels of income inequality, and welfare regimes that are strongly influenced by previous "radical" governments (i.e., military dictatorship and apartheid). The levels of inequality mean that low-income groups are likely to encounter difficulty in securing adequate housing without support.

Chile's housing regime has been strongly shaped by its adherence to the Washington Consensus approach of "enabling markets to work", providing a clear statement of neo-liberalism. The focus is, of course, on efficient markets, but the other side of the strategy is targeted demand-side assistance for low-income groups, although there are also construction subsidies. While an approach that emphasizes choice might be expected to be tenure neutral, there is a strong presumption in favour of homeownership. The promotion of homeownership is seen as part of an asset-based welfare strategy, in that lowered housing costs free up disposable income for spending on other necessities. South Africa's post-apartheid housing policies have also included subsidized homeownership aimed at lower income groups, often supported by urban land (owned by the state) being gifted to occupiers (although in practice the deeds may not be transferred). In both Chile and South Africa, cost restrictions have caused subsidized home-ownership housing to be located on cheaper peripheral land, which contributes to residential segregation, a form of stratification. There is a wider range of sources of housing welfare, including self-build, microfinance and sites-and-services.

Conclusions

This chapter has provided an analysis of the evolving macro-economic, wider welfare and housing regimes of countries from across the world, based on the more detailed preceding chapters.

It emphasizes the highly context-dependent nature of housing. To understand how housing systems operate, one must also understand the contexts in which they operate. The changing macroeconomic environment has had profound effects on housing systems. The relatively closed and regulated world economic system of the Bretton Woods era allowed countries to develop distinctive welfare and housing systems, derived from regimes that reflected national cultures and values. Globalization has changed that context, as economies were opened to competition and financial systems became integrated, linking housing markets across the world. So-called "deglobalization" – referring to the trade war and other tensions between China and the US – is relative.

Globalization impacted wider welfare regimes. Ideological preferences changed in some countries, but in others, pragmatism often spurred by economic crises or the general feeling that they were lagging prompted some governments to dilute the distinctiveness of some of the institutions of their welfare system. With degrees (and we stress the degrees varied) of liberalization, poverty and inequality rose. Yet, underlying organizational principles remained strong, and crucially distinctive distributional patterns produced by welfare regimes have remained intact. Consequently, the "classic" welfare regimes remain useful explanatory devices that are empirically founded, but the dilution of their distinctiveness merits the label "legacy" regimes.

Outside the "old" western world, other regimes have evolved: in the European post-socialist countries, in China and the Global South. Extending the welfare regime framework to cover these countries rests on identifying distinctive institutional features that reflect organizational principles and have distinctive distributional tendencies. In Russia and Hungary, imperfect markets and gaps in state welfare provide fertile ground for familialism; and if there is an organizing principle for the state it is a form of populist authoritarianism, designed to cement political support for the government. China has developed an urban welfare system that provides social protection with a system aimed at serving the economy. This may provide enough of an organizing principle to label China developmentalist. The system certainly has distinctive distributional tendencies. The economy has lifted millions of people out of absolute extreme poverty, but income inequality has soared. It is also highly stratified between urban residents, migrants and rural residents. Chile and South Africa's welfare regimes are more problematic. They have institutions that may suggest hybridity and reflect legacies from former regimes, but it is difficult to discern clear organizing principles. The distributional tendencies are clear: high levels of inequality.

The welfare regimes set the context for housing regimes. For the countries whose welfare regimes now produce higher levels of poverty and inequality, the boundaries of possibility have shrunk. For all, the context of globalization has changed.

Housing systems may be a distinctive "sector" of a welfare regime, but they tend to reflect the same underlying organizational principles. This proposition is given credence from the way in which alternative regime approaches (Varieties of Residential Capitalism and Growth Regime theory) tend to lead to the same country groupings as long-standing welfare regimes. However, the same process of the dilution of distinctive institutional features can be observed. This has had the effect of leading to the breakdown or weakening of the kind of "unitary" or "integrated" rental systems that Kemeny observed in the 1990s. The embrace of financial deregulation within the new world economic order by the Nordic countries created financialized homeownership sectors, while the era of ultra-low interest rates during the "long" post-GFC decade exacerbated housing inequality as affordability declined. Rising income inequality put Sweden's "housing for all" policy under greater strain, a situation that has been exacerbated by a diminished supply of cost rental housing – a policy choice. The pressures seem to have been less where supply has been supported, for example in Finland, which puts in doubt the continued relevance of the country's classification as a "dualist" rental system. The division with the corporatist/conservative countries has become clearer because they have remained resistant to mortgaged homeownership. Yet policy choices matter, and the German retreat from social housing in the 1990s and the rise of equity investment have weakened the integrated rental market with acute affordability issues leading back down the path of regulation. The contrast with Austria, where support for social housing has continued, is instructive in this respect. The divide between the social democratic and conservative regimes now lies in the way in which mortgaged ownership is overlaid (or not) on formerly unitary systems, while intra-regime differences may in large part be attributed to policy approaches to supply.

The response to the changing world economy, combined with the legacy of subsidized housebuilding, caused Spain's housing system to diverge from other familialist countries as it adopted a model of financialized property-driven growth in the run-up to the GFC. Subsequent definancialization has seen something of a reconvergence of the countries of southern Europe around a model characterized by high levels of homeownership combined with below-average levels of mortgage debt. In this respect the underlying (self) organizing principle of these systems unmortgaged homeownership remains a bulwark against the high levels of poverty and inequality that are produced by the wider welfare regime. Much the same phenomenon can be observed in formerly socialist countries in central and eastern Europe, the principal difference being that these countries have even higher levels of unmortgaged homeownership (and less mortgage debt). While housing policy interventions in Hungary and Russia may reflect the same authoritarian populist motivation as those in the wider welfare regimes, the familialist approach to housing (albeit the legacy of state housing that has been privatized) remains a strong organizing principle.

An important theoretical signal from these countries of southern, central and eastern Europe is that if the state does not define the system, other actors will. Where states are weak and markets underdeveloped, regimes may be shaped by household actors. In China, of course, the state does actively seek to define the housing system, and as we have seen the developmentalist nature of the wider welfare regime is replicated in the housing system which is deeply embedded within the growth model. The other countries of the Global South – Chile and South Africa – are more difficult to read. In both these countries, the state seeks to promote a form of social homeownership, which seems in part to be a response to high levels of inequality.

Governments operate within the macro-economic system, and housing operates within this and the wider welfare regime. Understanding how these operate is key to understanding how housing systems operate, and what policy can – and cannot – achieve. Reading welfare regimes and housing regimes, and the relationship between them has become more complex as the world has become more integrated. It has become simply impossible to understand housing systems from within. Describing institutional frameworks and policies will not explain how they operate. It is the dynamic relationship between actors within the contexts of the wider welfare regime and macro-level context that drives systems. Identifying systems on the basis of the role of the state, market and households may allow for categorization; but only understanding how they interrelate with one another provides an explanation.

Notes

1 Unless stated otherwise, the poverty line is set at 50 per cent of median disposable income after taxes and transfers, and adjusted ("equivalized") for household composition. See OECD (2023), Income inequality (indicator). doi: 10.1787/459aa7f1-en

2 Within the UK, Scotland is an exception having greatly extended tenure security and is developing forms of rent regulation.

References

Aalbers, M. B. (2016). *The Financialization of Housing: A Political Economy Approach*. Routledge.

EMF. (2022). *Hypostat 2022*.

Esping-Andersen, G. (1990). *The Three Worlds of Welfare Capitalism*. Princeton University Press.

Housing Europe. (2022). *State of Housing in Europe 2021*.

Kemeny, J. (1995). *From Public Housing to the Social Market: Rental Policy Strategies in Comparative Perspective*. Routledge.

Kemeny, J. (2006). Corporatism and Housing Regimes. *Housing, Theory and Society*, *23*(1), 1–18. https://doi.org/10.1080/14036090500375423

Reisenbichler, A. (2021). Housing Finance between Social Welfare and Growth Strategies. In A. Hassel & B. Palier (Eds.), *Growth and Welfare in Advanced Capitalist Economies* (pp. 320–347). Oxford University Press.

Schwartz, H., & Seabrooke. L. (2008). Varieties of Residential Capitalism in the International Political Economy: Old Welfare States and the New Politics of Housing. *Comparative European Politics*, *6*, 237–261. https://doi.org/10.1057/cep.2008.10

Stephens, M. (2009). The Role of the Social Rented Sector. In S. Fitzpatrick & M. Stephens (Eds.), *The Future of Social Housing* (pp. 27–38). Shelter.

Stephens, M. (2020b). Towards a Multi-layered Housing Regime Framework: Responses to Commentators. *Housing, Theory and Society*, *37*(5), 584–596. https://doi.org/10.1080/14036096.2020.1816573

Stephens, M., Lux, M., & Sunega, P. (2015). Post-Socialist Housing Systems in Europe: Housing Welfare Regimes by Default? *Housing Studies*, *30*(8), 1210–1234. https://doi.org/10.1080/02673037.2015.1013090

INDEX